# STUDENT SOLUTIONS MANUAL
## Richard Semmler

# BASIC COLLEGE MATHEMATICS

### FOURTH EDITION

## John Tobey    Jeffrey Slater

Prentice Hall

Upper Saddle River, NJ 07458

Executive Editor: Karin E. Wagner
Supplements Editor: Elizabeth Covello
Editorial Assistant: Rudy Leon
Assistant Managing Editor, Math Media Production: John Matthews
Production Editor: Wendy A. Perez
Supplement Cover Manager: Paul Gourhan
Supplement Cover Designer: PM Workshop Inc.
Manufacturing Buyer: Alan Fischer

© 2002 by Prentice-Hall, Inc.
Upper Saddle River, NJ 07458

Printed in the United States of America

10 9 8 7 6 5 4 3 2 1

ISBN    0-13-092419-9

Prentice-Hall International (UK) Limited, London
Prentice-Hall of Australia Pty. Limited, Sydney
Prentice-Hall Canada, Inc., Toronto
Prentice-Hall Hispanoamericana, S.A., Mexico City
Prentice-Hall of India Private Limited, New Delhi
Pearson Education Asia Pte. Ltd., Singapore
Prentice-Hall of Japan, Inc., Tokyo
Editora Prentice-Hall do Brazil, Ltda., Rio de Janeiro

# CONTENTS

## DIAGNOSTIC PRETEST

**1.**
$$\begin{array}{r} 3846 \\ +\ 527 \\ \hline 4373 \end{array}$$

**2.**
$$\begin{array}{r} 26 \\ 58\overline{)1508} \\ \underline{116} \\ 348 \\ \underline{348} \\ 0 \end{array}$$

**3.**
$$\begin{array}{r} {\scriptstyle 7\ 10\ 17} \\ 12,807 \\ -\ 11,679 \\ \hline 1,128 \end{array}$$

**4.**
$$\begin{array}{r} 115 \\ \times\ 8 \\ \hline 920 \end{array}$$

920 tons of sand

**5.** $\dfrac{3}{7} + \dfrac{2}{5} = \dfrac{3}{7} \times \dfrac{5}{5} + \dfrac{2}{5} \times \dfrac{7}{7}$

$\qquad = \dfrac{15}{35} + \dfrac{14}{35}$

$\qquad = \dfrac{29}{35}$

**6.** $3\dfrac{3}{4} \times 2\dfrac{1}{5} = \dfrac{15}{4} \times \dfrac{11}{5}$

$\qquad = \dfrac{5 \times 3 \times 11}{2 \times 2 \times 5}$

$\qquad = \dfrac{33}{4}$

$\qquad = 8\dfrac{1}{4}$

**7.**
$$\begin{array}{r} 2\frac{1}{6} \\ -\ 7\frac{1}{3} \end{array} \qquad \begin{array}{r} 2\frac{1}{6} \\ -\ 1\frac{2}{6} \end{array} \qquad \begin{array}{r} 1\frac{7}{6} \\ -\ 1\frac{2}{6} \\ \hline \frac{5}{6} \end{array}$$

**8.** $237 \div 7\dfrac{9}{10} = 237 \div \dfrac{79}{10}$

$\qquad = \dfrac{237}{1} \times \dfrac{10}{79}$

$\qquad = \dfrac{79 \times 3 \times 10}{79}$

$\qquad = 30$

30 miles per gallon

**9.**
$$\begin{array}{r} 51.06 \\ \times\ 0.307 \\ \hline 35742 \\ 0000 \\ 15\ 318 \\ \hline 15.67542 \end{array}$$

**10.**
$$\begin{array}{r} 3.4 \\ 0.026\overline{)0.0884} \\ \underline{78} \\ 104 \\ \underline{104} \\ 0 \end{array}$$

**11.**
$$\begin{array}{r} 24.375 \\ -\ 1.750 \\ \hline 22.625 \end{array}$$

22.625 cm

**12.**
$$\begin{array}{r} 20.5 \\ 5.8 \\ +\ 14.9 \\ \hline 41.2 \end{array}$$

41.2 miles

**13.**
$$\frac{3}{7} = \frac{n}{24}$$
$$3 \times 24 = 7 \times n$$
$$72 = 7 \times n$$
$$\frac{72}{7} = \frac{7 \times n}{7}$$
$$10.3 \approx n$$

**14.**
$$\frac{0.5}{0.8} = \frac{220}{n}$$
$$0.5 \times n = 0.8 \times 220$$
$$0.5 \times n = 176$$
$$\frac{0.5 \times n}{0.5} = \frac{176}{0.5}$$
$$n = 352$$

**15.**
$$\frac{n}{45} = \frac{600}{25}$$
$$n \times 25 = 45 \times 600$$
$$n \times 25 = 27,000$$
$$\frac{n \times 25}{25} = \frac{27,000}{25}$$
$$n = 1080$$

$1080

**16.**
$$\frac{n}{6} = \frac{300}{8}$$
$$n \times 8 = 6 \times 300$$
$$n \times 8 = 1800$$
$$\frac{n \times 25}{25} = \frac{1800}{8}$$
$$n = 225$$

225 miles apart

**17.**
$$\begin{array}{r} 0.375 \\ 8\overline{)3.000} \\ \underline{2\,4}\phantom{00} \\ 60 \\ \underline{56} \\ 40 \\ \underline{40} \\ 0 \end{array}$$

$$\frac{3}{8} = 0.375 = 37.5\%$$

**18.** $n = 138\%$ of $5600$
$$= 138\% \times 5600$$
$$= 1.38 \times 5600$$
$$= 7728$$

**19.**
$$\frac{53}{100} = \frac{2067}{b}$$
$$53 \times b = 100 \times 2067$$
$$53 \times b = 206,700$$
$$\frac{53 \times b}{53} = \frac{206,700}{53}$$
$$b = 3900$$

3900 students

**20.** percent: $\dfrac{9}{3000} = 0.003$
$$= 0.3\%$$

**21.** $15 \text{ qt} \times \dfrac{1 \text{ gal}}{4 \text{ qt}} = 3.75 \text{ gal}$

**22.** $3 \text{ cm} = 0.03 \text{ m}$

**23.** $1.56 \text{ tons} \times \dfrac{2000 \text{ lb}}{1 \text{ ton}} = 3120 \text{ lb}$

**24.** $4900 \text{ kg} = 4,900,000 \text{ g}$
$$= 4,900,000,000 \text{ mg}$$

**25.** $A = \dfrac{bh}{2}$
$$= \frac{(34 \text{ m})\,(23 \text{ m})}{2}$$
$$= \frac{782 \text{ m}^2}{2}$$
$$= 391 \text{ m}^2$$

**26.** 
$$A = \pi r^2$$
$$= 3.14 (5 \text{ yd})^2$$
$$= 3.14(25 \text{ yd}^2)$$
$$= 78.5 \text{ yd}^2$$

$$\text{Cost} = 78.5 \text{ yd}^2 \times \frac{\$35}{1 \text{ yd}^2}$$
$$= \$2747.50$$

**27.** 
$$\text{leg} = \sqrt{15^2 - 9^2}$$
$$= \sqrt{225 - 81}$$
$$= \sqrt{144}$$
$$= 12$$

12 meters

**28.** 
$$V = \pi r^2 h$$
$$= 3.14(5 \text{ ft})^2 (4 \text{ ft})$$
$$= 3.14(25 \text{ ft}^2)(4 \text{ ft})$$
$$= 314 \text{ ft}^3$$

$$\text{Pounds} = 314 \text{ ft}^3 \times \frac{70 \text{ lb}}{1 \text{ ft}^3}$$
$$= 21{,}980 \text{ pounds}$$

**29.** 500

**30.** $800 - 700 = 100$

**31.** 2000: $800 + 700 + 300 + 700 = 2500$
2001: $900 + 500 + 400 + 800 = 2600$
More were sold in 2001.

**32.** 
$$\text{Mean} = \frac{800 + 700 + 300 + 700}{4}$$
$$= \frac{2500}{4}$$
$$= 625$$

625 cars per quarter

**33.** 
$$-5 + (-2) + (-8) = -7 + (-8)$$
$$= -15$$

**34.** 
$$-8 - (-20) = -8 + 20$$
$$= 12$$

**35.** 
$$\left(-\frac{3}{4}\right) \div \left(\frac{5}{6}\right) = \left(-\frac{3}{4}\right) \times \left(\frac{6}{5}\right)$$
$$= \frac{-3 \times 3 \times 2}{2 \times 2 \times 5}$$
$$= -\frac{9}{10}$$

**36.** 
$$(-3)(2)(-1)(-3) = -6(-1)(-3)$$
$$= 6(-3)$$
$$= -18$$

**37.** 
$$9(x + y) - 3(2x - 5y)$$
$$= 9x + 9y - 6x + 15y$$
$$= 9x - 6x + 9y + 15y$$
$$= 3x + 24y$$

**38.** 
$$3x - 7 = 5x - 19$$
$$3x + (-3x) - 7 = 5x + (-3x) - 19$$
$$-7 = 2x - 19$$
$$-7 + 19 = 2x - 19 + 19$$
$$12 = 2x$$
$$\frac{12}{2} = \frac{2x}{2}$$
$$6 = x$$

**39.** $2(x-3)+4x=-2(3x+1)$

$$2x-6+4x=-6x-2$$

$$6x-6=-6x-2$$

$$6x+6x-6=-6x+6x-2$$

$$12x-6=-2$$

$$12x-6+6=-2+6$$

$$12x=4$$

$$\frac{12x}{12}=\frac{4}{12}$$

$$x=\frac{4}{12}=\frac{1}{3}$$

**40.**      $x=\text{width}$

$2x+4=\text{length}$

$$2(x)+2(2x+4)=134$$

$$2x+4x+8=134$$

$$6x+8=134$$

$$6x=126$$

$$\frac{6x}{6}=\frac{126}{6}$$

$$x=21$$

$\text{width}=21\text{ meters}$

$\text{length}=2(21)+4$

$\qquad=42+4$

$\qquad=46\text{ meters}$

# WHOLE NUMBERS

## Pretest Chapter 1

**1.** 78,310,436
= Seventy-eight million, three hundred ten thousand, four hundred thirty-six

**2.** $38,247 = 30,000 + 8000 + 200 + 40 + 7$

**3.** 5,064,122

**4.** 2,747,000

**5.** 2,583,000

**6.**
$$\begin{array}{r} 13 \\ 31 \\ 88 \\ 43 \\ + 69 \\ \hline 244 \end{array}$$

**7.**
$$\begin{array}{r} 28,318 \\ 5,039 \\ + 17,213 \\ \hline 50,570 \end{array}$$

**8.**
$$\begin{array}{r} 7148 \\ 500 \\ 19 \\ + 7062 \\ \hline 14,729 \end{array}$$

**9.**
$$\begin{array}{r} 6439 \\ - 4328 \\ \hline 2111 \end{array}$$

**10.**
$$\begin{array}{r} 100,450 \\ + 24,139 \\ \hline 76,311 \end{array}$$

**11.**
$$\begin{array}{r} 45,861,413 \\ - 43,879,761 \\ \hline 1,981,652 \end{array}$$

**12.** $9 \times 6 \times 1 \times 2 = 54 \times 1 \times 2$
$$= 54 \times 2$$
$$= 108$$

**13.**
$$\begin{array}{r} 2658 \\ \times 7 \\ \hline 18,606 \end{array}$$

**14.**
$$\begin{array}{r} 91 \\ \times 74 \\ \hline 364 \\ 637 \\ \hline 6734 \end{array}$$

**15.**
$$\begin{array}{r} 365 \\ \times 908 \\ \hline 2\,920 \\ 0\,00 \\ 328\,5 \\ \hline 331,420 \end{array}$$

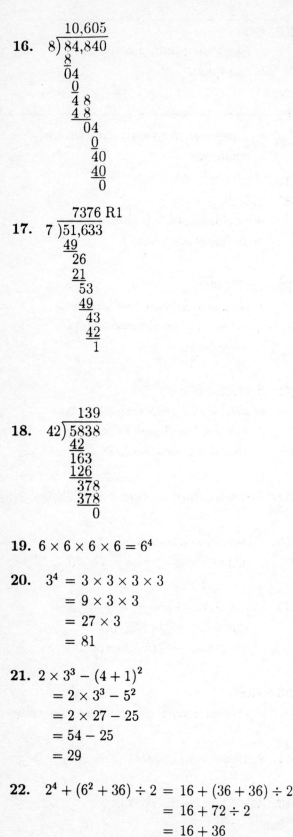

**16.**
$$
\begin{array}{r}
10{,}605 \\
8\,)\overline{84{,}840} \\
\underline{8}\phantom{0000} \\
04\phantom{00} \\
\underline{0}\phantom{00} \\
4\ 8\phantom{0} \\
\underline{4\ 8}\phantom{0} \\
04 \\
\underline{0} \\
40 \\
\underline{40} \\
0
\end{array}
$$

**17.**
$$
\begin{array}{r}
7376\ \text{R}1 \\
7\,)\overline{51{,}633} \\
\underline{49}\phantom{000} \\
26\phantom{00} \\
\underline{21}\phantom{00} \\
53\phantom{0} \\
\underline{49}\phantom{0} \\
43 \\
\underline{42} \\
1
\end{array}
$$

**18.**
$$
\begin{array}{r}
139 \\
42\,)\overline{5838} \\
\underline{42}\phantom{00} \\
163\phantom{0} \\
\underline{126}\phantom{0} \\
378 \\
\underline{378} \\
0
\end{array}
$$

**19.** $6 \times 6 \times 6 \times 6 = 6^4$

**20.** $3^4 = 3 \times 3 \times 3 \times 3$
$\phantom{3^4} = 9 \times 3 \times 3$
$\phantom{3^4} = 27 \times 3$
$\phantom{3^4} = 81$

**21.** $2 \times 3^3 - (4 + 1)^2$
$\phantom{2} = 2 \times 3^3 - 5^2$
$\phantom{2} = 2 \times 27 - 25$
$\phantom{2} = 54 - 25$
$\phantom{2} = 29$

**22.** $2^4 + (6^2 + 36) \div 2 = 16 + (36 + 36) \div 2$
$\phantom{2^4 + (6^2 + 36) \div 2} = 16 + 72 \div 2$
$\phantom{2^4 + (6^2 + 36) \div 2} = 16 + 36$
$\phantom{2^4 + (6^2 + 36) \div 2} = 52$

**23.** $6 \times 7 - 2 \times 6 - 2^3 + 17 \div 17$
$\phantom{2} = 6 \times 7 - 2 \times 6 - 8 + 17 \div 17$
$\phantom{2} = 42 - 12 - 8 + 17 \div 17$
$\phantom{2} = 42 - 12 - 8 + 1$
$\phantom{2} = 30 - 8 + 1$
$\phantom{2} = 22 + 1$
$\phantom{2} = 23$

**24.** 270,612 rounds to 271,000 since 6 is greater than 5.

**25.** 26,539 rounds to 26,500 since 3 is less than 5.

**26.** 59,540,000 rounds to 60,000,000 since 5 is greater than or equal to 5.

**27.**
$$
\begin{array}{r}
200 \\
500 \\
400 \\
800 \\
+\ 700 \\
\hline
2600
\end{array}
$$

**28.** $56{,}784 \times 459{,}202$ is approximately $60{,}000 \times 500{,}000 = 30{,}000{,}000{,}000.$

**29.**
$$
\begin{array}{r}
1483 \\
-\ 317 \\
\hline
1166\ \text{miles}
\end{array}
$$

**30.**
$$
\begin{array}{r}
22 \\
84\,)\overline{1848} \\
\underline{168}\phantom{0} \\
168 \\
\underline{168} \\
0
\end{array}
$$

$22 per share

**31.** $3 \times \$46 = \$138$     138
$\phantom{}3 \times \$29 = \$87$      87
$\phantom{}2 \times \$37 = \$74$   $+\ \ 74$
                                                 $\$299$

**32.**   **a.**      550,043
           −   401,851
           ───────
              148,192

148,192 people

**b.**      653,000
        +   134,890
        ───────
           787,890

787,890 people

## 1.1    Exercises

**1.** $6731 = 6000 + 700 + 30 + 1$

**3.** $108,276 = 100,000 + 8,000 + 200 + 70 + 6$

**5.** $23,761,345$
$= 20,000,000 + 3,000,000 + 700,000$
$+ 60,000 + 1,000 + 300 + 40 + 5$

**7.** $103,260,768$
$= 100,000,000 + 3,000,000 + 200,000$
$+ 60,000 + 700 + 60 + 8$

**9.** $600 + 70 + 1 = 671$

**11.** $9000 + 800 + 60 + 3 = 9863$

**13.** $30,000 + 9000 + 700 + 30 + 3 = 39,733$

**15.** $700,000 + 6000 + 200 = 706,200$

**17. a.** 7
**b.** 50,000

**19. a.** 4
**b.** 40,000

**21.** 53 = Fifty-three

**23.** 8936
= Eight thousand, nine hundred thirty-six.

**25.** 36,118
= Thirty-six thousand, one hundred eighteen.

**27.** 105,261
= One hundred five thousand, two hundred sixty-one.

**29.** 14,203,326
= Fourteen million, two hundred three thousand, three hundred twenty-six.

**31.** 4,302,156,200
= Four billion, three hundred two million, one hundred fifty-six thousand, two hundred.

**33.** Three hundred seventy-five = 375.

**35.** Twenty-seven thousand, three hundred eighty-two = 27,382

**37.** One hundred million, seventy-nine thousand, eight hundred twenty-six = 100,079,826.

**39.** $1965
= One thousand, nine hundred sixty-five.

**41.** 9 million or 9,000,000

**43.** 33 million or 33,000,000

**45.** $281 billion or $281,000,000,000

**47.** $574 billion or $574,000,000,000

**49. a.** 5
  **b.** 2

**51. a.** 7
  **b.** 8

**53.** 613,001,033,208,003

**55.** 6E22 = 60,000,000,000,000,000,000,000

**57.** 195

_____

## 1.2   Exercises

**1.** Answers may vary. Samples are below.
  **a.**  The order of the addends can be changed without changing the sum.
  **b.**  The addends can be grouped together in any way without changing the sum.

**3.**

| +  | 3  | 5  | 4  | 8  | 0 | 6  | 7  | 2  | 9  | 1  |
|----|----|----|----|----|---|----|----|----|----|----|
| 2  | 5  | 7  | 6  | 10 | 2 | 8  | 9  | 4  | 11 | 3  |
| 7  | 10 | 12 | 11 | 15 | 7 | 13 | 14 | 9  | 16 | 8  |
| 5  | 8  | 10 | 9  | 13 | 5 | 11 | 12 | 7  | 14 | 6  |
| 3  | 6  | 8  | 7  | 11 | 3 | 9  | 10 | 5  | 12 | 4  |
| 0  | 3  | 5  | 4  | 8  | 0 | 6  | 7  | 2  | 9  | 1  |
| 4  | 7  | 9  | 8  | 12 | 4 | 10 | 11 | 6  | 13 | 5  |
| 1  | 4  | 6  | 5  | 9  | 1 | 7  | 8  | 3  | 10 | 2  |
| 8  | 11 | 13 | 12 | 16 | 8 | 14 | 15 | 10 | 17 | 9  |
| 6  | 9  | 11 | 10 | 14 | 6 | 12 | 13 | 8  | 15 | 7  |
| 9  | 12 | 14 | 13 | 17 | 9 | 15 | 16 | 11 | 18 | 10 |

**5.**
```
    4
    2
    8
 +  9
   23
```

**7.**
```
    3
    8
    9
 +  6
   26
```

**9.**
```
   10
   36
 +  3
   49
```

**11.**
```
   63
   24
 + 12
   99
```

**13.**
```
   2847
   1634
 +   98
   4579
```

**15.**
```
    7738
    1363
 +  3255
   12,356
```

**17.**
```
    8235
 +  5626
   13,861
```

**19.**
```
   26,108
 + 16,371
   42,479
```

**21.**
```
   36
   41
   25
    6
 + 13
  121
```

**23.**
```
   106
    13
     4
    28
 + 981
  1132
```

**25.**
```
      32
     106
      47
  + 7193
    7378
```

**27.**
```
   1,362,214
   7,002,316
 + 3,214,896
  11,579,426
```

**29.**
```
    837,241,000
  + 298,039,240
  1,135,280,240
```

**31.**
```
      516,208
       24,317
  + 1,763,295
    2,303,820
```

**33.** $12 + 8 + 156 + 72$

$\quad = 20 + 156 + 72$

$\quad = 176 + 72 = 248$

**35.** $15,216 + 485 + 5208$

$\quad = 15,701 + 5208$

$\quad = 20,909$

**37.**
```
    224
    387
  + 183
    794
```
She spent $794 on gifts.

**39.**
```
    1184
    1632
  + 1395
    4211
```
The total amount earned was $4211.

**41.**
```
    124
    105
    147
  +  92
    468
```
468 feet of fencing are needed.

**43.**
```
    64,000,000
    31,800,000
  + 25,300,000
   121,100,000  square miles
```

**45.**
```
   2,144,856
     307,244
  +  470,239
   2,922,339  votes
```

**47. a.**
```
    415
    364
    159
  + 196
   1134  students
```

**b.**
```
   1134
     27
     68
    102
  +  61
   1392  students
```

**49.**
```
     87
     17
  +  98
    202  miles
```

**51.**
```
    568
    682
  + 703
   1953
```
Then,
```
    2387
  - 1953
    434  feet
```

**53. a.** Out-of-state
```
    5276
    2437
  + 1840
   $9553
```

**b.** In-state
```
    3640
    1926
  + 1753
   $7319
```

**c.** Foreign

$$
\begin{array}{r}
8352 \\
2855 \\
+\ 1840 \\
\hline
\$13{,}047
\end{array}
$$

**55.** $89 + 166 + 23 + 45 + 72 + 190 + 203 + 77 + 18 + 93 + 46 + 73 + 66 = 1161$

**57.** Answers may vary. A sample is.
You could not group the addends in groups that sum to 10s to make column addition easier.

**59.** $121{,}000{,}374 = $ One hundred twenty-one million, three hundred seventy-four.

**61.** Nine million, fifty-one thousand, seven hundred nineteen $= 9{,}051{,}719$.

---

## 1.3   Exercises

**1.** In subtraction, the minuend minus the subtrahend equals the difference. To check, we add the subtrahend and the difference to see if we get the minuend. If we do, the answer is correct.

**3.** We know that $1683 + 1592 = 32?5$. Therefore, if we add 8 tens and 9 tens, we get 17 tens which is 1 hundred and 7 tens. Thus, the ? should be replaced by 7.

**5.**
$$
\begin{array}{r}
6 \\
-\ 5 \\
\hline
1
\end{array}
$$

**7.**
$$
\begin{array}{r}
15 \\
-\ 6 \\
\hline
9
\end{array}
$$

**9.**
$$
\begin{array}{r}
16 \\
-\ 0 \\
\hline
16
\end{array}
$$

**11.**
$$
\begin{array}{r}
18 \\
-\ 9 \\
\hline
9
\end{array}
$$

**13.**
$$
\begin{array}{r}
11 \\
-\ 4 \\
\hline
7
\end{array}
$$

**15.**
$$
\begin{array}{r}
13 \\
-\ 7 \\
\hline
6
\end{array}
$$

**17.**
$$
\begin{array}{r}
11 \\
-\ 8 \\
\hline
3
\end{array}
$$

**19.**
$$
\begin{array}{r}
15 \\
-\ 6 \\
\hline
9
\end{array}
$$

**21.**
$$
\begin{array}{r}
47 \\
-\ 26 \\
\hline
21
\end{array}
\qquad
\begin{array}{r}
26 \\
+\ 21 \\
\hline
47
\end{array}
$$

**23.**
$$
\begin{array}{r}
85 \\
-\ 73 \\
\hline
12
\end{array}
\qquad
\begin{array}{r}
73 \\
+\ 12 \\
\hline
85
\end{array}
$$

**25.**
$$
\begin{array}{r}
126 \\
-\ 95 \\
\hline
31
\end{array}
\qquad
\begin{array}{r}
95 \\
+\ 31 \\
\hline
126
\end{array}
$$

**27.**
$$
\begin{array}{r}
768 \\
-\ 143 \\
\hline
625
\end{array}
\qquad
\begin{array}{r}
143 \\
+\ 625 \\
\hline
768
\end{array}
$$

**29.**
$$
\begin{array}{r}
2893 \\
-\ 572 \\
\hline
2321
\end{array}
\qquad
\begin{array}{r}
572 \\
+\ 2321 \\
\hline
2893
\end{array}
$$

**31.**
$$\begin{array}{r} 24{,}396 \\ -\ 13{,}205 \\ \hline 11{,}191 \end{array} \qquad \begin{array}{r} 13{,}205 \\ +\ 11{,}191 \\ \hline 24{,}396 \end{array}$$

**33.**
$$\begin{array}{r} 986{,}302 \\ -\ 433{,}201 \\ \hline 553{,}101 \end{array} \qquad \begin{array}{r} 433{,}201 \\ +\ 553{,}101 \\ \hline 986{,}302 \end{array}$$

**35.**
$$\begin{array}{r} 19 \\ +\ 110 \\ \hline 129 \end{array} \quad \text{Correct}$$

**37.**
$$\begin{array}{r} 1213 \\ +\ 6134 \\ \hline 7347 \end{array} \quad \text{Correct}$$

**39.**
$$\begin{array}{r} 5020 \\ +\ 1020 \\ \hline 6040 \end{array} \quad \text{Incorrect} \qquad \begin{array}{r} 6030 \\ -\ 5020 \\ \hline 1010 \end{array}$$

**41.**
$$\begin{array}{r} 41{,}181 \\ +\ 58{,}402 \\ \hline 99{,}583 \end{array} \quad \text{Correct}$$

**43.**
$$\begin{array}{r} \overset{8\ \ 13}{\cancel{9}\cancel{3}} \\ -\ 47 \\ \hline 46 \end{array}$$

**45.**
$$\begin{array}{r} \overset{1\ 15}{1\cancel{2}\cancel{5}} \\ -\ 88 \\ \hline 37 \end{array}$$

**47.**
$$\begin{array}{r} \overset{3\ \overset{14}{\cancel{4}}\ 11}{\cancel{4}\cancel{5}\cancel{1}} \\ -\ 376 \\ \hline 75 \end{array}$$

**49.**
$$\begin{array}{r} \overset{8\ 10}{\cancel{9}\cancel{0}5} \\ -\ 324 \\ \hline 581 \end{array}$$

**51.**
$$\begin{array}{r} \overset{1\ 9\ 9\ 9\ 10}{\cancel{2}\cancel{0}{,}\cancel{0}\cancel{0}\cancel{0}} \\ -\ 9{,}308 \\ \hline 10{,}692 \end{array}$$

**53.**
$$\begin{array}{r} \overset{\overset{11}{4}\ \cancel{1}\ 9\ 9\ 10}{1\cancel{5}2{,}\cancel{0}\cancel{0}\cancel{0}} \\ -117{,}908 \\ \hline 34{,}092 \end{array}$$

**55.**
$$\begin{array}{r} \overset{14\ 12\ 10}{\overset{3\ 4\ 8}{4}012} \\ \cancel{4}\cancel{5}{,}\cancel{3}\cancel{1}\cancel{2} \\ -37{,}865 \\ \hline 7{,}447 \end{array}$$

**57.**
$$\begin{array}{r} \overset{5\ 10\ 0\ 10}{\overset{9\ 10}{7}6\cancel{0}{,}\cancel{1}\cancel{0}8} \\ -536{,}992 \\ \hline 223{,}116 \end{array}$$

**59.**  $x + 14 = 19$
$$5 + 14 = 19$$
$$x = 5$$

**61.**  $34 = x + 13$
$$34 = 21 + 13$$
$$x = 21$$

**63.**  $100 + x = 127$
$$100 + 27 = 127$$
$$x = 27$$

**65.**
$$\begin{array}{r} 960 \\ -\ 778 \\ \hline 182 \end{array} \ \text{votes}$$

**67.**
$$\begin{array}{r} 9{,}918{,}040 \\ -\ 3{,}632{,}944 \\ \hline 6{,}285{,}096 \end{array}$$

**69.**  Total earned = \$475

Total paid out = \$142 + \$85 = \$227

$475 - 227 = 248$

He received \$248.

**71.**   $\begin{array}{r} 4,830,784 \\ -\ 3,413,864 \\ \hline 1,416,920 \end{array}$ people

**73.**   $\begin{array}{r} 8,881,826 \\ -\ 5,195,392 \\ \hline 3,686,434 \end{array}$ people

**75.**   $\begin{array}{r} 11,430,602 \\ -\ 11,110,285 \\ \hline 320,317 \end{array}$ people

**77.**   $\begin{array}{r} 413,471 \\ -\ 320,317 \\ \hline 93,154 \end{array}$ people

**79.**   $\begin{array}{r} 125 \\ -\ 96 \\ \hline 29 \end{array}$ homes

**81.**   $\begin{array}{r} 219 \\ -\ 139 \\ \hline 80 \end{array}$ homes

**83.**   $\begin{array}{r} 63 \\ -\ 45 \\ \hline 18 \end{array}$   $\begin{array}{r} 63 \\ -\ 55 \\ \hline 8 \end{array}$

The greatest change was between 1999 and 2000.

**85.** She should choose Willow Creek and Irving because the smallest difference was 9 homes

**87.** $a - b = b - a$ if $a$ and $b$ represent the same number.

**89.**   $\begin{array}{r} 276 \\ -\ 216 \\ \hline 60 \end{array}$   Now, $\dfrac{60}{12} = 5$ and   $\begin{array}{r} 110 \\ \times\ 5 \\ \hline \$550 \end{array}$

---

## Cumulative Review Problems

**91.** Eight million, four hundred sixty-six thousand, eight-four $= 8,466,084$

**93.** $16 + 27 + 82 + 34 + 9 = 168$

---

## 1.4   Exercises

**1.** Answers may vary. Samples are below.

    **a.** You can change the order of the factors without changing the product.

    **b.** You can group the factors in any way without changing the product.

**3.**

| +  | 6  | 2  | 3  | 8  | 0 | 5  | 7  | 9  | 12  | 4  |
|----|----|----|----|----|---|----|----|----|-----|----|
| 5  | 30 | 10 | 15 | 40 | 0 | 25 | 35 | 45 | 60  | 20 |
| 7  | 42 | 14 | 21 | 56 | 0 | 35 | 49 | 63 | 84  | 28 |
| 1  | 6  | 2  | 3  | 8  | 0 | 5  | 7  | 9  | 12  | 4  |
| 0  | 0  | 0  | 0  | 0  | 0 | 0  | 0  | 0  | 0   | 0  |
| 6  | 36 | 12 | 18 | 48 | 0 | 30 | 42 | 54 | 72  | 24 |
| 2  | 12 | 4  | 6  | 16 | 0 | 10 | 14 | 18 | 24  | 8  |
| 3  | 18 | 6  | 9  | 24 | 0 | 15 | 21 | 27 | 36  | 12 |
| 8  | 48 | 16 | 24 | 64 | 0 | 40 | 56 | 72 | 96  | 32 |
| 4  | 24 | 8  | 12 | 32 | 0 | 20 | 28 | 36 | 48  | 16 |
| 9  | 54 | 18 | 27 | 72 | 0 | 45 | 63 | 81 | 108 | 36 |

**5.**   $\begin{array}{r} 32 \\ \times\ 3 \\ \hline 96 \end{array}$

**7.**   $\begin{array}{r} 203 \\ \times\ 3 \\ \hline 609 \end{array}$

**9.**   $\begin{array}{r} 6102 \\ \times\ 3 \\ \hline 18,306 \end{array}$

**11.**   20,103
          ×     4
          ‾‾‾‾‾‾‾‾
          80,412

**13.**   14
          × 5
          ‾‾‾
          70

**15.**   87
          × 6
          ‾‾‾
          522

**17.**   326
          ×   5
          ‾‾‾‾‾
          1630

**19.**   2036
          ×    6
          ‾‾‾‾‾‾
          12,216

**21.**   12,526
          ×     8
          ‾‾‾‾‾‾‾
          100,208

**23.**   235,702
          ×      4
          ‾‾‾‾‾‾‾‾
          942,808

**25.**   156
          × 10
          ‾‾‾‾
          1560

**27.**   27,158
          ×     100
          ‾‾‾‾‾‾‾
          2,715,800

**29.**   482
          × 1000
          ‾‾‾‾‾‾
          482,000

**31.**   37,256
          ×   10,000
          ‾‾‾‾‾‾‾‾‾
          372,560,000

**33.**   423
          × 20
          ‾‾‾‾
          8460

**35.**   2120
          × 30
          ‾‾‾‾
          63,600

**37.**   14,000
          ×    4,000
          ‾‾‾‾‾‾‾‾
          56,000,000

**39.**   2103
          ×   32
          ‾‾‾‾‾
          4 206
          63 09
          ‾‾‾‾‾
          67,296

**41.**   146
          × 54
          ‾‾‾‾
          584
          730
          ‾‾‾
          7884

**43.**   89
          × 64
          ‾‾‾‾
          356
          534
          ‾‾‾‾
          5696

**45.**   607
          ×  25
          ‾‾‾‾‾
          3 035
          12 14
          ‾‾‾‾‾
          15,175

**47.**   569
          ×  73
          ‾‾‾‾‾
          1 707
          39 83
          ‾‾‾‾‾
          41,437

**49.**   912
          ×  76
          ‾‾‾‾‾
          5 472
          63 84
          ‾‾‾‾‾
          69,312

**51.**
$$
\begin{array}{r}
5123 \\
\times\ \ \ \ 29 \\
\hline
46\ 107 \\
102\ 46\ \ \\
\hline
148{,}567
\end{array}
$$

**53.**
$$
\begin{array}{r}
9053 \\
\times\ \ \ \ 91 \\
\hline
9\ 053 \\
814\ 77\ \ \\
\hline
823{,}823
\end{array}
$$

**55.**
$$
\begin{array}{r}
5536 \\
\times\ \ \ 224 \\
\hline
22\ 144 \\
110\ 72\ \ \\
1\ 107\ 2\ \ \ \\
\hline
1{,}240{,}064
\end{array}
$$

**57.**
$$
\begin{array}{r}
678 \\
\times\ \ \ 132 \\
\hline
1\ 356 \\
20\ 34\ \ \\
67\ 8\ \ \ \\
\hline
89{,}496
\end{array}
$$

**59.**
$$
\begin{array}{r}
2076 \\
\times\ \ \ 105 \\
\hline
10\ 380 \\
00\ 00\ \ \\
207\ 6\ \ \ \\
\hline
217{,}980
\end{array}
$$

**61.**
$$
\begin{array}{r}
3561 \\
\times\ \ \ 403 \\
\hline
10\ 683 \\
00\ 00\ \ \\
1\ 424\ 4\ \ \ \\
\hline
1{,}435{,}083
\end{array}
$$

**63.**
$$
\begin{array}{r}
6035 \\
\times\ \ 5006 \\
\hline
36\ 210 \\
30\ 175\ \ \ \ \\
\hline
30{,}211{,}210
\end{array}
$$

**65.**
$$
\begin{array}{r}
260 \\
\times\ \ \ \ 40 \\
\hline
10{,}400
\end{array}
$$

**67.**
$$
\begin{array}{r}
403 \\
\times\ \ 200 \\
\hline
80{,}600
\end{array}
$$

**69.** $7 \cdot 2 \cdot 5 = 7 \cdot 10 = 70$

**71.** $11 \cdot 7 \cdot 4 = 77 \cdot 4 = 308$

**73.** $10 \cdot 7 \cdot 10 = 700$

**75.**
$$
\begin{aligned}
12 \cdot 3 \cdot 5 \cdot 2 &= 12 \cdot 3 \cdot 10 \\
&= 36 \cdot 10 \\
&= 360
\end{aligned}
$$

**77.** $x = 8 \cdot 7 \cdot 6 \cdot 0$
$x = 0$

**79.**
$$
\begin{array}{r}
48 \\
\times\ 67 \\
\hline
336 \\
288\ \ \\
\hline
3216
\end{array}
$$

3216 square feet

**81.**
$$
\begin{array}{r}
12 \\
\times 14 \\
\hline
48 \\
12\ \ \\
\hline
168
\end{array}
\qquad
\begin{array}{r}
9 \\
\times 3 \\
\hline
27
\end{array}
\qquad
\begin{array}{r}
168 \\
+\ 27 \\
\hline
195
\end{array}
\text{ square feet}
$$

**83.**
$$
\begin{array}{r}
240 \\
\times\ \ 5 \\
\hline
1200
\end{array}
$$

$1200

**85.**
$$
\begin{array}{r}
266 \\
\times 12 \\
\hline
532 \\
266\ \ \\
\hline
3192
\end{array}
$$

$3192

**87.**
$$\begin{array}{r} 34 \\ \times 18 \\ \hline 272 \\ 34 \\ \hline 612 \end{array}$$

612 miles

**89.**
$$\begin{array}{r} 95 \\ \times 25 \\ \hline 475 \\ 190 \\ \hline 2375 \end{array}$$

2375 books

**91.**
$$\begin{array}{r} 6{,}890{,}000 \\ + \qquad 1070 \\ \hline \$7{,}372{,}300{,}000 \end{array}$$

**93.**

$$\begin{array}{r} 18 \\ \times\,2 \\ \hline 36 \end{array} \qquad \begin{array}{r} 26 \\ \times\,0 \\ \hline 0 \end{array} \qquad \begin{array}{r} 54 \\ \times\,3 \\ \hline 162 \end{array} \qquad \begin{array}{r} 36 \\ 0 \\ +\,162 \\ \hline 198 \end{array} \text{ black paws}$$

**95.**

$$\begin{array}{r} 18 \\ \times\,2 \\ \hline 36 \end{array} \qquad \begin{array}{r} 26 \\ \times\,1 \\ \hline 26 \end{array} \qquad \begin{array}{r} 54 \\ \times\,0 \\ \hline 0 \end{array} \qquad \begin{array}{r} 36 \\ 26 \\ +\,0 \\ \hline 62 \end{array} \text{ black ears}$$

**97.** $5\,(x) = 40$
$\quad\;\, 5\,(8) = 40$
$\qquad\quad x = 8$

**99.** $72 = 8\,(x)$
$\quad\;\, 72 = 8\,(9)$
$\qquad\;\; x = 9$

**101.** No, it would not always be true. In our number system $62 = 60 + 2$. But in roman numerals, $IV \neq I \cdot V$ and $IV \neq I + V$. The digit system in roman numerals involves subtraction.
$(XII) / (IV) \neq (XII / I) + (XII / V)$

## Cumulative Review Problems

**103.**
$$\begin{array}{r} {\scriptstyle 2\ 13107\ 14} \\ 34{,}084 \\ -27{,}328 \\ \hline 6{,}756 \end{array}$$

**105.**
$$\begin{array}{r} 12 \\ 2 \\ +\,3 \\ \hline \$17 \end{array} \qquad \begin{array}{r} {\scriptstyle 4\ 16} \\ 156 \\ -\;17 \\ \hline \$139 \end{array}$$

## 1.5   Exercises

**1. a.** When you divide a nonzero number by itself, the result is 1.

**b.** When you divide a number by 1, the result is that number.

**c.** When you divide zero by a nonzero number, the result is zero.

**d.** You cannot divide a number by 0.

**3.** $6\overline{)42}$ → $7$

**5.** $8\overline{)24}$ → $3$

**7.** $8\overline{)40}$ → $5$

**9.** $9\overline{)36}$ → $4$

**11.** $7\overline{)21}$ → $3$

**13.** $8\overline{)56}$ → $7$

**15.** $7\overline{)63}$ → $9$

**17.**  $8\overline{)72}$  with quotient $9$

**19.**  $9\overline{)63}$  with quotient $7$

**21.**  $6\overline{)24}$  with quotient $4$

**23.**  $1\overline{)7}$  with quotient $7$

**25.**  $6\overline{)0}$  with quotient $0$

**27.**  $54 \div 6 = 9$

**29.**  $6 \div 6 = 1$

**31.**  $29 \div 6$

$$
\begin{array}{r}
4 \text{ R5} \\
6\overline{)29} \\
\underline{24} \\
5
\end{array}
\qquad
\begin{array}{r}
\text{Check} \\
4 \\
\times\ 6 \\
\hline
24 \\
+\ 5 \\
\hline
29
\end{array}
$$

**33.**  $76 \div 8$

$$
\begin{array}{r}
9 \text{ R4} \\
8\overline{)76} \\
\underline{72} \\
4
\end{array}
\qquad
\begin{array}{r}
\text{Check} \\
9 \\
\times\ 8 \\
\hline
72 \\
+\ 4 \\
\hline
76
\end{array}
$$

**35.**  $128 \div 5$

$$
\begin{array}{r}
25 \text{ R3} \\
5\overline{)128} \\
\underline{10} \\
28 \\
\underline{25} \\
3
\end{array}
\qquad
\begin{array}{r}
\text{Check} \\
25 \\
\times\ 5 \\
\hline
125 \\
+\ 3 \\
\hline
128
\end{array}
$$

**37.**  $9\overline{)196}$

$$
\begin{array}{r}
21 \text{ R7} \\
9\overline{)196} \\
\underline{18} \\
16 \\
\underline{9} \\
7
\end{array}
\qquad
\begin{array}{r}
\text{Check} \\
21 \\
\times\ 8 \\
\hline
189 \\
+\ 7 \\
\hline
196
\end{array}
$$

**39.**  $9\overline{)288}$

$$
\begin{array}{r}
32 \\
9\overline{)288} \\
\underline{27} \\
18 \\
\underline{8} \\
0
\end{array}
\qquad
\begin{array}{r}
\text{Check} \\
32 \\
\times\ 9 \\
\hline
288
\end{array}
$$

**41.**  $5\overline{)185}$

$$
\begin{array}{r}
37 \\
5\overline{)185} \\
\underline{15} \\
35 \\
\underline{35} \\
0
\end{array}
\qquad
\begin{array}{r}
\text{Check} \\
37 \\
\times\ 5 \\
\hline
185
\end{array}
$$

**43.**  $4\overline{)1289}$

$$
\begin{array}{r}
322 \text{ R1} \\
4\overline{)1289} \\
\underline{12} \\
8 \\
\underline{8} \\
9 \\
\underline{8} \\
1
\end{array}
\qquad
\begin{array}{r}
\text{Check} \\
322 \\
\times\ 4 \\
\hline
1288 \\
+\ 1 \\
\hline
1289
\end{array}
$$

**45.**

```
    127 R1
 6)763        Check
   6
  ──            127
  16          ×   6
  12          ─────
  ──            762
  43          +   1
  42          ─────
  ──            763
   1
```

**47.**

```
    753
 8)6024       Check
  56
  ──            753
  42          ×   8
  40          ─────
  ──           6024
  24
  24
  ──
   0
```

**49.**

```
    1122 R1
 3)3367       Check
  3
  ─            1122
  3          ×   3
  3          ─────
  ─            3366
  6          +   1
  6          ─────
  ─            3367
   7
   6
   ─
   1
```

**51.**

```
   1757 R5
 7)12304      Check
   7
  ──           1757
  53         ×    7
  49         ──────
  ──          12,299
  40         +    5
  35         ──────
  ──          12,304
  54
  49
  ──
   5
```

**53.**

```
   1601 R5
 8)12813      Check
   8
  ──           1601
  48         ×    8
  48         ──────
  ──          12,808
   1         +    5
   0         ──────
  ──          12,813
  13
   8
   ─
   5
```

**55.**  $185 \div 36$

```
    5 R5
36)185
   180
   ───
     5
```

**57.**  $267 \div 52$

```
     5 R7
52)267
   260
   ───
     7
```

**59.**  $427 \div 61$

```
     7
61)427
   427
   ───
     0
```

**61.**

```
    160 R10
12)1930
   12
   ──
   73
   72
   ──
   10
    0
   ──
   10
```

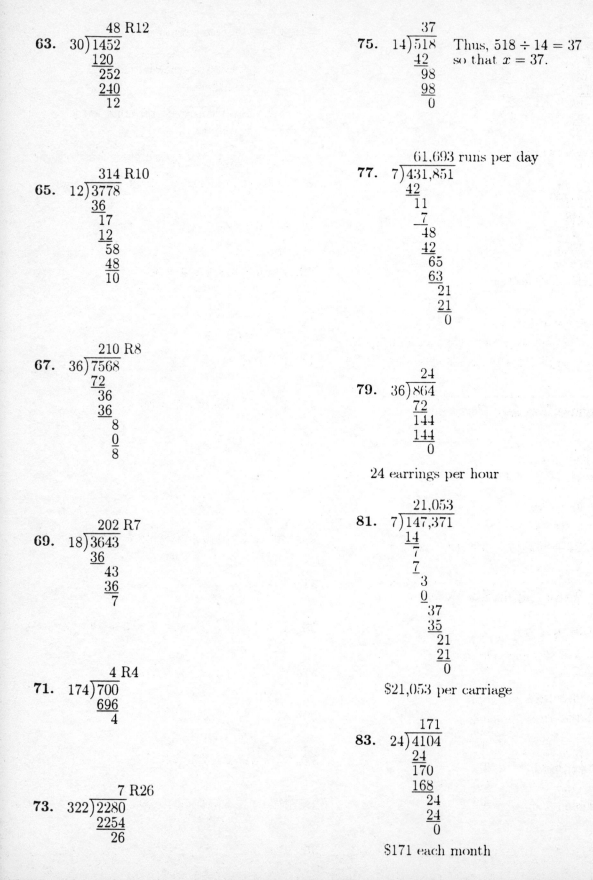

```
         48 R12
63.  30)1452
         120
         252
         240
          12
```

```
         314 R10
65.  12)3778
         36
         17
         12
         58
         48
         10
```

```
         210 R8
67.  36)7568
         72
         36
         36
          8
          0
          8
```

```
         202 R7
69.  18)3643
         36
         43
         36
          7
```

```
          4 R4
71.  174)700
          696
            4
```

```
           7 R26
73.  322)2280
          2254
            26
```

```
          37
75.  14)518      Thus, 518 ÷ 14 = 37
         42       so that x = 37.
         98
         98
          0
```

```
        61,693 runs per day
77.  7)431,851
        42
        11
         7
        48
        42
        65
        63
         21
         21
          0
```

```
         24
79.  36)864
         72
        144
        144
          0
```

24 earrings per hour

```
        21,053
81.  7)147,371
        14
         7
         7
          3
          0
         37
         35
         21
         21
          0
```

$21,053 per carriage

```
         171
83.  24)4104
         24
        170
        168
          24
          24
           0
```

$171 each month

**85.**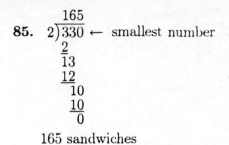

165 sandwiches

**87.**
$$34\overline{)2652}$$
$$\phantom{34)}238$$
$$\phantom{34)}272$$
$$\phantom{34)}272$$
$$\phantom{34)}\phantom{27}0$$

with quotient 78

Length is 78 feet

**89.** If $a \div b = b \div a$
$a$ and $b$ must represent the same number.
For example, if $a = 12$, then $b = 12$.

## Cumulative Review Problems

**91.**
$$\begin{array}{r} 128 \\ \times\ 43 \\ \hline 384 \\ 512\phantom{0} \\ \hline 5504 \end{array}$$

**93.** $316{,}214 + 89{,}981 = 406{,}195$

## Putting Your Skills To Work

**1.** $248 - 156 = \$92$
$92 \div 4 = \$23$ per person

**3.** $70 - 23 = \$47$ per night
$47 \times 4 = \$188$ for package

## 1.6    Exercises

**1.** $5^3$ means $5 \times 5 \times 5 = 125$

**3.** base

**5.** To insure consistency we

1. perform operations inside parentheses

2. simplify any expressions with exponents

3. multiply or divide from left to right

4. add or subtract from left to right

**7.** $6 \times 6 \times 6 \times 6 = 6^4$

**9.** $4 \times 4 \times 4 \times 4 \times 4 = 4^5$

**11.** $8 \times 8 \times 8 \times 8 = 8^4$

**13.** $9 = 9^1$

**15.** $2^4 = 2 \times 2 \times 2 \times 2 = 16$

**17.** $4^2 = 4 \times 4 = 16$

**19.** $6^3 = 6 \times 6 \times 6 = 216$

**21.** $10^4 = 10 \times 10 \times 10 \times 10 = 10{,}000$

**23.** $1^{17} = 1$

**25.** $2^6 = 2 \times 2 \times 2 \times 2 \times 2 \times 2 = 64$

**27.** $3^5 = 3 \times 3 \times 3 \times 3 \times 3 = 243$

**29.** $15^2 = 15 \times 15 = 225$

**31.** $7^3 = 7 \times 7 \times 7 = 343$

**33.** $4^4 = 4 \times 4 \times 4 \times 4 = 256$

**35.** $9^0 = 1$

**37.** $25^2 = 25 \times 25 = 625$

**39.** $10^6 = 10 \times 10 \times 10 \times 10 \times 10 \times 10$
$\phantom{10^6} = 1,000,000$

**41.** $4^5 = 4 \times 4 \times 4 \times 4 \times 4 = 1024$

**43.** $9^1 = 9$

**45.** $7^4 = 7 \times 7 \times 7 \times 7 = 2401$

**47.** $2^4 + 1^8 = 2 \times 2 \times 2 \times 2 + 1$
$\phantom{2^4 + 1^8} = 16 + 1$
$\phantom{2^4 + 1^8} = 17$

**49.** $6^3 + 3^2 = 6 \times 6 \times 6 + 3 \times 3$
$\phantom{6^3 + 3^2} = 216 + 9$
$\phantom{6^3 + 3^2} = 225$

**51.** $8^3 + 8 = 8 \times 8 \times 8 + 8$
$\phantom{8^3 + 8} = 512 + 8$
$\phantom{8^3 + 8} = 520$

**53.** $5 \times 6 - 3 + 10 = 30 - 3 + 10$
$\phantom{5 \times 6 - 3 + 10} = 27 + 10$
$\phantom{5 \times 6 - 3 + 10} = 37$

**55.** $3 \times 9 - 10 \div 2 = 27 - 5$
$\phantom{3 \times 9 - 10 \div 2} = 22$

**57.** $48 \div 2^3 + 4 = 48 \div 8 + 4$
$\phantom{48 \div 2^3 + 4} = 6 + 4$
$\phantom{48 \div 2^3 + 4} = 10$

**59.** $7 \times 3^2 + 4 - 8 = 7 \times 9 + 4 - 8$
$\phantom{7 \times 3^2 + 4 - 8} = 63 + 4 - 8$
$\phantom{7 \times 3^2 + 4 - 8} = 67 - 8$
$\phantom{7 \times 3^2 + 4 - 8} = 59$

**61.** $4 \times 9^2 - 4 \times (7 - 2)$
$\phantom{4 \times 9^2} = 4 \times 81 - 4 \times 5$
$\phantom{4 \times 9^2} = 324 - 20$
$\phantom{4 \times 9^2} = 304$

**63.** $(400 \div 20) \div 20 = 20 \div 20 = 1$

**65.** $950 \div (25 \div 5) = 950 \div 5 = 190$

**67.** $(12)(5) - (12 + 5) = (12)(5) - 17$
$\phantom{(12)(5) - (12 + 5)} = 60 - 17$
$\phantom{(12)(5) - (12 + 5)} = 43$

**69.** $3^2 + 4^2 \div 2^2 = 9 + 16 \div 4$
$\phantom{3^2 + 4^2 \div 2^2} = 9 + 4$
$\phantom{3^2 + 4^2 \div 2^2} = 13$

**71.** $(6)(7) - (12 - 8) \div 4 = 6(7) - 4 \div 4$
$\phantom{(6)(7) - (12 - 8) \div 4} = 42 - 1$
$\phantom{(6)(7) - (12 - 8) \div 4} = 41$

**73.** $100 - 3^2 \times 4 = 100 - 9 \times 4$
$\phantom{100 - 3^2 \times 4} = 100 - 36$
$\phantom{100 - 3^2 \times 4} = 64$

**75.** $5^2 + 2^2 + 3^3 = 25 + 4 + 27$
$\phantom{5^2 + 2^2 + 3^3} = 56$

**77.** $72 \div 9 \times 3 \times 1 \div 2$
$\phantom{72} = 8 \times 3 \times 1 \div 2$
$\phantom{72} = 24 \times 1 \div 2$
$\phantom{72} = 24 \div 2$
$\phantom{72} = 12$

**79.** $12^2 - 6 \times 3 \times 4 \times 0$
$\phantom{12^2} = 144 - 6 \times 3 \times 4 \times 0$
$\phantom{12^2} = 144 - 72 \times 0$
$\phantom{12^2} = 144 - 0$
$\phantom{12^2} = 144$

**81.** $4^2 \times 6 \div 3 = 16 \times 6 \div 3$
$$= 96 \div 3$$
$$= 32$$

**83.** $5 + 6 \times 2 \div 4 - 1 = 5 + 12 \div 4 - 1$
$$= 5 + 3 - 1$$
$$= 7$$

**85.** $3 + 3^2 \times 6 + 4 = 3 + 9 \times 6 + 4$
$$= 3 + 54 + 4$$
$$= 61$$

**87.** $12 \div 2 \times 3 - 2^4 = 12 \div 2 \times 3 - 16$
$$= 6 \times 3 - 16$$
$$= 18 - 16$$
$$= 2$$

**89.** $3^2 \times 6 \div 9 + 4 \times 3$
$$= 9 \times 6 \div 9 + 4 \times 3$$
$$= 54 \div 9 + 12$$
$$= 6 + 12$$
$$= 18$$

**91.** $4(10 - 6)^3 \div (2 + 6) = 4 \times 4^3 \div 8$
$$= 4 \times 64 \div 8$$
$$= 256 \div 8$$
$$= 32$$

**93.** $1200 - 2^3(3) \div 6$
$$= 1200 - 8(3) \div 6$$
$$= 1200 - 24 \div 6$$
$$= 1200 - 4$$
$$= 1196$$

**95.** $250 \div 5 + 20 - 3^2$
$$= 250 \div 5 + 20 - 9$$
$$= 50 + 20 - 9$$
$$= 61$$

**97.** $20 \div 4 \times 3 - 6 + 2 \times (1 + 4)$
$$= 20 \div 4 \times 3 - 6 + 2 \times 5$$
$$= 5 \times 3 - 6 + 10$$
$$= 15 - 6 + 10$$
$$= 19$$

**99.** $2 \times 3 + (11 + 5) - 4 \times 5$
$$= 2 \times 3 + 16 - 4 \times 5$$
$$= 6 + 16 - 20$$
$$= 2$$

**101.** 23 hours, 56 minute, 4 seconds
$$= 23(60)(60) + 56(60) + 4$$
$$= 82,800 + 3360 + 4$$
$$= 86,164 \text{ seconds}$$

## Cumulative Review Problems

**103.** 156,312
$$= 100,000 + 50,000 + 6000$$
$$+ 300 + 10 + 2$$

**105.** 261,763,002
Two hundred sixty-one million, seven hundred sixty-three thousand, two

## 1.7   Exercises

**1.** Locate the rounding place. If the digit to the right of the rounding place is 5 or greater than 5, round up. If the digit to the right of the rounding place is less than 5, round down.
Examples will vary.

**3.** 8$\underline{3}$ rounds to 80 since 3 is less than 5.

**5.** 6$\underline{5}$ rounds to 70 since 5 is equal to 5.

**7.** 9$\underline{2}$ rounds to 90 since 2 is less than 5.

**9.** 52<u>6</u> rounds to 530 since 6 is greater than 5.

**11.** 167<u>2</u> rounds to 1670 since 2 is equal to 5.

**13.** 2<u>4</u>7 rounds to 200 since 4 is less than 5.

**15.** 27<u>8</u>1 rounds to 2800 since 8 is greater than 5.

**17.** 76<u>9</u>2 rounds to 7700 since 9 is greater than 5.

**19.** 7<u>6</u>21 rounds to 8000 since 6 is greater than 5.

**21.** 1<u>6</u>72 rounds to 2000 since 6 is greater than 5.

**23.** 27,<u>8</u>63 rounds to 28,000 since 8 is greater than 5.

**25.** 8<u>3</u>2,400 rounds to 800,000 since 3 is less than 5.

**27.** 15,<u>1</u>69,873 rounds to 15,000,000 stars since 1 is less than 5.

**29. a.** 163,<u>2</u>98 rounds to 163,000 since 2 is less than 5.
   **b.** 163,2<u>9</u>8 rounds to 163,300 since 9 is greater than 5.

**31. a.** 3,7<u>0</u>5,392 rounds to 3,700,000 square miles since 0 is less than 5. 9,5<u>9</u>6,960 rounds to 9,600,000 square kilometers since 9 is greater than 5.

**b.** 3,70<u>5</u>,392 rounds to 3,710,000 square miles since 5 is equal to 5. 9,59<u>6</u>,960 rounds to 9,600,000 square kilometers since 6 is greater than 5.

**33.**
$$\begin{array}{r} 600 \\ 300 \\ +\ 100 \\ \hline 1000 \end{array}$$

**35.**
$$\begin{array}{r} 30 \\ 80 \\ 60 \\ +\ 30 \\ \hline 200 \end{array}$$

**37.**
$$\begin{array}{r} 100,000 \\ 50,000 \\ +\ 100,000 \\ \hline 250,000 \end{array}$$

**39.**
$$\begin{array}{r} 600,000 \\ -\ 100,000 \\ \hline 500,000 \end{array}$$

**41.**
$$\begin{array}{r} 80,000,000 \\ -\ 60,000,000 \\ \hline 20,000,000 \end{array}$$

**43.**
$$\begin{array}{r} 30,000,000 \\ -\ 20,000,000 \\ \hline 10,000,000 \end{array}$$

**45.**
$$\begin{array}{r} 60 \\ \times\ 50 \\ \hline 3000 \end{array}$$

**47.**
$$\begin{array}{r} 1000 \\ +\ 8 \\ \hline 8000 \end{array}$$

**49.**
$$\begin{array}{r} 600,000 \\ \times\ \ \ \ \ 300 \\ \hline 180,000,000 \end{array}$$

**51.** $20\overline{)20,000}$ gives $1,000$

**53.** $40\overline{)200,000}$ gives $5,000$

**55.** $880\overline{)4,000,000}$ gives $5,000$

**57.**
$$\begin{array}{r} 400 \\ 500 \\ 900 \\ +\ 200 \\ \hline 2000 \ \ \text{Incorrect} \end{array}$$

**59.**
$$\begin{array}{r} 100,000 \\ 50,000 \\ +\ 40,000 \\ \hline 190,000 \ \ \text{Incorrect} \end{array}$$

**61.**
$$\begin{array}{r} 300,000 \\ -\ 100,000 \\ \hline 200,000 \ \ \text{Correct} \end{array}$$

**63.**
$$\begin{array}{r} 80,000,000 \\ -\ 50,000,000 \\ \hline 30,000,000 \ \ \text{Incorrect} \end{array}$$

**65.**
$$\begin{array}{r} 200 \\ \times\ 20 \\ \hline 4000 \ \ \text{Incorrect} \end{array}$$

**67.**
$$\begin{array}{r} 6000 \\ \times\ \ \ \ 70 \\ \hline 420,000 \ \ \text{Correct} \end{array}$$

**69.** $80\overline{)400,000}$ gives $5000$ Correct

**71.** $400\overline{)200,000}$ gives $500$ Correct

**73.**
$$\begin{array}{r} 60 \\ \times\ 40 \\ \hline 2400 \ \ \text{square feet} \end{array}$$

**75.**
$$\begin{array}{r} 20\ 000 \\ 20\ 000 \\ 40\ 000 \\ +\ \ 60\ 000 \\ \hline \$140,000 \end{array}$$

**77.**
$$\begin{array}{r} 300 \\ \times\ \ \ 100 \\ \hline 30,000 \ \ \text{pizzas} \end{array}$$

**79.**
$$\begin{array}{r} 660,000\ 000 \\ -\ 460,000\ 000 \\ \hline 200,000,000 \ \ \text{passengers} \end{array}$$

**81.**
$$\begin{array}{r} 590,000 \\ -\ 270,000 \\ \hline 320,000 \ \ \text{square miles} \end{array}$$

**83. a.** $20,000\overline{)8,000,000,000}$ gives $400,000$ hours

**b.** $20\overline{)400,000}$ gives $20,000$ days

## Cumulative Review Problems

**85.** $26 \times 3 + 20 \div 4 = 78 + 5$
$$= 83$$

**87.** $3 \times (16 \div 4) + 8 \times 2 = 3 \times 4 + 8 \times 2$
$$= 12 + 16$$
$$= 28$$

## 1.8  Exercises

**1.**
```
  24,111
     327
+   793
  25,231
```
There is a total of 25,231

```
   793
 − 327
   466  more people
```

**3.**
```
  42,318
− 37,650
   4,668
```
She went over budget by $4668.

**5.**
```
    250
×    15
   3750
```
She ordered 3750 hors d'oeuvres.

**7.** 96 ÷ 16
```
       6
16) 96
    96
     0
```
The unit cost is 6¢ per ounce.

**9.** 322 ÷ 14
```
      23
14) 322
    28
    42
    42
     0
```
It cost $23 each time.

**11.**
```
       5
60) 300
    300
      0
```
It will take him 5 hours to fill the order.

**13.** 480 ÷ 60 = 8
100,000 × 8 = 800,000
800,000 people will be born.

**15.**
```
    7356
    3257
    4777
+   4992
   20,382
```
The gross income was $20,382

**17.**
```
  250      35      75      22
×   2    ×  2    ×  2    ×  3
  500      70     150      66
```
```
   500
    70
   150
+   66
   786
```
The total cost was $786.

**19.**
```
    61       223
   385        29
   945        98
   732    +  435
+  144       785
  2267
```
```
   2267
 −  785
   1482
```
The balance is $1482.

**21.**
```
     250       85
×     85    +  57
  21,250     4845
```
```
   21,250
 −  4845
   16,405
```
Her profit is $16,405.

**23.** $(15{,}276 - 14{,}926) \div 14$

$= 350 \div 14$

$= 25$

The car achieves 25 miles per gallon.

**25.** oaks $= 3 \times 18 = 54$

maples $= 2 \times 54 = 108$

pines $= 7 \times 108 = 756$

Total number of trees is

$$\begin{array}{r} 18 \\ 54 \\ 108 \\ + \ 756 \\ \hline 936 \ \text{trees} \end{array}$$

**27.** $53 + 44 + 21 = 118$ students

**29.** $(200 + 174) - 189$

$= 374 - 189$

185 students

**31.** 
$$\begin{array}{r} 33{,}867{,}000{,}000 \\ - \ 19{,}133{,}000{,}000 \\ \hline \$14{,}734{,}000{,}000 \end{array}$$

**33.** 
$$\begin{array}{r} 51{,}735{,}000{,}000 \\ - \ 42{,}561{,}000{,}000 \\ \hline 9{,}174{,}000{,}000 \end{array}$$

For year 2005,

$$\begin{array}{r} 51{,}735{,}000{,}000 \\ + \ 9{,}174{,}000{,}000 \\ \hline \$60{,}909{,}000{,}000 \end{array}$$

## Cumulative Review Problems

**35.** $7^3 = 7 \times 7 \times 7 = 343$

**37.** 
$$\begin{array}{r} 126 \\ \times \ 38 \\ \hline 1008 \\ 378 \\ \hline 4788 \end{array}$$

## Chapter 1 Review Problems

1. Three hundred seventy-six

2. Five thousand eighty-two

3. One hundred nine thousand, two hundred seventy-six

4. Four hundred twenty-three million, five hundred seventy-six thousand, fifty-five.

5. $4364 = 4000 + 300 + 60 + 4$

6. $27{,}986 = 20{,}000 + 7000 + 900 + 80 + 6$

7. $42{,}166{,}037 = 40{,}000{,}000 + 2{,}000{,}000 + 100{,}000 + 60{,}000 + 6{,}000 + 30 + 7$

8. $1{,}305{,}128 = 1{,}000{,}000 + 300{,}000 + 5000 + 100 + 20 + 8$

9. 924

10. 6095

11. 1,328,828

12. 45,092,651

13. 
$$\begin{array}{r} 76 \\ + \ 39 \\ \hline 115 \end{array}$$

14. 
$$\begin{array}{r} 36 \\ + \ 94 \\ \hline 130 \end{array}$$

15. 
$$\begin{array}{r} 127 \\ + \ 563 \\ \hline 690 \end{array}$$

16.
```
    12
    28
    34
  + 76
   150
```

17.
```
   123
    61
     9
    84
 + 123
   400
```

18.
```
   937
   405
 + 256
  1598
```

19.
```
   125
   364
 + 980
  1469
```

20.
```
   28,364
 + 97,059
  125,423
```

21.
```
    1356
    2892
     561
      89
 +  9805
   14,703
```

22.
```
      26
     503
     935
   1 257
 + 7 861
   10,582
```

23.
```
    36
  − 19
    17
```

24.
```
    54
  − 48
     6
```

25.
```
   126
  − 99
    27
```

26.
```
   543
 − 372
   171
```

27.
```
   1296
 − 1137
    159
```

28.
```
   9821
 − 4993
   4828
```

29.
```
   201,010
 − 137,864
    63,146
```

30.
```
   101,300
 −  98,274
     3,026
```

31.
```
   1,986,312
 − 1,761,555
     224,757
```

32.
```
   7,216,03
 − 5,985,312
   1,230,691
```

33.
```
    57
  ×  2
   114
```

34.
```
    12
  ×  3
    36
```

35.
```
    36
  ×  0
     0
```

**36.**
$$\begin{array}{r} 24 \\ \times\ 1 \\ \hline 24 \end{array}$$

**37.** $1 \times 3 \times 6 = 3 \times 6 = 18$

**38.** $2 \times 4 \times 8 = 8 \times 8 = 64$

**39.** $5 \times 7 \times 3 = 35 \times 3 = 105$

**40.** $4 \times 6 \times 5 = 24 \times 5 = 120$

**41.** $\begin{aligned} 8 \times 1 \times 9 \times 2 &= 8 \times 9 \times 2 \\ &= 72 \times 2 \\ &= 144 \end{aligned}$

**42.** $7 \times 6 \times 0 \times 4 = 0$

**43.** $\begin{aligned} 3 \cdot 4 \cdot 2 \cdot 2 \cdot 5 &= 12 \cdot 2 \cdot 2 \cdot 5 \\ &= 24 \cdot 2 \cdot 5 \\ &= 48 \cdot 5 \\ &= 240 \end{aligned}$

**44.** $\begin{aligned} 1 \cdot 2 \cdot 7 \cdot 3 \cdot 4 &= 2 \cdot 7 \cdot 3 \cdot 4 \\ &= 14 \cdot 3 \cdot 4 \\ &= 42 \cdot 4 = 168 \end{aligned}$

**45.** $26,121 \times 100 = 2,612,100$

**46.** $84,312 \times 1000 = 84,312,000$

**47.** $832 \times 100,000 = 83,200,000$

**48.** $563 \times 1,000,000 = 563,000,000$

**49.**
$$\begin{array}{r} 58 \\ \times\ 32 \\ \hline 116 \\ 174\ \ \\ \hline 1856 \end{array}$$

**50.**
$$\begin{array}{r} 36 \\ \times\ 24 \\ \hline 144 \\ 72\ \ \\ \hline 864 \end{array}$$

**51.**
$$\begin{array}{r} 150 \\ \times\ 27 \\ \hline 1050 \\ 300\ \ \\ \hline 4050 \end{array}$$

**52.**
$$\begin{array}{r} 360 \\ \times\ 38 \\ \hline 2\ 880 \\ 10\ 80\ \ \\ \hline 13,680 \end{array}$$

**53.**
$$\begin{array}{r} 709 \\ \times\ 36 \\ \hline 4\ 254 \\ 21\ 27\ \ \\ \hline 25,524 \end{array}$$

**54.**
$$\begin{array}{r} 502 \\ \times\ 48 \\ \hline 4\ 016 \\ 20\ 08\ \ \\ \hline 24,096 \end{array}$$

**55.**
$$\begin{array}{r} 123 \\ \times\ 714 \\ \hline 492 \\ 1\ 23\ \ \\ 86\ 1\ \ \ \\ \hline 87,822 \end{array}$$

**56.**
$$\begin{array}{r} 431 \\ \times\ 623 \\ \hline 1\ 293 \\ 8\ 62\ \ \\ 258\ 6\ \ \ \\ \hline 268,513 \end{array}$$

**57.**
$$\begin{array}{r} 1782 \\ \times\ 305 \\ \hline 8\ 910 \\ 534\ 60\ \ \\ \hline 543,510 \end{array}$$

**58.**
$$\begin{array}{r} 2057 \\ \times\ \ \ 124 \\ \hline 8\ 228 \\ 41\ 14 \\ 205\ 7 \\ \hline 255{,}068 \end{array}$$

**59.**
$$\begin{array}{r} 300 \\ \times\ \ \ 500 \\ \hline 150{,}000 \end{array}$$

**60.**
$$\begin{array}{r} 400 \\ \times\ \ \ 600 \\ \hline 240{,}000 \end{array}$$

**61.**
$$\begin{array}{r} 1200 \\ \times\ \ \ 6000 \\ \hline 7{,}200{,}000 \end{array}$$

**62.**
$$\begin{array}{r} 2500 \\ \times\ \ \ 3000 \\ \hline 7{,}500{,}000 \end{array}$$

**63.**
$$\begin{array}{r} 100{,}000 \\ \times\ \ \ 20{,}000 \\ \hline 2{,}000{,}000{,}000 \end{array}$$

**64.**
$$\begin{array}{r} 300{,}000 \\ \times\ \ \ 40{,}000 \\ \hline 12{,}000{,}000{,}000 \end{array}$$

**65.** $20 \div 10 = 2$

**66.** $40 \div 8 = 5$

**67.** $70 \div 5 = 14$

**68.** $36 \div 9 = 4$

**69.** $0 \div 8 = 0$

**70.** $12 \div 1 = 12$

**71.** $7 \div 1 = 7$

**72.** $0 \div 5 = 0$

**73.** $\dfrac{49}{7} = 7$

**74.** $\dfrac{42}{6} = 7$

**75.** $\dfrac{5}{0}$ is not possible.

**76.** $\dfrac{24}{6} = 4$

**77.** $\dfrac{56}{8} = 7$

**78.** $\dfrac{48}{8} = 6$

**79.** $\dfrac{72}{9} = 8$

**80.** Not possible

**81.**
$$\begin{array}{r} 125 \\ 6\overline{)750} \\ \underline{6}\phantom{50} \\ 15\phantom{0} \\ \underline{12}\phantom{0} \\ 30 \\ \underline{30} \\ 0 \end{array}$$

**82.**
$$\begin{array}{r} 125 \\ 7\overline{)875} \\ \underline{2}\phantom{50} \\ 17\phantom{0} \\ \underline{14}\phantom{0} \\ 35 \\ \underline{35} \\ 0 \end{array}$$

**83.**
$$\begin{array}{r} 258 \\ 5\overline{)1290} \\ \underline{10}\phantom{90} \\ 29\phantom{0} \\ \underline{25}\phantom{0} \\ 40 \\ \underline{40} \\ 0 \end{array}$$

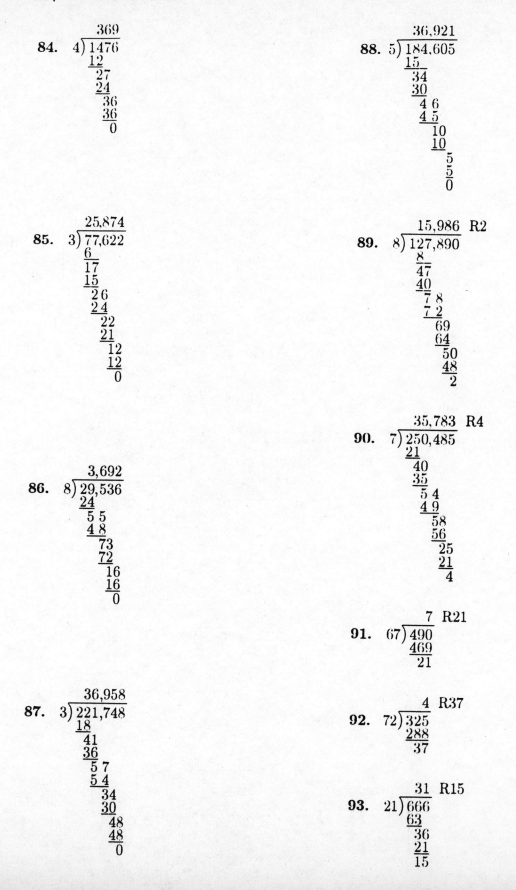

**84.**
$$
\begin{array}{r}
369 \\
4\overline{)1476} \\
12 \\
\overline{27} \\
24 \\
\overline{\phantom{0}36} \\
36 \\
\overline{\phantom{00}0}
\end{array}
$$

**85.**
$$
\begin{array}{r}
25{,}874 \\
3\overline{)77{,}622} \\
6 \\
\overline{17} \\
15 \\
\overline{26} \\
24 \\
\overline{22} \\
21 \\
\overline{12} \\
12 \\
\overline{0}
\end{array}
$$

**86.**
$$
\begin{array}{r}
3{,}692 \\
8\overline{)29{,}536} \\
24 \\
\overline{5\,5} \\
4\,8 \\
\overline{73} \\
72 \\
\overline{16} \\
16 \\
\overline{0}
\end{array}
$$

**87.**
$$
\begin{array}{r}
36{,}958 \\
3\overline{)221{,}748} \\
18 \\
\overline{41} \\
36 \\
\overline{5\,7} \\
5\,4 \\
\overline{34} \\
30 \\
\overline{48} \\
48 \\
\overline{0}
\end{array}
$$

**88.**
$$
\begin{array}{r}
36{,}921 \\
5\overline{)184{,}605} \\
15 \\
\overline{34} \\
30 \\
\overline{4\,6} \\
4\,5 \\
\overline{10} \\
10 \\
\overline{5} \\
5 \\
\overline{0}
\end{array}
$$

**89.**
$$
\begin{array}{r}
15{,}986 \ \text{R2} \\
8\overline{)127{,}890} \\
8 \\
\overline{47} \\
40 \\
\overline{7\,8} \\
7\,2 \\
\overline{69} \\
64 \\
\overline{50} \\
48 \\
\overline{2}
\end{array}
$$

**90.**
$$
\begin{array}{r}
35{,}783 \ \text{R4} \\
7\overline{)250{,}485} \\
21 \\
\overline{40} \\
35 \\
\overline{5\,4} \\
4\,9 \\
\overline{58} \\
56 \\
\overline{25} \\
21 \\
\overline{4}
\end{array}
$$

**91.**
$$
\begin{array}{r}
7 \ \text{R21} \\
67\overline{)490} \\
469 \\
\overline{21}
\end{array}
$$

**92.**
$$
\begin{array}{r}
4 \ \text{R37} \\
72\overline{)325} \\
288 \\
\overline{37}
\end{array}
$$

**93.**
$$
\begin{array}{r}
31 \ \text{R15} \\
21\overline{)666} \\
63 \\
\overline{36} \\
21 \\
\overline{15}
\end{array}
$$

**94.**
$$
\begin{array}{r}
14 \text{ R11} \\
22{\overline{)319}} \\
\underline{22}\phantom{0} \\
99 \\
\underline{88} \\
11
\end{array}
$$

**95.**
$$
\begin{array}{r}
38 \text{ R30} \\
68{\overline{)2614}} \\
\underline{204}\phantom{0} \\
574 \\
\underline{544} \\
30
\end{array}
$$

**96.**
$$
\begin{array}{r}
54 \text{ R38} \\
76{\overline{)4142}} \\
\underline{380}\phantom{0} \\
342 \\
\underline{304} \\
38
\end{array}
$$

**97.**
$$
\begin{array}{r}
95 \\
45{\overline{)4275}} \\
\underline{405}\phantom{0} \\
225 \\
\underline{225} \\
0
\end{array}
$$

**98.**
$$
\begin{array}{r}
258 \\
35{\overline{)9030}} \\
\underline{70}\phantom{00} \\
203 \\
\underline{175} \\
280 \\
\underline{280} \\
0
\end{array}
$$

**99.**
$$
\begin{array}{r}
54 \\
132{\overline{)7128}} \\
\underline{660}\phantom{0} \\
528 \\
\underline{528} \\
0
\end{array}
$$

**100.**
$$
\begin{array}{r}
19 \\
204{\overline{)3876}} \\
\underline{204}\phantom{0} \\
1836 \\
\underline{1836} \\
0
\end{array}
$$

**101.** $13 \times 13 = 13^2$

**102.** $24 \times 24 = 24^2$

**103.** $8 \times 8 \times 8 \times 8 \times 8 = 8^5$

**104.** $9 \times 9 \times 9 \times 9 \times 9 = 9^5$

**105.** $2^6 = 2 \times 2 \times 2 \times 2 \times 2 \times 2 = 64$

**106.** $3^4 = 3 \times 3 \times 3 \times 3 = 81$

**107.** $2^7 = 2 \times 2 \times 2 \times 2 \times 2 \times 2 \times 2 = 128$

**108.** $5^3 = 5 \times 5 \times 5 = 125$

**109.** $7^2 = 7 \times 7 = 49$

**110.** $9^2 = 9 \times 9 = 81$

**111.** $6^3 = 6 \times 6 \times 6 = 216$

**112.** $4^4 = 4 \times 4 \times 4 \times 4 = 256$

**113.** $\begin{aligned} 7 + 2 \times 3 - 5 &= 7 + 6 - 5 \\ &= 13 - 5 \\ &= 8 \end{aligned}$

**114.** $\begin{aligned} 6 \times 2 - 4 + 3 &= 12 - 4 + 3 = 8 + 3 \\ &= 11 \end{aligned}$

**115.** $\begin{aligned} 2^5 + 4 - (5 + 3^2) &= 32 + 4 - (5 + 9) \\ &= 32 + 4 - 14 \\ &= 36 - 14 \\ &= 22 \end{aligned}$

**116.** $\begin{aligned} 4^3 + 20 \div (2 + 2^3) &= 64 + 20 \div (2 + 8) \\ &= 64 + 20 \div 10 \\ &= 64 + 2 \\ &= 66 \end{aligned}$

**117.** $\begin{aligned} 3^3 \times 4 - 6 \div 6 &= 27 \times 4 - 6 \div 6 \\ &= 108 - 1 \\ &= 107 \end{aligned}$

**118.** $20 \div 20 + 5^3 \times 3 = 20 \div 20 + 25 \times 3$

$$= 1 + 75$$

$$= 76$$

**119.** $2^3 \times 5 \div 8 + 3 \times 4$

$$= 8 \times 5 \div 8 + 3 \times 4$$

$$= 40 \div 8 + 12$$

$$= 5 + 12$$

$$= 17$$

**120.** $3^2 \times 6 \div 3 + 5 \times 6 = 9 \times 6 \div 3 + 5 \times 6$

$$= 54 \div 3 + 5 \times 6$$

$$= 18 + 30$$

$$= 48$$

**121.** $9 \times 2^2 + 3 \times 4 - 36 \div (4 + 5)$

$$= 9 \times 4 + 3 \times 4 - 36 \div 9$$

$$= 36 + 12 - 4$$

$$= 44$$

**122.** $5 \times 3 + 5 \times 4^2 - 14 \div (6 + 1)$

$$= 5 \times 3 + 5 \times 16 - 14 \div 7$$

$$= 15 + 80 - 2$$

$$= 95 - 2$$

$$= 93$$

**123.** 127$\underline{5}$ rounds to 1280.

**124.** 567$\underline{3}$ rounds to 5670.

**125.** 15,30$\underline{5}$ rounds to 15,310.

**126.** 42,64$\underline{4}$ rounds to 42,640.

**127.** 12,3$\underline{5}$0 rounds to 12,000.

**128.** 22,$\underline{9}$86 rounds to 23,000.

**129.** 675,$\underline{8}$00 rounds to 676,000.

**130.** 202,$\underline{4}$98 rounds to 202,000.

**131.** 5,6$\underline{6}$8,243 rounds to 5,700,000.

**132.** 9,99$\underline{5}$,312 rounds to 10,000,000.

**133.**
$$
\begin{array}{r}
600 \\
600 \\
900 \\
+\ 900 \\
\hline
3000
\end{array}
$$

**134.**
$$
\begin{array}{r}
30,000 \\
40,000 \\
+\ 70,000 \\
\hline
140,000
\end{array}
$$

**135.**
$$
\begin{array}{r}
4,000,000 \\
-\ 3,000,000 \\
\hline
1,000,000
\end{array}
$$

**136.**
$$
\begin{array}{r}
30,000 \\
-\ 20,000 \\
\hline
10,000
\end{array}
$$

**137.**
$$
\begin{array}{r}
2000 \\
\times\quad 6000 \\
\hline
12,000,000
\end{array}
$$

**138.**
$$
\begin{array}{r}
3,000,000 \\
\times\quad\quad 9000 \\
\hline
2,700,000,000
\end{array}
$$

**139.**
$$
\begin{array}{r}
4,000 \\
20\overline{)80,000} \\
\underline{80}\phantom{,000} \\
0
\end{array}
$$

**140.**
$$
\begin{array}{r}
20,000 \\
400\overline{)8,000,000}
\end{array}
$$

**141.**
$$
\begin{array}{r}
90 \\
40 \\
90 \\
+\ 60 \\
\hline
280
\end{array}
\quad \text{Correct}
$$

142.
$$\begin{array}{r} 900,000 \\ -\ 400,000 \\ \hline 500,000 \end{array}$$  Correct

143.
$$\begin{array}{r} 200,000 \\ \times\qquad 5000 \\ \hline 1,000,000,000 \end{array}$$  Incorrect

144.
$$\begin{array}{r} 30,000 \\ 30\overline{)900,000} \end{array}$$

145.
$$\begin{array}{r} 34 \\ \times\ \ 6 \\ \hline 204 \end{array}$$

204 soft-drink cans

146.
$$\begin{array}{r} 25 \\ \times\ \ 7 \\ \hline 175 \end{array}$$ words

147.
$$\begin{array}{r} 1362 \\ 562 \\ +\ 473 \\ \hline 2397 \end{array}$$

2397 kilometers were driven.

148.
$$\begin{array}{r} 26,300 \\ 14,520 \\ +\ 18,650 \\ \hline \$59,470 \end{array}$$

149.
$$\begin{array}{r} 14,630 \\ -\ 4,329 \\ \hline 10,301 \end{array}$$

10,301 feet between them.

150.
$$\begin{array}{r} 11,658 \\ -\ 4,630 \\ \hline \$7,028 \end{array}$$

151.
$$\begin{array}{r} 1\,356 \\ 24\overline{)32,544} \\ \underline{24}\phantom{,544} \\ 8\,5\phantom{44} \\ \underline{7\,2}\phantom{44} \\ 1\,34\phantom{4} \\ \underline{1\,20}\phantom{4} \\ 144 \\ \underline{144} \\ 0 \end{array}$$

Cost per passenger was $1356.

152.
$$\begin{array}{r} 64 \\ 92\overline{)5888} \\ \underline{552}\phantom{8} \\ 368 \\ \underline{368} \\ 0 \end{array}$$

Cost per share was $64.

153.
| Deposits | Checks |
|---|---|
| 24 | 18 |
| 105 | 145 |
| 36 | 250 |
| + 177 | + 461 |
| 342 | 874 |

$810 + 342 - 874 = 278$

Her balance will be $278.

154.
| 436 | |
|---|---|
| 16 | 29 |
| 98 | 128 |
| 125 | 100 |
| + 318 | + 402 |
| $993 | 659 |

Balance is $993 - $659 = $334

155.
$$\begin{array}{r} 56,720 \\ -\ 56,320 \\ \hline 400 \end{array}$$ miles

$$\begin{array}{r} 25 \\ 16\overline{)400} \\ \underline{32}\phantom{0} \\ 80 \\ \underline{80} \\ 0 \end{array}$$

He got 25 miles per gallon.

**156.**

$$\begin{array}{r} 24,780 \\ -\ 24,396 \\ \hline 384 \end{array}$$

$$\begin{array}{r} 24 \\ 16\overline{)384} \\ 32 \\ \hline 64 \\ 64 \end{array}$$

24 miles per gallon

**157.** $3 \times 279 + 4 \times 61 + 2 \times 1980$

$= 837 + 244 + 3960$

$= 5041$

The total price was $5041.

**158.**

$$\begin{array}{rrrr} 118 & 120 & 24 & 2832 \\ \times\ 24 & \times\ 4 & \times\ 6 & 480 \\ \hline 472 & 480 & 144 & +\ 144 \\ 436 & & & \overline{3456} \\ \hline 2832 & & & \end{array}$$

$3456 for total purchase

**159.**

$$\begin{array}{r} 55,000,000 \\ -\ 14,500,000 \\ \hline 40,500,000 \end{array}$$

The difference is 40,500,000 tons.

**160.**

$$\begin{array}{r} 55,000,000 \\ -\ 33,600,000 \\ \hline 21,400,000 \end{array} \text{ tons between 1990 and 1995}$$

**161.**

$$\begin{array}{rr} 63,500,000 & 63,500,000 \\ -\ 55,000,000 & +\ 8,500,000 \\ \hline 8,500,000 & 72,000,000 \end{array}$$

72,000,000 tons in 2005.

**162.**

$$\begin{array}{r} 63,500,000 \\ -\ 55,000,000 \\ \hline 8,500,000 \end{array}$$

and $8,500,000 \div 5 = 1,700,000$ tons

**163.** $(360 + 50 + 48) \times 42,900,000$

$= 458 \times 42,900,000$

$= 19,648,200,000$

The cost is $19,648,200,000.

**164.** $1500 - 241 - 407 - 3 \times 117$

$= 1500 - 241 - 407 - 351$

$= 501$ pieces

## Chapter 1 Test

**1.** $44,007,635 = $ Forty-four million, seven thousand, six hundred thirty-five

**2.** $26,859 = 20,000 + 6000 + 800 + 50 + 9$

**3.** Three million, five hundred eighty-one thousand, seventy-six $= 3,581,076$

**4.**

$$\begin{array}{r} 156 \\ 93 \\ 8 \\ 127 \\ +\ 593 \\ \hline 977 \end{array}$$

**5.**

$$\begin{array}{r} 470 \\ 386 \\ +\ 189 \\ \hline 1045 \end{array}$$

**6.**

$$\begin{array}{r} 135,484 \\ 2,376 \\ 81,004 \\ +\ 100,113 \\ \hline 318,977 \end{array}$$

**7.**

$$\begin{array}{r} 8961 \\ -\ 894 \\ \hline 8067 \end{array}$$

**8.**

$$\begin{array}{r} 300,523 \\ -\ 262,182 \\ \hline 38,341 \end{array}$$

**9.**
$$
\begin{array}{r}
18,400,100 \\
-\ 13,174,332 \\
\hline
5,225,768
\end{array}
$$

**10.** $1 \times 6 \times 9 \times 7$

$\qquad = 6 \times 9 \times 7$

$\qquad = 54 \times 7$

$\qquad = 378$

**11.**
$$
\begin{array}{r}
45 \\
\times\ 96 \\
\hline
270 \\
405 \\
\hline
4320
\end{array}
$$

**12.**
$$
\begin{array}{r}
147 \\
\times\ \ \ 625 \\
\hline
735 \\
2\ 94 \\
88\ 2 \\
\hline
91,875
\end{array}
$$

**13.**
$$
\begin{array}{r}
18,491 \\
\times\ \ \ \ \ \ 7 \\
\hline
129,437
\end{array}
$$

**14.**
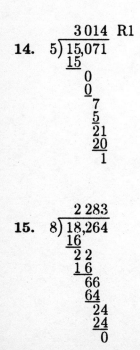

$$
\begin{array}{r}
3\,014\ \ \text{R1} \\
5\overline{)15{,}071} \\
\underline{15} \\
0 \\
\underline{0} \\
7 \\
\underline{5} \\
21 \\
\underline{20} \\
1
\end{array}
$$

**15.**
$$
\begin{array}{r}
2\,283 \\
8\overline{)18{,}264} \\
\underline{16} \\
2\ 2 \\
\underline{16} \\
66 \\
\underline{64} \\
24 \\
\underline{24} \\
0
\end{array}
$$

**16.**
$$
\begin{array}{r}
352 \\
37\overline{)13{,}024} \\
\underline{11\ 1} \\
1\ 92 \\
\underline{1\ 85} \\
74 \\
\underline{74} \\
0
\end{array}
$$

**17.** $11 \times 11 \times 11 = 11^3$

**18.** $2^6 = 2 \times 2 \times 2 \times 2 \times 2 \times 2 = 64$

**19.** $5 + 6^2 - 2 \times (9 - 6)^2$

$\qquad = 5 + 6^2 - 2 \times 3^2$

$\qquad = 5 + 36 - 2 \times 9$

$\qquad = 5 + 36 - 18$

$\qquad = 41 - 18$

$\qquad = 23$

**20.** $2^3 + 4^3 + 18 \div 3 = 8 + 64 + 18 \div 3$

$\qquad\qquad\qquad\qquad = 8 + 64 + 6$

$\qquad\qquad\qquad\qquad = 72 + 6$

$\qquad\qquad\qquad\qquad = 78$

**21.** $4 \times 6 + 3^3 \times 2 + 23 \div 23$

$\qquad = 4 \times 6 + 27 \times 2 + 23 \div 23$

$\qquad = 24 + 54 + 1$

$\qquad = 78 + 1$

$\qquad = 79$

**22.** 88,07<u>3</u> rounds to 88,070 since 3 is less than 5.

**23.** 6,46<u>2</u>,431 rounds to 6,460,000 since 2 is less than 5.

**24.** 4,7<u>8</u>2,163 rounds to 4,800,000 since 8 is greater than 5.

**25.** $5,000,000 \times 30,000 = 150,000,000,000$

**26.**

```
  1000
  4000
  4000
+ 9000
 18,000
```

**27.**

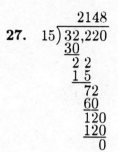

```
        2148
15) 32,220
    30
     2 2
     1 5
       72
       60
      120
      120
        0
```

Each person paid $2148.

**28.**

```
  602
- 135
  467
```

467 feet

**29.**  $3 \times 2 + 1 \times 45 + 2 \times 21 + 2 \times 17$

$= 6 + 45 + 42 + 34$

$= 127$

His total bill was $127.

**30.**

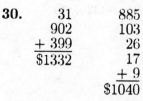

```
    31        885
   902        103
 + 399         26
 $1332         17
              + 9
            $1040
```

Balance is $1332 − $1040 = $292

**31.**

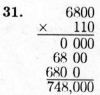

```
    6800
 ×   110
   0 000
  68 00
 680 0
 748,000
```

Area of runway is 748,000 square feet.

**32.** Perimeter is  $2 \times 8 + 2 \times 15 = 16 + 30$

$= 48$ feet

# FRACTIONS

## Pretest Chapter 2

**1.** $\dfrac{3}{8}$

**2.**

| ★ | ★ | ★ |
|---|---|---|
|   |   |   |

**3.** $\dfrac{5}{124}$

**4.** $\dfrac{3}{18} = \dfrac{3 \div 3}{18 \div 3} = \dfrac{1}{6}$

**5.** $\dfrac{13}{39} = \dfrac{13 \div 13}{39 \div 13} = \dfrac{1}{3}$

**6.** $\dfrac{16}{112} = \dfrac{16 \div 16}{112 \div 16} = \dfrac{1}{7}$

**7.** $\dfrac{175}{200} = \dfrac{175 \div 25}{200 \div 25} = \dfrac{7}{8}$

**8.** $\dfrac{44}{121} = \dfrac{44 \div 11}{121 \div 11} = \dfrac{4}{11}$

**9.** $3\dfrac{2}{3} = \dfrac{3 \times 3 + 2}{3} = \dfrac{11}{3}$

**10.** $6\dfrac{1}{9} = \dfrac{9 \times 6 + 1}{9} = \dfrac{55}{9}$

**11.** $\begin{array}{r} 24 \\ 4\overline{)97} \\ 8 \\ \hline 17 \\ 16 \\ \hline 1 \end{array}$ $\quad \dfrac{97}{4} = 24\dfrac{1}{4}$

**12.** $\begin{array}{r} 5 \\ 5\overline{)29} \\ 25 \\ \hline 4 \end{array}$ $\quad \dfrac{29}{5} = 5\dfrac{4}{5}$

**13.** $\begin{array}{r} 2 \\ 17\overline{)36} \\ 34 \\ \hline 2 \end{array}$ $\quad \dfrac{36}{17} = 2\dfrac{2}{17}$

**14.** $\dfrac{5}{11} \times \dfrac{1}{4} = \dfrac{5 \times 1}{11 \times 4}$
$= \dfrac{5}{44}$

**15.** $\dfrac{3}{7} \times \dfrac{14}{9} = \dfrac{3 \times 2 \times 7}{7 \times 3 \times 3} = \dfrac{2}{3}$

**16.** $12\dfrac{1}{3} \times 5\dfrac{1}{2} = \dfrac{37}{3} \times \dfrac{11}{2}$
$= \dfrac{37 \times 11}{3 \times 2}$
$= \dfrac{407}{6} = 67\dfrac{5}{6}$

**17.** $\dfrac{3}{7} \div \dfrac{3}{7} = \dfrac{3}{7} \times \dfrac{7}{3} = 1$

**18.** $\dfrac{7}{16} \div \dfrac{7}{8} = \dfrac{7}{16} \times \dfrac{8}{7}$
$= \dfrac{7 \times 8}{2 \times 8 \times 7}$
$= \dfrac{1}{2}$

**19.** $6\frac{4}{7} \div 1\frac{5}{21} = \frac{46}{7} \div \frac{26}{21}$

$= \frac{46}{7} \times \frac{21}{26}$

$= \frac{2 \times 23 \times 3 \times 7}{7 \times 2 \times 13}$

$= \frac{69}{13}$

$= 5\frac{4}{13}$

**20.** $8 \div \frac{12}{7} = \frac{8}{1} \div \frac{12}{7}$

$= \frac{8}{1} \times \frac{7}{12}$

$= \frac{2 \times 4 \times 7}{1 \times 3 \times 4}$

$= \frac{14}{3} = 4\frac{2}{3}$

**21.** $\frac{1}{8}, \frac{3}{4}, \frac{1}{2}$

$8 = 2 \times 2 \times 2$

$4 = 2 \times 2$

$2 = 2$

$\text{LCD} = 2 \times 2 \times 2 = 8$

**22.** $9 = 3 \times 3$

$45 = 3 \times 3 \times 5$

$\text{LCD} = 3 \times 3 \times 5 = 45$

**23.** $\frac{4}{11}, \frac{2}{55}$

$11 = 11$

$55 = 5 \times 11$

$\text{LCD} = 5 \times 11 = 55$

**24.** $24 = 2 \times 2 \times 2 \times 3$

$36 = 2 \times 2 \times 3 \times 3$

$\text{LCD} = 2 \times 2 \times 2 \times 3 \times 3$

$= 72$

**25.** $\frac{7}{18} - \frac{3}{24} = \frac{7}{18} \times \frac{4}{4} - \frac{3}{24} \times \frac{3}{3}$

$= \frac{28}{72} - \frac{9}{72}$

$= \frac{19}{72}$

**26.** $\frac{5}{24} + \frac{4}{9} + \frac{1}{36}$

$= \frac{15}{72} + \frac{32}{72} + \frac{2}{72}$

$= \frac{15 + 32 + 2}{72} = \frac{49}{72}$

**27.** $8 - 3\frac{2}{3} = 7\frac{3}{3} - 3\frac{2}{3}$

$= 4\frac{1}{3}$

**28.** $1\frac{5}{6} + 2\frac{1}{7} = 1\frac{35}{42} + 2\frac{6}{42}$

$= 3\frac{41}{42}$

**29.** $\frac{5}{9} \times \frac{9}{2} \div \frac{3}{2} = \frac{5}{2} \div \frac{3}{2}$

$= \frac{5}{2} \times \frac{2}{3}$

$= \frac{5}{3}$

$= 1\frac{2}{3}$

**30.** $13\frac{1}{2} - 6\frac{1}{3} = 13\frac{3}{6} - 6\frac{2}{6}$

$= 7\frac{1}{6}$

$7\frac{1}{6}$ miles

**31.** $4\frac{3}{12} + 3\frac{1}{18} = 4\frac{9}{36} + 3\frac{2}{36}$

$= 7\frac{11}{36}$ tons

**32.** $776 \div 43\frac{1}{9} = \frac{776}{1} \div \frac{388}{9}$

$\qquad = \frac{776}{1} \times \frac{9}{388}$

$\qquad = \frac{2 \times 388 \times 9}{1 \times 388}$

$\qquad = 18$

18 students

---

## 2.1  Exercises

**1.** fraction

**3.** denominator

**5.** 3 is the numerator.
5 is the denominator

**7.** 2 is the numerator.
3 is the denominator

**9.** 1 is the numerator.
17 is the denominator.

**11.** $\frac{1}{3}$

**13.** $\frac{5}{6}$

**15.** $\frac{3}{4}$

**17.** $\frac{5}{6}$

**19.** $\frac{1}{4}$

**21.** $\frac{3}{10}$

**23.** $\frac{5}{8}$

**25.** $\frac{4}{7}$

**27.** $\frac{7}{8}$

**29.** $\frac{2}{5}$

**31.** $\frac{4}{5}$

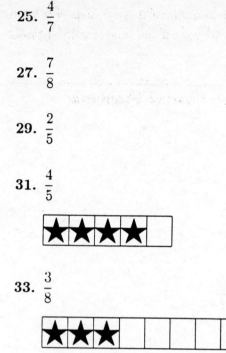

**33.** $\frac{3}{8}$

**35.** $\frac{7}{10}$

**37.** $\frac{\text{silver bells}}{\text{wreaths}} = \frac{31}{95}$

**39.** $\frac{\text{weekend earnings}}{\text{jukebox price}} = \frac{209}{750}$

**41.** $\frac{\text{roast}}{\text{total}} = \frac{89}{122 + 89} = \frac{89}{211}$

**43.** $\frac{\text{rowing}}{\text{total}} = \frac{7}{9 + 7 + 13} = \frac{7}{29}$

**45.** $\frac{\text{ribs or beans}}{\text{total bowls}} = \frac{5 + 4}{2 + 3 + 4 + 5} = \frac{9}{14}$

**47. a.** $\frac{50 + 40}{94 + 101} = \frac{90}{195}$

$\quad$ **b.** $\frac{3 + 19}{94 + 101} = \frac{22}{195}$

**49.** $\frac{0}{6}$ is the amount of money each of 6 business owners get if the business has a profit of $0.

---

## Cumulative Review Problems

**51.**
$$
\begin{array}{r}
18 \\
27 \\
34 \\
16 \\
125 \\
+\,21 \\
\hline
241
\end{array}
$$

**53.**
$$
\begin{array}{r}
4136 \\
\times\quad 29 \\
\hline
37\,224 \\
82\,72\phantom{0} \\
\hline
119,944
\end{array}
$$

**55.**
$$
\begin{array}{r}
282 \\
866 \\
42 \\
317 \\
102 \\
99 \\
+\,115 \\
\hline
1823
\end{array}
$$

$$
\begin{array}{r}
2004 \\
-\,1823 \\
\hline
181 \;\text{reference books}
\end{array}
$$

---

## 2.2  Exercises

**1.** 11, 19, 41, 5

**3.** composite number

**5.** Answers may vary. Example: $6 = 2 \times 3$

**7.** $15 = 3 \times 5$

**9.** $6 = 2 \times 3$

**11.** $49 = 7 \times 7 = 7^2$

**13.** $64 = 8 \times 8$
$\qquad = 2 \times 2 \times 2 \times 2 \times 2 \times 2$
$\qquad = 2^6$

**15.** $55 = 5 \times 11$

**17.** $63 = 7 \times 9$
$\qquad = 7 \times 3 \times 3$
$\qquad = 7 \times 3^2$

**19.** $75 = 3 \times 25$
$\qquad = 3 \times 5 \times 5$
$\qquad = 3 \times 5^2$

**21.** $54 = 6 \times 9$
$\qquad = 2 \times 3 \times 3 \times 3$
$\qquad = 2 \times 3^3$

**23.** $84 = 4 \times 21$
$\qquad = 2 \times 2 \times 3 \times 7$
$\qquad = 2^2 \times 3 \times 7$

**25.** $98 = 2 \times 49$
$\qquad = 2 \times 7 \times 7$
$\qquad = 2 \times 7^2$

**27.** Prime

**29.** $57 = 3 \times 19$

**31.** Prime

**33.** $62 = 2 \times 31$

**35.** Prime

**37.** Prime

**39.** $121 = 11 \times 11 = 11^2$

**41.** Prime

**43.** $\dfrac{18}{27} = \dfrac{18 \div 8}{27 \div 9} = \dfrac{2}{3}$

**45.** $\dfrac{32}{48} = \dfrac{32 \div 16}{48 \div 16} = \dfrac{2}{3}$

**47.** $\dfrac{30}{48} = \dfrac{30 \div 6}{48 \div 6} = \dfrac{5}{8}$

**49.** $\dfrac{35}{60} = \dfrac{35 \div 5}{60 \div 5} = \dfrac{7}{12}$

**51.** $\dfrac{3}{15} = \dfrac{3 \times 1}{3 \times 5} = \dfrac{1}{5}$

**53.** $\dfrac{66}{88} = \dfrac{3 \times 2 \times 11}{2 \times 2 \times 2 \times 11} = \dfrac{3}{4}$

**55.** $\dfrac{18}{24} = \dfrac{2 \times 3 \times 3}{2 \times 2 \times 2 \times 3} = \dfrac{3}{4}$

**57.** $\dfrac{27}{45} = \dfrac{3 \times 3 \times 3}{3 \times 3 \times 5} = \dfrac{3}{5}$

**59.** $\dfrac{33}{36} = \dfrac{3 \times 11}{3 \times 12} = \dfrac{11}{12}$

**61.** $\dfrac{63}{108} = \dfrac{3 \times 3 \times 7}{2 \times 2 \times 3 \times 3 \times 3} = \dfrac{7}{12}$

**63.** $\dfrac{88}{121} = \dfrac{11 \times 8}{11 \times 11} = \dfrac{8}{11}$

**65.** $\dfrac{150}{200} = \dfrac{3 \times 50}{4 \times 50} = \dfrac{3}{4}$

**67.** $\dfrac{220}{260} = \dfrac{11 \times 20}{13 \times 20} = \dfrac{11}{13}$

**69.** $\dfrac{3}{17} \overset{?}{=} \dfrac{15}{85}$

$3 \times 85 \overset{?}{=} 17 \times 15$

$255 = 255$

Yes

**71.** $\dfrac{12}{40} \overset{?}{=} \dfrac{3}{13}$

$12 \times 13 \overset{?}{=} 40 \times 3$

$156 \neq 120$

No

**73.** $\dfrac{23}{27} \overset{?}{=} \dfrac{92}{107}$

$23 \times 107 \overset{?}{=} 27 \times 92$

$2461 \neq 2484$

No

**75.** $\dfrac{27}{75} \overset{?}{=} \dfrac{45}{95}$

$27 \times 95 \overset{?}{=} 75 \times 45$

$2565 \neq 3375$

No

**77.** $\dfrac{65}{70} \overset{?}{=} \dfrac{13}{14}$

$65 \times 14 \overset{?}{=} 70 \times 13$

$910 = 910$

Yes

**79.** $\dfrac{6}{10} = \dfrac{2 \times 3}{2 \times 5} = \dfrac{3}{5}$

**81.** $\dfrac{95 - 15}{95} = \dfrac{80}{95} = \dfrac{16 \times 5}{19 \times 5} = \dfrac{16}{19}$ passed

**83.** $\dfrac{2000}{16,000 + 2000} = \dfrac{2000}{18,000}$

$\qquad\qquad = \dfrac{1 \times 2000}{9 \times 2000} = \dfrac{1}{9}$

**85.** Total student body is

$1100 + 1700 + 900 + 500 + 300 = 4500.$

Short commute is

$\dfrac{1700}{4500} = \dfrac{17 \times 100}{45 \times 100} = \dfrac{17}{45}$

**87.** $\dfrac{500 + 300}{4500} = \dfrac{800}{4500}$

$= \dfrac{8 \times 100}{45 \times 100} = \dfrac{8}{45}$

---

# Cumulative Review Problems

**89.**
$$\begin{array}{r} 386 \\ \times\ 425 \\ \hline 1\,930 \\ 7\,72\phantom{0} \\ 154\,4\phantom{00} \\ \hline 164{,}050 \end{array}$$

**91.**
$$\begin{array}{r} 3600 \\ \times\ \ \ 1700 \\ \hline 6{,}120{,}000 \end{array}$$

---

## 2.3   Exercises

**1. (a)** Multiply the whole number by the denominator of the fraction.
   **(b)** Add the numerator of the fraction to the product formed in step (a).
   **(c)** Write the sum found in step (b) over the denominator of the fraction.

**3.** $4\dfrac{2}{3} = \dfrac{3 \times 4 + 2}{3} = \dfrac{14}{3}$

**5.** $2\dfrac{3}{7} = \dfrac{7 \times 2 + 3}{7} = \dfrac{17}{7}$

**7.** $6\dfrac{1}{7} = \dfrac{6 \times 7 + 1}{7} = \dfrac{43}{7}$

**9.** $10\dfrac{2}{3} = \dfrac{3 \times 10 + 2}{3} = \dfrac{32}{3}$

**11.** $21\dfrac{2}{3} = \dfrac{3 \times 21 + 2}{3} = \dfrac{65}{3}$

**13.** $9\dfrac{1}{6} = \dfrac{6 \times 9 + 1}{6} = \dfrac{55}{6}$

**15.** $28\dfrac{1}{6} = \dfrac{6 \times 28 + 1}{6} = \dfrac{169}{6}$

**17.** $10\dfrac{11}{12} = \dfrac{12 \times 10 + 11}{12} = \dfrac{131}{12}$

**19.** $7\dfrac{9}{10} = \dfrac{10 \times 7 + 9}{10} = \dfrac{79}{10}$

**21.** $8\dfrac{1}{25} = \dfrac{25 \times 8 + 1}{25} = \dfrac{201}{25}$

**23.** $105\dfrac{1}{2} = \dfrac{2 \times 105 + 1}{2} = \dfrac{211}{2}$

**25.** $164\dfrac{2}{3} = \dfrac{3 \times 164 + 2}{3} = \dfrac{494}{3}$

**27.** $8\dfrac{11}{15} = \dfrac{15 \times 8 + 11}{15} = \dfrac{131}{15}$

**29.** $5\dfrac{13}{25} = \dfrac{25 \times 5 + 13}{25} = \dfrac{138}{25}$

**31.** $3\overline{\smash)4}$  $\phantom{x}$  $\dfrac{4}{3} = 1\dfrac{1}{3}$
$\phantom{3)}\dfrac{3}{1}$

**33.** $4\overline{\smash)11}$  $\phantom{x}$  $\dfrac{11}{4} = 2\dfrac{3}{4}$
$\phantom{4)}\dfrac{8}{3}$

**35.** $6\overline{\smash)15}$  $\phantom{x}$  $\dfrac{15}{6} = 2\dfrac{3}{6} = 2\dfrac{1}{2}$
$\phantom{6)}\dfrac{12}{3}$

**37.** $8\overline{\smash)27}$  $\phantom{x}$  $\dfrac{27}{8} = 3\dfrac{3}{8}$
$\phantom{8)}\dfrac{24}{3}$

**39.** $12\overline{\smash)60}$  $\phantom{x}$  $\dfrac{60}{12} = 5$
$\phantom{12)}\dfrac{60}{0}$

**41.**  $9\overline{)86}$  $\dfrac{9}{\underline{81}}$  $\dfrac{5}{}$     $\dfrac{86}{9} = 9\dfrac{5}{9}$

**43.**  $13\overline{)28}$  $\dfrac{2}{\underline{26}}$  $\dfrac{2}{}$     $\dfrac{28}{13} = 2\dfrac{2}{13}$

**45.**  $16\overline{)51}$  $\dfrac{3}{\underline{48}}$  $\dfrac{3}{}$     $\dfrac{51}{16} = 3\dfrac{3}{16}$

**47.**  $3\overline{)28}$  $\dfrac{9}{\underline{27}}$  $\dfrac{1}{}$     $\dfrac{28}{3} = 9\dfrac{1}{3}$

**49.**  $2\overline{)35}$  $\dfrac{17}{\underline{2}}$  $\dfrac{15}{\underline{14}}$  $\dfrac{1}{}$     $\dfrac{35}{2} = 17\dfrac{1}{2}$

**51.**  $7\overline{)91}$  $\dfrac{13}{\underline{7}}$  $\dfrac{21}{\underline{21}}$  $\dfrac{0}{}$     $\dfrac{91}{7} = 13$

**53.**  $15\overline{)210}$  $\dfrac{14}{\underline{15}}$  $\dfrac{60}{\underline{60}}$  $\dfrac{0}{}$     $\dfrac{210}{15} = 14$

**55.**  $17\overline{)102}$  $\dfrac{6}{\underline{102}}$  $\dfrac{0}{}$     $\dfrac{102}{17} = 6$

**57.**  $11\overline{)403}$  $\dfrac{36}{\underline{33}}$  $\dfrac{73}{\underline{66}}$  $\dfrac{7}{}$     $\dfrac{403}{11} = 36\dfrac{7}{11}$

**59.**  $2\dfrac{9}{12} = 2\dfrac{3 \times 3}{3 \times 4} = 2\dfrac{3}{4}$

**61.**  $4\dfrac{11}{66} = 4\dfrac{11 \times 1}{11 \times 6} = 4\dfrac{1}{6}$

**63.**  $12\dfrac{15}{40} = 12\dfrac{5 \times 3}{5 \times 8} = 12\dfrac{3}{8}$

**65.**  $\dfrac{24}{6} = \dfrac{6 \times 4}{6 \times 1} = 4$

**67.**  $\dfrac{36}{15} = \dfrac{12 \times 3}{5 \times 3} = \dfrac{12}{5}$

**69.**  $\dfrac{78}{9} = \dfrac{26 \times 3}{3 \times 3} = \dfrac{26}{3}$

**71.**  $126\overline{)340}$  $\dfrac{2}{\underline{252}}$  $\dfrac{88}{}$

$\dfrac{340}{126} = 2\dfrac{88}{126} = 2\dfrac{44 \times 2}{63 \times 2} = 2\dfrac{44}{63}$

**73.**  $424\overline{)986}$  $\dfrac{2}{\underline{848}}$  $\dfrac{138}{}$

$\dfrac{986}{424} = 2\dfrac{138}{424} = 2\dfrac{69 \times 2}{212 \times 2} = 2\dfrac{69}{212}$

**75.**  $296\overline{)508}$  $\dfrac{1}{\underline{296}}$  $\dfrac{212}{}$

$\dfrac{508}{296} = 1\dfrac{212}{296} = 1\dfrac{53 \times 4}{74 \times 4} = 1\dfrac{53}{74}$

**77.**  $360\dfrac{2}{3} = \dfrac{3 \times 360 + 2}{3} = \dfrac{1082}{3}$ yards

**79.**

$$3\overline{)151} \\ \phantom{3)}\underline{15} \\ \phantom{3)1}1 \\ \phantom{3)1}\underline{0} \\ \phantom{3)1}1$$

$$\frac{50}{\phantom{0}}$$

$$\frac{151}{3} = 50\frac{1}{3} \text{ acres}$$

**81.** No. 101 is prime and is not a factor of 5687.

---

## Cumulative Review Problems

**83.**
$$\begin{array}{r} {}^{7\ 11\ 10\ 10}\\ 1,39\cancel{8},\cancel{2}\cancel{1}\cancel{0} \\ -\ 1,137,963 \\ \hline 260,247 \end{array}$$

**85.** $\dfrac{900,000}{3000} = 300$

---

## 2.4   Exercises

**1.** $\dfrac{3}{5} \times \dfrac{7}{11} = \dfrac{21}{55}$

**3.** $\dfrac{3}{4} \times \dfrac{5}{13} = \dfrac{15}{52}$

**5.** $\dfrac{\overset{1}{\cancel{6}}}{\underset{1}{\cancel{3}}} \times \dfrac{\overset{2}{\cancel{10}}}{\underset{2}{\cancel{12}}} = \dfrac{1}{1} \times \dfrac{2}{2} = 1$

**7.** $\dfrac{7}{\underset{6}{\cancel{36}}} \times \dfrac{\overset{5}{\cancel{30}}}{9} = \dfrac{7}{6} \times \dfrac{5}{9} = \dfrac{35}{54}$

**9.** $\dfrac{\overset{5}{\cancel{15}}}{\underset{4}{\cancel{28}}} \times \dfrac{7}{\underset{3}{\cancel{9}}} = \dfrac{5}{4} \times \dfrac{1}{3} = \dfrac{5}{12}$

**11.** $\dfrac{\overset{3}{\cancel{9}}}{\underset{2}{\cancel{10}}} \times \dfrac{\overset{7}{\cancel{35}}}{\underset{4}{\cancel{12}}} = \dfrac{3}{2} \times \dfrac{7}{4} = \dfrac{21}{8} = 2\dfrac{5}{8}$

**13.** $8 \times \dfrac{3}{7} = \dfrac{8}{1} \times \dfrac{3}{7} = \dfrac{24}{7} = 3\dfrac{3}{7}$

**15.** $\dfrac{5}{16} \times 8 = \dfrac{5}{\underset{2}{\cancel{16}}} \times \dfrac{\overset{1}{\cancel{8}}}{1} = \dfrac{5}{2} = 2\dfrac{1}{2}$

**17.** $\dfrac{\overset{1}{\cancel{4}}}{\underset{3}{\cancel{9}}} \times \dfrac{\overset{1}{\cancel{3}}}{\underset{1}{\cancel{7}}} \times \dfrac{\overset{1}{\cancel{7}}}{\underset{2}{\cancel{8}}} = \dfrac{1}{3} \times \dfrac{1}{1} \times \dfrac{1}{2} = \dfrac{1}{6}$

**19.** $\dfrac{\overset{1}{\cancel{4}}}{\underset{1}{\cancel{3}}} \times \dfrac{1}{\underset{2}{\cancel{8}}} \times \dfrac{\overset{1}{\cancel{35}}}{7} = \dfrac{1}{1} \times \dfrac{1}{2} \times \dfrac{1}{1} = \dfrac{1}{2}$

**21.** $2\dfrac{3}{4} \times \dfrac{8}{9} = \dfrac{11}{\underset{1}{\cancel{4}}} \times \dfrac{\overset{1}{\cancel{8}}}{9} = \dfrac{22}{9} = 2\dfrac{4}{9}$

**23.** $1\dfrac{1}{4} \times 3\dfrac{2}{3} = \dfrac{5}{4} \times \dfrac{11}{3} = \dfrac{55}{12} = 4\dfrac{7}{12}$

**25.** $2\dfrac{1}{2} \times 6 = \dfrac{5}{\underset{1}{\cancel{2}}} \times \dfrac{\overset{3}{\cancel{6}}}{1} = \dfrac{15}{1} = 15$

**27.** $2\dfrac{3}{10} \times \dfrac{3}{5} = \dfrac{23}{10} \times \dfrac{3}{5} = \dfrac{69}{50} = 1\dfrac{19}{50}$

**29.** $1\dfrac{3}{16} \times 0 = \dfrac{19}{16} \times 0 = 0$

**31.** $4\dfrac{1}{5} \times 12\dfrac{2}{9} = \dfrac{\overset{7}{\cancel{21}}}{\underset{1}{\cancel{5}}} \times \dfrac{\overset{22}{\cancel{110}}}{\underset{3}{\cancel{9}}} = \dfrac{154}{3} = 51\dfrac{1}{3}$

**33.** $6\dfrac{2}{5} \times \dfrac{1}{4} = \dfrac{\overset{8}{\cancel{32}}}{5} \times \dfrac{1}{\underset{1}{\cancel{4}}} = \dfrac{8}{5} = 1\dfrac{3}{5}$

**35.** $\dfrac{5}{5} \times 11\dfrac{5}{7} = 1 \times 11\dfrac{5}{7} = 11\dfrac{5}{7}$

**37.** $\dfrac{2}{7} \cdot x = \dfrac{18}{35}$

$\dfrac{2 \cdot 9}{7 \cdot 5} = \dfrac{18}{35}$

$x = \dfrac{9}{5}$

**39.** $\dfrac{7}{13} \cdot x = \dfrac{56}{117}$

$\dfrac{7 \cdot 8}{13 \cdot 9} = \dfrac{56}{117}$

$x = \dfrac{8}{9}$

**41.** $8\dfrac{3}{4} \times 4\dfrac{1}{3} = \dfrac{4 \times 8 + 3}{4} \times \dfrac{3 \times 4 + 1}{3}$

$= \dfrac{35}{4} \times \dfrac{13}{3}$

$= \dfrac{455}{12}$

$= 37\dfrac{11}{12}$ square miles

**43.** $360 \times 4\dfrac{1}{3} = \dfrac{\overset{120}{\cancel{360}}}{1} \times \dfrac{13}{\underset{1}{\cancel{3}}}$

$= 120 \times 13$

$= 1560$ miles

**45.** $90\dfrac{1}{2} \times 18 = \dfrac{181}{\underset{1}{\cancel{2}}} \times \dfrac{\overset{9}{\cancel{18}}}{1}$

$= 181 \times 9$

$= 1629$ grams

**47.** $12\dfrac{1}{4} \times \dfrac{3}{4} = \dfrac{49}{4} \times \dfrac{3}{4}$

$= \dfrac{147}{16}$

$= 9\dfrac{3}{16}$ ounces

**49.** $12{,}064 \times \dfrac{1}{32} = \dfrac{12{,}064}{32}$

$= 377$ companies

**51.** $3\dfrac{1}{3} \times 1\dfrac{1}{4} \times \dfrac{1}{2} = \dfrac{7}{2} \times \dfrac{5}{4} \times \dfrac{1}{2}$

$= \dfrac{35}{16}$

$= 2\dfrac{3}{16}$ miles

**53.** **a.** $\dfrac{3}{4} \times \dfrac{5}{6} = \dfrac{3 \times 5}{4 \times 3 \times 2}$

$= \dfrac{5}{8}$

$\dfrac{5}{8}$ of the students

**b.** $8600 \times \dfrac{5}{8} = \dfrac{8600}{1} \times \dfrac{5}{8}$

$= \dfrac{1075 \times 8 \times 5}{1 \times 8}$

$= 5375$

5375 students

**55.** The step of dividing the numerator and denominator by the same number allows us to work with smaller numbers when we do the multiplication. Also, this allows us to avoid the step of having to simplify the fraction in the final answer.

## Cumulative Review Problems

**57.**
$$\begin{array}{r} 529 \text{ cars} \\ 31)\overline{16399} \\ \underline{155}\phantom{00} \\ 89\phantom{0} \\ \underline{62}\phantom{0} \\ 279 \\ \underline{279} \\ 0 \end{array}$$

**59.**
$$\begin{array}{r} 146 \\ \times\ 12 \\ \hline 292 \\ 146\phantom{0} \\ \hline 1752 \text{ lines} \end{array}$$

## 2.5   Exercises

1. Think of a simple problem like $3 \div \frac{1}{2}$. One way to think of it is how many $\frac{1}{2}$'s can be placed in 3? For example, how many $\frac{1}{2}$ pound rocks could be put in a bag that holds 3 pounds of rocks? The answer is 6. If we inverted the first fraction by mistake, we would have $\frac{1}{3} \times \frac{1}{2} = \frac{1}{6}$. We know that is wrong since there are obviously several $\frac{1}{2}$ pound rocks in a bag that holds 3 pounds of rocks. The answer $\frac{1}{6}$ would make no sense.

3. $\frac{7}{8} \div \frac{2}{3} = \frac{7}{8} \times \frac{3}{2} = \frac{21}{16} = 1\frac{5}{16}$

5. $\frac{2}{3} \div \frac{4}{27} = \frac{\overset{1}{\cancel{2}}}{\underset{1}{\cancel{3}}} \times \frac{\overset{9}{\cancel{27}}}{\underset{2}{\cancel{4}}} = \frac{9}{2} = 4\frac{1}{2}$

7. $\frac{5}{9} \div \frac{10}{27} = \frac{\overset{1}{\cancel{5}}}{\underset{1}{\cancel{9}}} \times \frac{\overset{3}{\cancel{27}}}{\underset{2}{\cancel{10}}} = \frac{3}{2} = 1\frac{1}{2}$

9. $\frac{4}{5} \div 1 = \frac{4}{5} \times \frac{1}{1} = \frac{4}{5}$

11. $\frac{3}{11} \div 4 = \frac{3}{11} \times \frac{1}{4} = \frac{3}{44}$

13. $\frac{2}{9} \div \frac{1}{6} = \frac{2}{\underset{3}{\cancel{9}}} \times \frac{\overset{2}{\cancel{6}}}{1} = \frac{4}{3} = 1\frac{1}{3}$

15. $\frac{4}{15} \div \frac{4}{15} = \frac{\overset{1}{\cancel{4}}}{\underset{1}{\cancel{15}}} \times \frac{\overset{1}{\cancel{15}}}{\underset{1}{\cancel{4}}} = 1$

17. $\frac{3}{7} \div \frac{7}{3} = \frac{3}{7} \times \frac{3}{7} = \frac{9}{49}$

19. $\frac{4}{3} \div \frac{7}{27} = \frac{4}{\underset{1}{\cancel{3}}} \times \frac{\overset{9}{\cancel{27}}}{7} = \frac{36}{7} = 5\frac{1}{7}$

21. $0 \div \frac{3}{17} = 0 \times \frac{17}{3} = 0$

23. $\frac{18}{19} \div 0$   Cannot be done

25. $\frac{9}{16} \div \frac{3}{4} = \frac{9}{16} \times \frac{4}{3} = \frac{3}{4}$

27. $\frac{3}{7} \div \frac{15}{28} = \frac{3}{7} \times \frac{28}{15} = \frac{4}{5}$

29. $\frac{10}{25} \div \frac{20}{50} = \frac{10}{25} \times \frac{50}{20} = 1$

31. $8 \div \frac{4}{5} = \frac{8}{1} \times \frac{5}{4} = \frac{10}{1} = 10$

33. $\frac{7}{8} \div 4 = \frac{7}{8} \times \frac{1}{4} = \frac{7}{32}$

35. $6000 \div \frac{6}{5} = \frac{6000}{1} \times \frac{5}{6} = 5000$

37. $\frac{\frac{3}{5}}{6} = \frac{3}{5} \times \frac{1}{6} = \frac{1}{10}$

39. $\frac{\frac{5}{8}}{\frac{25}{7}} = \frac{5}{8} \times \frac{7}{25} = \frac{7}{40}$

41. $3\frac{1}{5} \div \frac{3}{10} = \frac{16}{5} \times \frac{10}{3} = \frac{32}{3} = 10\frac{2}{3}$

43. $2\frac{1}{3} \div 6 = \frac{7}{3} \times \frac{1}{6} = \frac{7}{18}$

45. $5\frac{1}{4} \div 2\frac{5}{8} = \frac{21}{4} \div \frac{21}{8} = \frac{21}{4} \times \frac{8}{21} = 2$

47. $5 \div 1\frac{1}{4} = \frac{5}{1} \div \frac{5}{4} = \frac{5}{1} \times \frac{4}{5} = 4$

**49.** $12\frac{1}{2} \div 5\frac{5}{6} = \frac{25}{2} \div \frac{35}{6} = \frac{25}{2} \times \frac{6}{35} = \frac{15}{7} = 2\frac{1}{7}$

**51.** $8\frac{1}{4} \div 2\frac{3}{4} = \frac{33}{4} \div \frac{11}{4} = \frac{33}{4} \times \frac{4}{11} = 3$

**53.** $3\frac{1}{2} \times \frac{9}{16} = \frac{7}{2} \times \frac{9}{16} = \frac{63}{32} = 1\frac{31}{32}$

**55.** $3\frac{3}{4} \div 9 = \frac{15}{4} \times \frac{1}{9} = \frac{15}{36}$

**57.** $\dfrac{3\frac{1}{2}}{\frac{1}{4}} = \frac{7}{2} \div \frac{1}{4} = \frac{7}{2} \times \frac{4}{1} = 14$

**59.** $\dfrac{1\frac{1}{4}}{1\frac{7}{8}} = \frac{5}{4} \div \frac{15}{8} = \frac{5}{4} \times \frac{8}{15} = \frac{2}{3}$

**61.** $\dfrac{\frac{7}{12}}{3\frac{2}{3}} = \frac{7}{12} \div \frac{11}{3} = \frac{7}{12} \times \frac{3}{11} = \frac{7}{44}$

**63.** $\dfrac{5\frac{1}{3}}{2\frac{1}{2}} = \frac{16}{3} \div \frac{5}{2} = \frac{16}{3} \times \frac{2}{5} = \frac{32}{15} = 2\frac{2}{15}$

**65.** $x \div \frac{4}{3} = \frac{21}{20}$

$x \times \frac{3}{4} = \frac{21}{20}$

$\frac{7 \cdot 3}{5 \cdot 4} = \frac{21}{20}$

$x = \frac{7}{5}$

**67.** $x \div \frac{9}{5} = \frac{20}{63}$

$x \times \frac{5}{9} = \frac{20}{63}$

$\frac{4 \cdot 5}{7 \cdot 9} = \frac{20}{63}$

$x = \frac{4}{7}$

**69.** $20\frac{1}{4} \div 9 = \frac{81}{4} \div \frac{9}{1}$

$= \frac{81}{4} \times \frac{1}{9}$

$= \frac{9 \times 9 \times 1}{4 \times 9}$

$= \frac{9}{4}$

$= 2\frac{1}{4}$ gallons

**71.** $125 \div 3\frac{1}{3} = \frac{125}{1} \div \frac{10}{3}$

$= \frac{125}{1} \times \frac{3}{10}$

$= \frac{5 \times 25 \times 3}{5 \times 2}$

$= \frac{75}{2}$

$= 37\frac{1}{2}$ miles per hour

**73.** $87\frac{1}{2} \div 3\frac{1}{8} = \frac{175}{2} \div \frac{25}{8}$

$= \frac{175}{2} \times \frac{8}{25}$

$= \frac{25 \times 7 \times 2 \times 4}{2 \times 25}$

$= 28$ flags

**75.** $10 \div 1\frac{2}{3} = \frac{10}{1} \div \frac{5}{3}$

$= \frac{10}{1} \times \frac{3}{5}$

$= \frac{2 \times 5 \times 3}{1 \times 5}$

$= 6$ dresses

**77.** $4\frac{3}{4} \div \frac{5}{6} = \frac{19}{4} \times \frac{6}{5} = \frac{19 \times 2 \times 3}{2 \times 2 \times 5}$

$= \frac{57}{10} = 5\frac{7}{10}$ attempts

In reality, it is 6 drill attempts.

**79.** $15 \div 5 = 3$

Exact is

$$14\frac{2}{3} \div 5\frac{1}{6} = \frac{44}{3} \div \frac{31}{6} = \frac{44}{3} \times \frac{6}{31}$$

$$= \frac{88}{31} = 2\frac{26}{31}$$

## Cumulative Review Problems

**81.** $39,576,304 =$ Thirty-nine million, five hundred seventy-six thousand, three hundred four.

**83.** $126 + 34 + 9 + 891 + 12 + 27 = 1099$

## Test on Sections 2.1–2.5

**1.** $\dfrac{23}{32}$

**2.** $\dfrac{112}{340}$

**3.** $\dfrac{19}{38} = \dfrac{19 \div 19}{38 \div 19} = \dfrac{1}{2}$

**4.** $\dfrac{12}{15} = \dfrac{12 \div 3}{15 \div 3} = \dfrac{4}{5}$

**5.** $\dfrac{24}{66} = \dfrac{24 \div 6}{66 \div 6} = \dfrac{4}{11}$

**6.** $\dfrac{125}{155} = \dfrac{125 \div 5}{155 \div 5} = \dfrac{25}{31}$

**7.** $\dfrac{39}{52} = \dfrac{39 \div 13}{52 \div 13} = \dfrac{3}{4}$

**8.** $\dfrac{51}{34} = \dfrac{51 \div 17}{34 \div 17} = \dfrac{3}{2} = 1\frac{1}{2}$

**9.** $3\frac{7}{12} = \dfrac{12 \times 3 + 7}{12} = \dfrac{43}{12}$

**10.** $4\frac{1}{8} = \dfrac{8 \times 4 + 1}{8} = \dfrac{33}{8}$

**11.**
$$\begin{array}{r} 6 \\ 7{\overline{)45}} \\ \underline{42} \\ 3 \end{array} \qquad \frac{45}{7} = 6\frac{3}{7}$$

**12.**
$$\begin{array}{r} 8 \\ 4{\overline{)33}} \\ \underline{32} \\ 1 \end{array} \qquad \frac{33}{4} = 8\frac{1}{4}$$

**13.** $\dfrac{4}{5} \times \dfrac{2}{7} = \dfrac{4 \times 2}{5 \times 7} = \dfrac{8}{35}$

**14.** $\dfrac{15}{7} \times \dfrac{3}{5} = \dfrac{15 \times 3}{7 \times 5} = \dfrac{3 \times 5 \times 3}{7 \times 5}$

$$= \dfrac{3 \times 3}{7} = \dfrac{9}{7} = 1\frac{2}{7}$$

**15.** $18 \times \dfrac{5}{6} = \dfrac{18}{1} \times \dfrac{5}{6} = \dfrac{3 \times 6 \times 5}{6} = 3 \times 5 = 15$

**16.** $\dfrac{3}{8} \times 44 = \dfrac{3}{8} \times \dfrac{44}{1} = \dfrac{3 \times 4 \times 11}{2 \times 4}$

$$= \dfrac{3 \times 11}{2} = \dfrac{33}{2} = 16\frac{1}{2}$$

**17.** $2\frac{1}{3} \times 5\frac{3}{4} = \dfrac{7}{3} \times \dfrac{23}{4} = \dfrac{161}{12} = 13\frac{5}{12}$

**18.** $1\frac{3}{7} \times 3\frac{1}{3} = \dfrac{10}{7} \times \dfrac{10}{3} = \dfrac{10 \times 10}{7 \times 3}$

$$= \dfrac{100}{21} = 4\frac{16}{21}$$

**19.** $\dfrac{4}{7} \div \dfrac{3}{4} = \dfrac{4}{7} \times \dfrac{4}{3} = \dfrac{4 \times 4}{7 \times 3} = \dfrac{16}{21}$

**20.** $\dfrac{8}{9} \div \dfrac{1}{6} = \dfrac{8}{9} \times \dfrac{6}{1} = \dfrac{8 \times 3 \times 2}{3 \times 3}$

$$= \dfrac{8 \times 2}{3} = \dfrac{16}{3} = 5\frac{1}{3}$$

**21.** $5\frac{1}{4} \div \dfrac{3}{4} = \dfrac{21}{4} \times \dfrac{4}{3} = \dfrac{7 \times 3 \times 4}{4 \times 3} = 7$

**22.** $5\dfrac{3}{5} \div \dfrac{1}{2} = \dfrac{28}{5} \div \dfrac{1}{2} = \dfrac{28}{5} \times \dfrac{2}{1}$

$\qquad = \dfrac{28 \times 2}{5} = \dfrac{56}{5} = 11\dfrac{1}{5}$

**23.** $2\dfrac{1}{4} \times 3\dfrac{1}{2} = \dfrac{9}{4} \times \dfrac{7}{2} = \dfrac{63}{8} = 7\dfrac{7}{8}$

**24.** $6 \times 2\dfrac{1}{3} = \dfrac{6}{1} \times \dfrac{7}{3} = \dfrac{2 \times 3 \times 7}{3}$

$\qquad = 2 \times 7 = 14$

**25.** $5 \div 1\dfrac{7}{8} = 5 \div \dfrac{15}{8} = \dfrac{5}{1} \times \dfrac{8}{15}$

$\qquad = \dfrac{5 \times 8}{5 \times 3} = \dfrac{8}{3} = 2\dfrac{2}{3}$

**26.** $5\dfrac{3}{4} \div 2 = \dfrac{23}{4} \div 2 = \dfrac{23}{4} \times \dfrac{1}{2}$

$\qquad = \dfrac{23}{4 \times 2} = \dfrac{23}{8} = 2\dfrac{7}{8}$

**27.** $\dfrac{13}{20} \div \dfrac{4}{5} = \dfrac{13}{20} \times \dfrac{5}{4} = \dfrac{13 \times 5}{5 \times 4 \times 4} = \dfrac{13}{16}$

**28.** $\dfrac{4}{7} \div 8 = \dfrac{4}{7} \times \dfrac{1}{8} = \dfrac{4}{7 \times 2 \times 4}$

$\qquad = \dfrac{1}{14}$

**29.** $\dfrac{9}{22} \times \dfrac{11}{16} = \dfrac{9 \times 11}{2 \times 11 \times 16} = \dfrac{9}{32}$

**30.** $\dfrac{14}{25} \times \dfrac{65}{42} = \dfrac{7 \times 2 \times 13 \times 5}{5 \times 5 \times 2 \times 3 \times 7}$

$\qquad = \dfrac{13}{5 \times 3} = \dfrac{13}{15}$

**31.** $5\dfrac{1}{4} \times 8\dfrac{3}{4} = \dfrac{21}{4} \times \dfrac{35}{4}$

$\qquad = \dfrac{735}{16} = 45\dfrac{15}{16}$ square feet

**32.** $2\dfrac{2}{3} \times 1\dfrac{1}{2} = \dfrac{8}{3} \times \dfrac{3}{2} = \dfrac{4 \times 2 \times 3}{3 \times 2}$

$\qquad = 4$ cups

**33.** $4\dfrac{1}{2} \times \dfrac{1}{3} = \dfrac{9}{2} \times \dfrac{1}{3} = \dfrac{3 \times 3}{2 \times 3} = \dfrac{3}{2} = 1\dfrac{1}{2}$ miles

**34.** $12\dfrac{3}{8} \div \dfrac{3}{4} = \dfrac{99}{8} \times \dfrac{4}{3}$

$\qquad = \dfrac{3 \times 33 \times 4}{4 \times 2 \times 3}$

$\qquad = \dfrac{33}{2} = 16\dfrac{1}{2}$ packages

He had 16 full packages with $\dfrac{3}{4} \times \dfrac{1}{2} = \dfrac{3}{8}$ pound left over.

**35.** $136 \times \dfrac{3}{8} = \dfrac{136}{1} \times \dfrac{3}{8} = \dfrac{17 \times 8 \times 3}{8}$

$\qquad = 17 \times 3 = 51$ computers

**36.** $12{,}000 \div \dfrac{3}{5} = \dfrac{12{,}000}{1} \times \dfrac{5}{3}$

$\qquad = 4000 \times 5$

$\qquad = 20{,}000$ homes

$20{,}000 - 12{,}000 = 8000$ homes to be inspected

**37.** $132 \div 8\dfrac{1}{4} = 132 \div \dfrac{33}{4} = \dfrac{132}{1} \times \dfrac{4}{33}$

$\qquad = \dfrac{33 \times 4 \times 4}{33} = 4 \times 4$

$\qquad = 16$ hours

**38.** $56\dfrac{1}{2} \div 8\dfrac{1}{4} = \dfrac{113}{2} \div \dfrac{33}{4} = \dfrac{113}{2} \times \dfrac{4}{33}$

$\qquad = \dfrac{113 \times 2 \times 2}{2 \times 33}$

$\qquad = \dfrac{226}{33} = 6\dfrac{28}{33}$

He can make 6 full tents, with

$8\dfrac{1}{4} \times \dfrac{28}{33} = \dfrac{33}{4} \times \dfrac{28}{33} = 7$ yards left over

**39.** $12 \div \dfrac{9}{120} = \dfrac{12}{1} \times \dfrac{120}{9} = \dfrac{3 \times 4 \times 3 \times 40}{3 \times 3}$

$\qquad = 4 \times 40 = 160$ days

## 2.6   Exercises

1. $8 = 2 \times 2 \times 2$
$14 = 2 \times 7$
$LCM = 2 \times 2 \times 2 \times 7 = 56$

3. $20 = 2 \times 2 \times 5$
$50 = 2 \times 5 \times 5$
$LCM = 2 \times 2 \times 5 \times 5 = 100$

5. $12 = 2 \times 2 \times 3$
$15 = 3 \times 5$
$LCM = 2 \times 2 \times 3 \times 5 = 60$

7. $9 = 3 \times 3$
$36 = 2 \times 2 \times 3 \times 3$
$LCM = 2 \times 2 \times 3 \times 3 = 36$

9. $21 = 3 \times 7$
$49 = 7 \times 7$
$LCM = 3 \times 7 \times 7 = 147$

11. $5 = 5$
$10 = 2 \times 5$
$LCD = 2 \times 5 = 10$

13. $7 = 7$
$4 = 2 \times 2$
$LCD = 2 \times 2 \times 7 = 28$

15. $5 = 5$
$7 = 7$
$LCD = 5 \times 7 = 35$

17. $9 = 3 \times 3$
$6 = 2 \times 3$
$LCD = 2 \times 3 \times 3 = 18$

19. $12 = 2 \times 2 \times 3$
$15 = 3 \times 5$
$LCD = 2 \times 2 \times 3 \times 5 = 60$

21. $16 = 2 \times 2 \times 2 \times 2$
$4 = 2 \times 2$
$LCD = 2 \times 2 \times 2 \times 2 = 16$

23. $10 = 2 \times 5$
$45 = 3 \times 3 \times 5$
$LCD = 2 \times 3 \times 3 \times 5 = 90$

25. $12 = 2 \times 2 \times 3$
$30 = 2 \times 3 \times 5$
$LCD = 2 \times 2 \times 3 \times 5 = 60$

27. $21 = 3 \times 7$
$35 = 5 \times 7$
$LCD = 3 \times 5 \times 7 = 105$

29. $18 = 2 \times 3 \times 3$
$12 = 2 \times 2 \times 3$
$LCD = 2 \times 2 \times 3 \times 3 = 36$

31. $3 = 3$
$2 = 2$
$6 = 2 \times 3$
$LCD = 2 \times 3 = 6$

33. $24 = 2 \times 2 \times 2 \times 3$
$15 = 3 \times 5$
$30 = 2 \times 3 \times 5$
$LCD = 2 \times 2 \times 2 \times 3 \times 5 = 120$

35. $11 = 11$
$12 = 2 \times 2 \times 3$
$6 = 2 \times 3$
$LCD = 2 \times 2 \times 3 \times 11 = 132$

**37.** $12 = 2 \times 2 \times 3$
$21 = 3 \times 7$
$14 = 2 \times 7$
$LCD = 2 \times 2 \times 3 \times 7 = 84$

**39.** $15 = 3 \times 5$
$12 = 2 \times 2 \times 3$
$8 = 2 \times 2 \times 2$
$LCD = 2 \times 2 \times 2 \times 3 \times 5 = 120$

**41.** $\dfrac{1}{3} = \dfrac{1}{3} \times \dfrac{3}{3} = \dfrac{3}{9}$
$3$

**43.** $\dfrac{5}{6} = \dfrac{5}{6} \times \dfrac{9}{9} = \dfrac{45}{54}$
$45$

**45.** $\dfrac{4}{11} = \dfrac{4}{11} \times \dfrac{5}{5} = \dfrac{20}{55}$
$20$

**47.** $\dfrac{7}{24} = \dfrac{7}{24} \times \dfrac{2}{2} = \dfrac{14}{48}$
$14$

**49.** $\dfrac{8}{9} = \dfrac{8}{9} \times \dfrac{12}{12} = \dfrac{96}{108}$
$96$

**51.** $\dfrac{7}{20} = \dfrac{7}{20} \times \dfrac{9}{9} = \dfrac{63}{180}$
$63$

**53.** $\quad\dfrac{7}{12} \quad$ and $\quad \dfrac{5}{9}$
$\dfrac{7}{12} \times \dfrac{3}{3} \quad$ and $\quad \dfrac{5}{9} \times \dfrac{4}{4}$
$\quad\dfrac{21}{36} \quad$ and $\quad \dfrac{20}{36}$

**55.** $\quad\dfrac{3}{25} \quad$ and $\quad \dfrac{7}{40}$
$\dfrac{3}{25} \times \dfrac{8}{8} \quad$ and $\quad \dfrac{7}{40} \times \dfrac{5}{5}$
$\quad\dfrac{24}{200} \quad$ and $\quad \dfrac{35}{200}$

**57.** $\quad\dfrac{9}{10} \quad$ and $\quad \dfrac{19}{20}$
$\dfrac{9}{10} \times \dfrac{16}{16} \quad$ and $\quad \dfrac{19}{20} \times \dfrac{8}{8}$
$\quad\dfrac{144}{160} \quad$ and $\quad \dfrac{152}{160}$

**59.** $\quad 5 = 5$
$35 = 5 \times 7$
$LCD = 5 \times 7 = 35$
$\quad\dfrac{2}{5} \quad$ and $\quad \dfrac{9}{35}$
$\dfrac{2}{5} \times \dfrac{7}{7} \quad$ and $\quad \dfrac{9}{35}$
$\quad\dfrac{14}{35} \quad$ and $\quad \dfrac{9}{35}$

**61.** $\quad 12 = 2 \times 2 \times 3$
$16 = 2 \times 2 \times 2 \times 2$
$LCD = 2 \times 2 \times 2 \times 2 \times 3 = 48$
$\quad\dfrac{5}{12} \quad$ and $\quad \dfrac{1}{16}$
$\dfrac{5}{12} \times \dfrac{4}{4} \quad$ and $\quad \dfrac{1}{16} \times \dfrac{3}{3}$
$\quad\dfrac{20}{48} \quad$ and $\quad \dfrac{3}{48}$

**63.** $\quad 20 = 2 \times 2 \times 5$
$16 = 2 \times 2 \times 2 \times 2$
$LCD = 2 \times 2 \times 2 \times 2 \times 5 = 80$
$\quad\dfrac{13}{20} \quad$ and $\quad \dfrac{11}{16}$
$\dfrac{13}{20} \times \dfrac{4}{4} \quad$ and $\quad \dfrac{11}{16} \times \dfrac{5}{5}$
$\quad\dfrac{52}{80} \quad$ and $\quad \dfrac{55}{80}$

**65.**  $12 = 2 \times 2 \times 3$
$15 = 3 \times 5$
$LCD = 2 \times 2 \times 3 \times 5 = 60$

$$\frac{4}{15} \quad \text{and} \quad \frac{5}{12}$$

$$\frac{4}{15} \times \frac{4}{4} \quad \text{and} \quad \frac{5}{12} \times \frac{5}{5}$$

$$\frac{16}{20} \quad \text{and} \quad \frac{25}{60}$$

**67.**  $24 = 2 \times 2 \times 2 \times 3$
$36 = 2 \times 2 \times 3 \times 3$
$72 = 2 \times 2 \times 2 \times 3 \times 3$
$LCD = 2 \times 2 \times 2 \times 3 \times 3 = 72$

$$\frac{5}{24} \times \frac{3}{3}, \frac{11}{36} \times \frac{2}{2}, \frac{3}{72}$$

$$\frac{15}{72}, \frac{22}{72}, \frac{3}{72}$$

**69.**  $56 = 2 \times 2 \times 2 \times 7$
$8 = 2 \times 2 \times 2$
$7 = 7$
$LCD = 2 \times 2 \times 2 \times 7 = 56$

$$\frac{3}{56}, \frac{7}{8} \times \frac{7}{7}, \frac{5}{7} \times \frac{8}{8}$$

$$\frac{3}{56}, \frac{49}{56}, \frac{40}{56}$$

**71.**  $63 = 3 \times 3 \times 7$
$21 = 3 \times 7$
$9 = 3 \times 3$
$LCD = 3 \times 3 \times 7 = 63$

$$\frac{5}{63}, \frac{4}{21} \times \frac{3}{3}, \frac{8}{9} \times \frac{7}{7}$$

$$\frac{5}{63}, \frac{12}{63}, \frac{56}{63}$$

**73.**  a.   $16 = 2 \times 2 \times 2 \times 2$
$4 = 2 \times 2$
$8 = 2 \times 2 \times 2$
$LCD = 2 \times 2 \times 2 \times 2 = 16$

b.  $\dfrac{3}{16}, \dfrac{3}{4} \times \dfrac{4}{4}, \dfrac{3}{8} \times \dfrac{2}{2}$

$$\frac{3}{16}, \frac{2}{16}, \frac{6}{16}$$

---

## Cumulative Review Problems

**75.**
$$\begin{array}{r} 178 \\ 32\overline{)5699} \\ 32 \phantom{00} \\ \hline 249 \phantom{0} \\ 224 \phantom{0} \\ \hline 259 \\ 256 \\ \hline 3 \end{array}$$

178 R3

**77.**
$$\begin{array}{r} 369 \\ \times\ 27 \\ \hline 2583 \\ 738 \phantom{0} \\ \hline 9963 \end{array}$$

**79.**

| $11,205$ | $13,359$ | $15,569$ | $18,340$ |
|---|---|---|---|
| $-\ 8,414$ | $-\ 11,205$ | $-\ 13,359$ | $-\ 15,569$ |
| $2,791$ | $2,154$ | $2,210$ | $2,771$ |

Smallest increase is $2,154 between 1985 and 1990.

**81.**
$$\begin{array}{r} 18,340 \\ -\ 9,700 \\ \hline 8,640 \end{array}$$

Income must be at least $8640 per year.

**83.**
$$\begin{array}{r} 18,340 \\ -\ 13,359 \\ \hline 4,981 \end{array}$$

For year 2010, the poverty level is

$$\begin{array}{r} 18,340 \\ +\ 4,981 \\ \hline \$23,321 \end{array}$$

---

## 2.7   Exercises

1. $\dfrac{5}{9} + \dfrac{2}{9} = \dfrac{5+2}{9} = \dfrac{7}{9}$

3. $\dfrac{5}{18} + \dfrac{7}{18} = \dfrac{12}{18} = \dfrac{2}{3}$

5. $\dfrac{5}{24} - \dfrac{3}{24} = \dfrac{2}{24} = \dfrac{1}{12}$

7. $\dfrac{53}{88} - \dfrac{19}{88} = \dfrac{34}{88} = \dfrac{17}{44}$

9. $\dfrac{2}{7} + \dfrac{1}{2} = \dfrac{2}{7} \times \dfrac{2}{2} + \dfrac{1}{2} \times \dfrac{7}{7}$
   $\phantom{\dfrac{2}{7} + \dfrac{1}{2}} = \dfrac{4}{14} + \dfrac{7}{14}$
   $\phantom{\dfrac{2}{7} + \dfrac{1}{2}} = \dfrac{11}{14}$

11. $\dfrac{3}{10} + \dfrac{3}{20} = \dfrac{3}{10} \times \dfrac{2}{2} + \dfrac{3}{20}$
   $\phantom{\dfrac{3}{10} + \dfrac{3}{20}} = \dfrac{6}{20} + \dfrac{3}{20}$
   $\phantom{\dfrac{3}{10} + \dfrac{3}{20}} = \dfrac{9}{20}$

13. $\dfrac{1}{8} + \dfrac{3}{4} = \dfrac{1}{8} + \dfrac{3}{4} \times \dfrac{2}{2}$
   $\phantom{\dfrac{1}{8} + \dfrac{3}{4}} = \dfrac{1}{8} + \dfrac{6}{6}$
   $\phantom{\dfrac{1}{8} + \dfrac{3}{4}} = \dfrac{7}{8}$

15. $\dfrac{4}{5} + \dfrac{7}{20} = \dfrac{4}{5} \times \dfrac{4}{4} + \dfrac{7}{20}$
   $\phantom{\dfrac{4}{5} + \dfrac{7}{20}} = \dfrac{16}{20} + \dfrac{7}{20} = \dfrac{23}{20} = 1\dfrac{3}{20}$

17. $\dfrac{3}{10} + \dfrac{7}{100} = \dfrac{3}{10} \times \dfrac{10}{10} + \dfrac{7}{100}$
   $\phantom{\dfrac{3}{10} + \dfrac{7}{100}} = \dfrac{30}{100} + \dfrac{7}{100}$
   $\phantom{\dfrac{3}{10} + \dfrac{7}{100}} = \dfrac{37}{100}$

19. $\dfrac{3}{25} + \dfrac{1}{35} = \dfrac{3}{25} \times \dfrac{7}{7} + \dfrac{1}{35} \times \dfrac{5}{5}$
   $\phantom{\dfrac{3}{25} + \dfrac{1}{35}} = \dfrac{21}{175} + \dfrac{5}{175}$
   $\phantom{\dfrac{3}{25} + \dfrac{1}{35}} = \dfrac{26}{175}$

21. $\dfrac{7}{8} + \dfrac{5}{12} = \dfrac{7}{8} \times \dfrac{3}{3} + \dfrac{5}{12} \times \dfrac{2}{2}$
   $\phantom{\dfrac{7}{8} + \dfrac{5}{12}} = \dfrac{21}{24} + \dfrac{10}{24} = \dfrac{31}{24}$
   $\phantom{\dfrac{7}{8} + \dfrac{5}{12}} = 1\dfrac{7}{24}$

23. $\dfrac{3}{8} + \dfrac{3}{10} = \dfrac{3}{8} \times \dfrac{5}{5} + \dfrac{3}{10} \times \dfrac{4}{4}$
   $\phantom{\dfrac{3}{8} + \dfrac{3}{10}} = \dfrac{15}{40} + \dfrac{12}{40}$
   $\phantom{\dfrac{3}{8} + \dfrac{3}{10}} = \dfrac{27}{40}$

25. $\dfrac{5}{12} - \dfrac{1}{6} = \dfrac{5}{12} - \dfrac{1}{6} \times \dfrac{2}{2}$
   $\phantom{\dfrac{5}{12} - \dfrac{1}{6}} = \dfrac{5}{12} - \dfrac{2}{12} = \dfrac{3}{12}$
   $\phantom{\dfrac{5}{12} - \dfrac{1}{6}} = \dfrac{1}{4}$

27. $\dfrac{3}{7} - \dfrac{1}{5} = \dfrac{3}{7} \times \dfrac{5}{5} - \dfrac{1}{5} \times \dfrac{7}{7}$
   $\phantom{\dfrac{3}{7} - \dfrac{1}{5}} = \dfrac{15}{35} - \dfrac{7}{35}$
   $\phantom{\dfrac{3}{7} - \dfrac{1}{5}} = \dfrac{8}{35}$

**29.** $\dfrac{7}{9} - \dfrac{2}{27} = \dfrac{7}{9} \times \dfrac{3}{3} - \dfrac{2}{27}$

$\qquad\qquad = \dfrac{21}{27} - \dfrac{2}{27}$

$\qquad\qquad = \dfrac{19}{27}$

**31.** $\dfrac{5}{12} - \dfrac{7}{30} = \dfrac{5}{12} \times \dfrac{5}{5} - \dfrac{7}{30} \times \dfrac{2}{2}$

$\qquad\qquad = \dfrac{25}{60} - \dfrac{14}{60}$

$\qquad\qquad = \dfrac{11}{60}$

**33.** $\dfrac{11}{12} - \dfrac{2}{3} = \dfrac{11}{12} - \dfrac{2}{3} \times \dfrac{4}{4}$

$\qquad\qquad = \dfrac{11}{12} - \dfrac{8}{12}$

$\qquad\qquad = \dfrac{3}{12} = \dfrac{1}{4}$

**35.** $\dfrac{16}{25} - \dfrac{2}{5} = \dfrac{16}{25} - \dfrac{2}{5} \times \dfrac{5}{5}$

$\qquad\qquad = \dfrac{16}{25} - \dfrac{10}{25}$

$\qquad\qquad = \dfrac{6}{25}$

**37.** $\dfrac{5}{12} - \dfrac{5}{16} = \dfrac{5}{12} \times \dfrac{4}{4} - \dfrac{5}{16} \times \dfrac{3}{3}$

$\qquad\qquad = \dfrac{20}{48} - \dfrac{15}{48}$

$\qquad\qquad = \dfrac{5}{48}$

**39.** $\dfrac{10}{16} - \dfrac{5}{8} = \dfrac{10}{16} - \dfrac{5}{8} \times \dfrac{2}{2}$

$\qquad\qquad = \dfrac{10}{16} - \dfrac{10}{16} = 0$

**41.** $\dfrac{23}{36} - \dfrac{2}{9} = \dfrac{23}{36} - \dfrac{2}{9} \times \dfrac{4}{4}$

$\qquad\qquad = \dfrac{23}{36} - \dfrac{8}{36} = \dfrac{15}{36}$

$\qquad\qquad = \dfrac{5}{12}$

**43.** $\dfrac{4}{5} + \dfrac{1}{20} + \dfrac{3}{4} = \dfrac{4}{5} \times \dfrac{4}{4} + \dfrac{1}{20} + \dfrac{3}{4} \times \dfrac{5}{5}$

$\qquad\qquad = \dfrac{16}{20} + \dfrac{1}{20} + \dfrac{15}{20} = \dfrac{32}{20}$

$\qquad\qquad = \dfrac{8}{5} = 1\dfrac{3}{5}$

**45.** $\dfrac{5}{30} + \dfrac{3}{40} + \dfrac{1}{8}$

$\qquad = \dfrac{5}{30} \times \dfrac{4}{4} + \dfrac{3}{40} \times \dfrac{3}{3} + \dfrac{1}{8} \times \dfrac{15}{15}$

$\qquad = \dfrac{20}{120} + \dfrac{9}{120} + \dfrac{15}{120}$

$\qquad = \dfrac{44}{120} = \dfrac{11}{30}$

**47.** $\dfrac{7}{30} + \dfrac{2}{5} + \dfrac{5}{6}$

$\qquad = \dfrac{7}{30} + \dfrac{2}{5} \times \dfrac{6}{6} + \dfrac{5}{6} \times \dfrac{5}{5}$

$\qquad = \dfrac{7}{30} + \dfrac{12}{30} + \dfrac{25}{30}$

$\qquad = \dfrac{44}{30} = \dfrac{22}{15} = 1\dfrac{7}{15}$

**49.** $x + \dfrac{1}{7} = \dfrac{5}{14}$

$\qquad x + \dfrac{1}{7} \times \dfrac{2}{2} = \dfrac{5}{14}$

$\qquad x + \dfrac{2}{14} = \dfrac{5}{14}$

$\qquad \dfrac{3}{14} + \dfrac{2}{14} = \dfrac{5}{14}$

$\qquad x = \dfrac{3}{14}$

**51.**
$$x + \frac{2}{3} = \frac{9}{11}$$
$$x + \frac{2}{3} \times \frac{11}{11} = \frac{9}{11} \times \frac{3}{3}$$
$$x + \frac{22}{33} = \frac{27}{33}$$
$$\frac{5}{33} + \frac{22}{33} = \frac{27}{33}$$
$$x = \frac{5}{33}$$

**53.**
$$x - \frac{1}{5} = \frac{4}{12}$$
$$x - \frac{1}{5} \times \frac{12}{12} = \frac{4}{12} \times \frac{5}{5}$$
$$x - \frac{12}{60} = \frac{20}{60}$$
$$\frac{32}{60} - \frac{12}{60} = \frac{20}{60}$$
$$x = \frac{32}{60} = \frac{8}{15}$$

**55.**
$$\frac{3}{4} + \frac{2}{3} = \frac{3}{4} \times \frac{3}{3} + \frac{2}{3} \times \frac{4}{4}$$
$$= \frac{9}{12} + \frac{8}{12}$$
$$= \frac{17}{12}$$
$$= 1\frac{5}{12} \text{ cups}$$

**57.**
$$\frac{2}{3} + \frac{5}{6} = \frac{2}{3} \times \frac{2}{2} + \frac{5}{6}$$
$$= \frac{4}{6} + \frac{5}{6}$$
$$= \frac{9}{6} = \frac{3}{2} = 1\frac{1}{2} \text{ pounds}$$

**59.**
$$\frac{11}{12} - \frac{3}{5} = \frac{11}{12} \times \frac{5}{5} - \frac{3}{5} \times \frac{12}{12}$$
$$= \frac{55}{60} - \frac{36}{60}$$
$$= \frac{19}{60} \text{ lost}$$

**61.** Before he ate half, there were
$$6 \div \frac{1}{2} = 6 \times 2 = 12$$
chocolates. While walking, he ate $\frac{1}{4}$ of the chocolates, leaving
$$1 - \frac{1}{4} = \frac{3}{4}$$
in the box. The box had
$$12 \div \frac{3}{4} = \frac{12}{1} \times \frac{4}{3} = 16 \text{ chocolates.}$$

**63.**
$$\frac{5}{6} - \frac{9}{14} = \frac{35}{42} - \frac{27}{42}$$
$$= \frac{8}{42}$$
$$= \frac{4}{21}$$
$\frac{4}{21}$ of the membership

## Cumulative Review Problems

**65.** $\dfrac{15}{85} = \dfrac{15 \div 5}{85 \div 5} = \dfrac{3}{17}$

**67.**
$$16)\overline{\phantom{1}123}\phantom{xx} \quad \frac{123}{16} = 7\frac{11}{16}$$

with work:
$$\begin{array}{r} 7 \\ 16)\overline{123} \\ \underline{112} \\ 11 \end{array} \qquad \frac{123}{16} = 7\frac{11}{16}$$

**69.**
$$2\frac{1}{2} \times 3\frac{3}{4} = \frac{5}{2} \times \frac{15}{4}$$
$$= \frac{75}{8} = 9\frac{3}{8}$$

## 2.8   Exercises

**1.**
$$\begin{array}{r} 7\frac{1}{8} \\ + 2\frac{5}{8} \\ \hline 9\frac{6}{8} = 9\frac{3}{4} \end{array}$$

**3.**
$$\begin{array}{r} 15\frac{3}{14} \\ - 11\frac{1}{14} \\ \hline 4\frac{2}{14} = 4\frac{1}{7} \end{array}$$

**5.**  $12\frac{1}{3}$       $12\frac{2}{6}$

$\phantom{12}+ 5\frac{1}{6}$      $\phantom{12}+ 5\frac{1}{6}$

$\phantom{12+5}\overline{17\frac{3}{6}} = 17\frac{1}{2}$

**7.**  $\phantom{1}5\frac{4}{5}$       $\phantom{1}5\frac{8}{10}$

$+ 10\frac{3}{10}$      $\phantom{1}+ 10\frac{3}{10}$

$\phantom{+10}\overline{15\frac{11}{10}} = 16\frac{1}{10}$

**9.**  $1$       $\frac{7}{7}$

$- \frac{3}{7}$      $- \frac{3}{7}$

$\phantom{-}\overline{\frac{4}{7}}$

**11.**  $1\frac{5}{6}$       $1\frac{20}{24}$

$+ \phantom{1}\frac{7}{8}$      $+ \phantom{1}\frac{21}{24}$

$\phantom{+1}\overline{1\frac{41}{24}} = 2\frac{17}{24}$

**13.**  $7\frac{1}{2}$       $7\frac{2}{4}$

$+ 8\frac{3}{4}$      $+ 8\frac{3}{4}$

$\phantom{+8}\overline{15\frac{5}{4}} = 16\frac{1}{4}$

**15.**  $9\frac{2}{3}$       $9\frac{16}{24}$

$- 7\frac{1}{8}$      $- 7\frac{3}{24}$

$\phantom{-7}\overline{2\frac{13}{24}}$

**17.**  $12\frac{1}{3}$       $12\frac{5}{15}$       $11\frac{20}{15}$

$- 7\frac{2}{5}$      $- 7\frac{6}{15}$      $- 7\frac{6}{15}$

$\phantom{-7-7}\overline{4\frac{14}{15}}$

**19.**  $30$       $29\frac{7}{7}$

$- 15\frac{3}{7}$      $- 15\frac{3}{7}$

$\phantom{-15}\overline{14\frac{4}{7}}$

**21.**  $15\frac{4}{15}$

$+ 26\frac{8}{15}$

$\overline{41\frac{12}{15}} = 41\frac{4}{5}$

**23.**  $4\frac{1}{3}$       $4\frac{4}{12}$

$+ 2\frac{1}{4}$      $+ 2\frac{3}{12}$

$\phantom{+2}\overline{6\frac{7}{12}}$

**25.**  $3\frac{3}{4}$       $3\frac{9}{12}$

$+ 4\frac{5}{12}$      $+ 4\frac{5}{12}$

$\phantom{+4}\overline{7\frac{14}{12}} = 8\frac{2}{12} = 8\frac{1}{6}$

**27.**  $47\frac{3}{10}$       $47\frac{12}{40}$

$+ 26\frac{5}{8}$      $+ 26\frac{25}{40}$

$\phantom{+26}\overline{73\frac{37}{40}}$

**29.**  $19\frac{5}{6}$       $19\frac{5}{6}$

$- 14\frac{1}{3}$      $- 14\frac{2}{6}$

$\phantom{-14}\overline{5\frac{3}{6}} = 5\frac{1}{2}$

**31.**  $5\frac{2}{12}$       $5\frac{10}{24}$

$- 5\frac{10}{24}$      $- 5\frac{10}{24}$

$\phantom{-5}\overline{0}$

**33.**  $12\frac{3}{20}$       $12\frac{9}{60}$       $11\frac{69}{60}$

$- 7\frac{7}{15}$      $- 7\frac{28}{60}$      $- 7\frac{28}{60}$

$\phantom{-7-7}\overline{4\frac{41}{60}}$

**35.**  $12$       $11\frac{15}{15}$

$- 3\frac{7}{15}$      $- 3\frac{7}{15}$

$\phantom{-3}\overline{8\frac{8}{15}}$

**37.**  $120$       $119\frac{8}{8}$

$- 17\frac{3}{8}$      $- 17\frac{3}{8}$

$\phantom{-17}\overline{102\frac{5}{8}}$

**39.**  $3\frac{1}{8}$       $3\frac{3}{24}$

$2\frac{1}{3}$      $2\frac{8}{24}$

$+ 7\frac{3}{4}$      $+ 7\frac{18}{24}$

$\phantom{+7}\overline{12\frac{29}{24}} = 13\frac{5}{24}$

**41.**  $20\frac{3}{4}$       $20\frac{6}{8}$

$+ 22\frac{3}{8}$      $+ 22\frac{3}{8}$

$\phantom{+22}\overline{42\frac{9}{8}} = 43\frac{1}{8}$

$43\frac{1}{8}$ miles

**43.**

$$
\begin{array}{r} 6\frac{3}{8} \\ -\ 4\frac{1}{3} \\ \hline \end{array}
\qquad
\begin{array}{r} 6\frac{9}{24} \\ -\ 4\frac{8}{24} \\ \hline 2\frac{1}{24} \end{array}
$$

$2\frac{1}{24}$ pounds

**45.**

$$
\begin{array}{r} 72\frac{1}{2} \\ -\ 69\frac{3}{4} \\ \hline \end{array}
\qquad
\begin{array}{r} 72\frac{2}{4} \\ -\ 69\frac{3}{4} \\ \hline \end{array}
\qquad
\begin{array}{r} 71\frac{6}{4} \\ -\ 69\frac{3}{4} \\ \hline 2\frac{3}{4} \end{array}
$$

$2\frac{3}{4}$ inches

**47.**

**a.**

$$
\begin{array}{r} 5\frac{1}{8} \\ +\ 4\frac{2}{3} \\ \hline \end{array}
\qquad
\begin{array}{r} 5\frac{3}{24} \\ +\ 4\frac{16}{24} \\ \hline 9\frac{19}{24} \end{array}
$$

$9\frac{19}{24}$ pounds

**b.**

$$
\begin{array}{r} 16 \\ -\ 9\frac{19}{24} \\ \hline \end{array}
\qquad
\begin{array}{r} 15\frac{24}{24} \\ -\ 9\frac{19}{24} \\ \hline 6\frac{5}{24} \end{array}
$$

$6\frac{5}{24}$ pounds

**49.** $\dfrac{379}{8} + \dfrac{89}{5} = \dfrac{1895}{40} + \dfrac{712}{40}$

$$= \dfrac{2607}{40}$$

$$= 65\dfrac{7}{40}$$

**51.** Estimate: $35 + 24 = 59$

Exact: $\quad \begin{array}{r} 35\frac{1}{6} \\ +\ 24\frac{5}{12} \\ \hline \end{array} \qquad \begin{array}{r} 35\frac{2}{12} \\ +\ 24\frac{5}{12} \\ \hline 59\frac{7}{12} \end{array}$

Difference: $\quad \begin{array}{r} 59\frac{7}{12} \\ -\ 59 \\ \hline \frac{7}{12} \end{array}$

**53.** $\dfrac{6}{7} - \dfrac{4}{7} \times \dfrac{1}{3} = \dfrac{6}{7} - \dfrac{4}{21}$

$$= \dfrac{6}{7} \times \dfrac{3}{3} - \dfrac{4}{21}$$

$$= \dfrac{18}{21} - \dfrac{4}{21} = \dfrac{14}{21}$$

$$= \dfrac{2}{3}$$

**55.** $\dfrac{1}{2} + \dfrac{3}{8} \div \dfrac{3}{4} = \dfrac{1}{2} + \dfrac{3}{8} \times \dfrac{4}{3}$

$$= \dfrac{1}{2} + \dfrac{1}{2} = \dfrac{2}{2}$$

$$= 1$$

**57.** $\dfrac{5}{7} \times \dfrac{7}{2} \div \dfrac{3}{2} = \dfrac{5}{7} \times \dfrac{7}{2} \times \dfrac{2}{3} = \dfrac{5}{3} = 1\dfrac{2}{3}$

**59.** $\dfrac{3}{5} \times \dfrac{1}{2} + \dfrac{1}{5} \div \dfrac{2}{3} = \dfrac{3}{5} \times \dfrac{1}{2} + \dfrac{1}{5} \times \dfrac{3}{2}$

$$= \dfrac{3}{10} + \dfrac{3}{10} = \dfrac{6}{10}$$

$$= \dfrac{3}{5}$$

**61.** $\left(\dfrac{3}{5} - \dfrac{3}{20}\right) \times \dfrac{4}{5} = \left(\dfrac{12}{20} - \dfrac{3}{20}\right) \times \dfrac{4}{5}$

$$= \dfrac{9}{20} \times \dfrac{4}{5}$$

$$= \dfrac{9}{25}$$

**63.** $\dfrac{8}{7} \div \left(\dfrac{2}{3} + \dfrac{1}{12}\right) = \dfrac{8}{7} \div \left(\dfrac{8}{12} + \dfrac{1}{12}\right)$

$$= \dfrac{8}{7} \div \dfrac{9}{12}$$

$$= \dfrac{8}{7} \times \dfrac{12}{9} = \dfrac{32}{21}$$

$$= 1\dfrac{11}{21}$$

**65.** $\dfrac{1}{4} \times \left(\dfrac{2}{3}\right)^2 = \dfrac{1}{4} \times \dfrac{4}{9} = \dfrac{1}{9}$

**67.** $\left(\dfrac{4}{3}\right)^2 \times \dfrac{9}{11} = \dfrac{16}{9} \times \dfrac{9}{11} = \dfrac{16}{11} = 1\dfrac{5}{11}$

---

## Cumulative Review Problems

**69.**
$$\begin{array}{r} 6737 \\ \times\ \ \ \ 76 \\ \hline 40\ 422 \\ 471\ 59 \\ \hline 512{,}012 \end{array}$$

**71.** Changes are

$$25 \times 27 + 520 + 30 \times 8 + 2972$$
$$= 675 + 520 + 240 + 2972$$
$$= \$4407$$

Amount left over is

$$6300 - 4407 = \$1893$$

---

## Putting Your Skills To Work

**1.** Eggs: $20 \times \dfrac{1}{2} = 10$

Milk: $5 \times \dfrac{1}{2} = \dfrac{5}{2} = 2\dfrac{1}{2}$

Shortening: $4\dfrac{1}{3} \times \dfrac{1}{2} = \dfrac{13}{3} \times \dfrac{1}{2} = \dfrac{13}{6} = 2\dfrac{1}{6}$

Orange Peel: $6 \times \dfrac{1}{2} = 3$

Lemon Peel: $9\dfrac{1}{2} \times \dfrac{1}{2} = \dfrac{19}{2} \times \dfrac{1}{2} = \dfrac{19}{4} = 4\dfrac{3}{4}$

Flour: $16\dfrac{1}{3} \times \dfrac{1}{2} = \dfrac{49}{3} \times \dfrac{1}{2} = \dfrac{49}{6} = 8\dfrac{1}{6}$

Baking Powder: $15\dfrac{3}{4} \times \dfrac{1}{2} = \dfrac{63}{4} \times \dfrac{1}{2} = \dfrac{63}{8} = 7\dfrac{7}{8}$

Salt: $4\dfrac{1}{2} \times \dfrac{1}{2} = \dfrac{9}{2} \times \dfrac{1}{2} = \dfrac{9}{4} = 2\dfrac{1}{4}$

Lemon Juice: $12 \times \dfrac{1}{2} = 6$

Sugar: $7\dfrac{1}{4} \times \dfrac{1}{2} = \dfrac{29}{4} \times \dfrac{1}{2} = \dfrac{29}{8} = 3\dfrac{5}{8}$

**3.** Eggs: $20 \times \dfrac{5}{8} = \dfrac{100}{8} = \dfrac{25}{2} = 12\dfrac{1}{2}$

Milk: $5 \times \dfrac{5}{8} = \dfrac{25}{8} = 3\dfrac{1}{8}$

Shortening: $4\dfrac{1}{3} \times \dfrac{5}{8} = \dfrac{13}{3} \times \dfrac{5}{8} = \dfrac{65}{24} = 2\dfrac{17}{24}$

Orange Peel: $6 \times \dfrac{5}{8} = \dfrac{30}{8} = \dfrac{15}{4} = 3\dfrac{3}{4}$

Lemon Peel: $9\dfrac{1}{2} \times \dfrac{5}{8} = \dfrac{19}{2} \times \dfrac{5}{8} = \dfrac{95}{16} = 5\dfrac{15}{16}$

Flour: $16\dfrac{1}{3} \times \dfrac{5}{8} = \dfrac{49}{3} \times \dfrac{5}{8} = \dfrac{245}{24} = 10\dfrac{5}{24}$

Baking Powder: $15\dfrac{3}{4} \times \dfrac{5}{8} = \dfrac{63}{4} \times \dfrac{5}{8} = \dfrac{315}{32} = 9\dfrac{27}{32}$

Salt: $4\dfrac{1}{2} \times \dfrac{5}{8} = \dfrac{9}{2} \times \dfrac{5}{8} = \dfrac{45}{16} = 2\dfrac{13}{16}$

Lemon Juice: $12 \times \dfrac{5}{8} = \dfrac{60}{8} = \dfrac{15}{2} = 7\dfrac{1}{2}$

Sugar: $7\dfrac{1}{4} \times \dfrac{5}{8} = \dfrac{29}{4} \times \dfrac{5}{8} = \dfrac{145}{32} = 4\dfrac{17}{32}$

---

## 2.9   Exercises

**1.**
$$\begin{array}{r} 5\frac{1}{4} \\ 2\frac{5}{6} \\ +\ 8\frac{7}{12} \\ \hline \end{array} \qquad \begin{array}{r} 5\frac{3}{12} \\ 2\frac{10}{12} \\ +\ 8\frac{7}{12} \\ \hline 15\frac{20}{12} = 16\frac{8}{12} = 16\frac{2}{3}\ \text{feet} \end{array}$$

**3.**
$$\begin{array}{r} 7\frac{5}{6} \\ 8\frac{1}{8} \\ +\ 9\frac{1}{2} \\ \hline \end{array} \qquad \begin{array}{r} 7\frac{20}{24} \\ 8\frac{3}{24} \\ +\ 9\frac{12}{24} \\ \hline 24\frac{35}{24} = 25\frac{11}{24}\ \text{tons} \end{array}$$

**5.** $\dfrac{1}{16} + \dfrac{3}{4} + \dfrac{1}{16} + \dfrac{3}{16} + \dfrac{1}{2}$

$$= \dfrac{1}{16} + \dfrac{12}{16} + \dfrac{1}{16} + \dfrac{3}{16} + \dfrac{8}{16}$$

$$= \dfrac{25}{16}$$

$$= 1\dfrac{9}{16}\ \text{inches}$$

**7.** Sum:

$$
\begin{array}{r} 6\frac{3}{4} \\ + 9\frac{1}{2} \\ \hline \end{array}
\qquad
\begin{array}{r} 6\frac{3}{4} \\ + 9\frac{2}{4} \\ \hline 15\frac{5}{4} = 16\frac{1}{4} \end{array}
$$

Miles to end:

$$
\begin{array}{r} 26\frac{1}{5} \\ - 16\frac{1}{4} \\ \hline \end{array}
\qquad
\begin{array}{r} 26\frac{4}{20} \\ - 16\frac{5}{20} \\ \hline \end{array}
\qquad
\begin{array}{r} 25\frac{24}{20} \\ - 16\frac{5}{20} \\ \hline 9\frac{19}{20} \text{ miles} \end{array}
$$

**9.** Sum:

$$
\begin{array}{r} 3\frac{1}{2} \\ 1\frac{5}{6} \\ + 1\frac{1}{5} \\ \hline \end{array}
\qquad
\begin{array}{r} 3\frac{15}{30} \\ 1\frac{25}{30} \\ + 1\frac{6}{30} \\ \hline 5\frac{46}{30} = 6\frac{16}{30} = 6\frac{8}{15} \end{array}
$$

Difference:

$$
\begin{array}{r} 7 \\ - 6\frac{8}{15} \\ \hline \end{array}
\qquad
\begin{array}{r} 6\frac{15}{15} \\ - 6\frac{8}{15} \\ \hline \frac{7}{15} \text{ of a foot} \end{array}
$$

**11.** $36\dfrac{3}{4} \times 7\dfrac{1}{2} = \dfrac{147}{4} \times \dfrac{15}{2}$

$$= \frac{2205}{8}$$

$$= 275\frac{5}{8} \text{ gallons}$$

**13.** $22\dfrac{1}{2} \times 4\dfrac{3}{4}$

$$= \frac{45}{2} \times \frac{19}{4}$$

$$= \frac{855}{8}$$

$$= 106\frac{7}{8}$$

$106\dfrac{7}{8}$ nautical miles

**15.** $\dfrac{1}{5} + \dfrac{1}{15} + \dfrac{1}{20} = \dfrac{12}{60} + \dfrac{4}{60} + \dfrac{3}{60}$

$$= \frac{12+4+3}{60}$$

$$= \frac{19}{60}$$

$\dfrac{19}{60}(660) = 209$, so \$209 is deducted.

$660 - 209 = 451$

She has \$451 per week left.

**17. a.** $36\dfrac{3}{4} \div \dfrac{7}{8} = \dfrac{147}{4} \div \dfrac{7}{8}$

$$= \frac{147}{4} \times \frac{8}{7}$$

$$= \frac{7 \times 21 \times 4 \times 2}{4 \times 7}$$

$$= 42$$

42 bags

**b.** $42 \times 3\dfrac{1}{2} = \dfrac{42}{1} \times \dfrac{7}{2} = 147$

\$147

**c.** $147 - 2 = 145$

\$145

**19. a.** $18\dfrac{1}{2} - 1\dfrac{1}{4} - 3\dfrac{1}{8}$

$$= \frac{37}{2} - \frac{5}{4} - \frac{25}{8} = \frac{148}{8} - \frac{10}{8} - \frac{25}{8}$$

$$= \frac{148 - 10 - 25}{8}$$

$$= \frac{113}{8}$$

$$= 14\frac{1}{8}$$

$14\dfrac{1}{8}$ ounces of bread

**b.**

$$
\begin{array}{r} 14\frac{3}{4} \\ - 14\frac{1}{8} \\ \hline \end{array}
\qquad
\begin{array}{r} 14\frac{6}{8} \\ - 14\frac{1}{8} \\ \hline \frac{5}{8} \end{array}
$$

$\dfrac{5}{8}$ of an ounce

**21. a.** $160\frac{1}{8} \div 5\frac{1}{4} = \frac{1281}{8} \div \frac{21}{4}$

$$= \frac{1281}{8} \times \frac{4}{21}$$

$$= \frac{61}{2}$$

$$= 30\frac{1}{2} \text{ knots}$$

**b.** $213\frac{1}{2} \div \frac{61}{2} = \frac{427}{2} \div \frac{61}{2}$

$$= \frac{427}{2} \times \frac{2}{61}$$

$$= \frac{427}{61}$$

$$= 7 \text{ hours}$$

**23. a.** $6856\frac{1}{4} \div 1\frac{1}{4} = \frac{27,425}{4} \div \frac{5}{4}$

$$= \frac{27,425}{4} \times \frac{4}{5}$$

$$= 5485 \text{ bushels}$$

**b.** $6856\frac{1}{4} \times 1\frac{3}{4} = \frac{27,425}{4} \times \frac{7}{4}$

$$= \frac{191,975}{16}$$

$$= 11,998\frac{7}{16} \text{ cubic feet}$$

**c.** $11,998\frac{7}{16} \div 1\frac{1}{4} = \frac{191,975}{16} \times \frac{4}{5}$

$$= \frac{38,395}{4}$$

$$= 9598\frac{3}{4} \text{ bushels}$$

**25.** Salad this year:

$$
\begin{array}{ll}
12\frac{3}{4} & 12\frac{27}{36} \\
10\frac{1}{3} & 10\frac{12}{36} \\
+\ 6\frac{8}{9} & +\ 6\frac{32}{36} \\
\hline
& 28\frac{71}{36} = 29\frac{35}{36}
\end{array}
$$

Salad last year:

$$29\frac{35}{36} \times \frac{1}{2} = \frac{1079}{36} \times \frac{1}{2} = \frac{1079}{72}$$

$$= 14\frac{71}{72} \text{ lbs}$$

## Cumulative Review Problems

**27.**
$$
\begin{array}{r}
16,846 \\
19,321 \\
+\ 8,078 \\
\hline
44,245
\end{array}
$$

**29.**
$$
\begin{array}{r}
1683 \\
\times\ \ \ 27 \\
\hline
11\ 781 \\
33\ 66\ \ \\
\hline
45,441
\end{array}
$$

## Chapter 2 Review Problems

**1.** $\frac{5}{12}$

**2.** $\frac{3}{8}$

**3.** $\frac{4}{7}$

**4.** $\frac{7}{9}$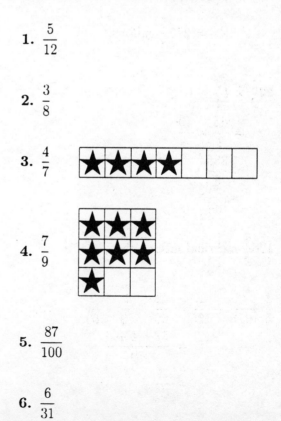

**5.** $\frac{87}{100}$

**6.** $\frac{6}{31}$

**7.** $54 = 2 \times 27$
$= 2 \times 3 \times 9$
$= 2 \times 3 \times 3 \times 3$
$= 2 \times 3^3$

**8.** $42 = 6 \times 7$
$= 2 \times 3 \times 7$

**9.** $168 = 8 \times 21$
$= 2 \times 2 \times 2 \times 3 \times 7$
$= 2^3 \times 3 \times 7$

**10.** Prime

**11.** $78 = 2 \times 39$
$= 2 \times 3 \times 13$

**12.** Prime

**13.** $\dfrac{12}{42} = \dfrac{12 \div 6}{42 \div 6} = \dfrac{2}{7}$

**14.** $\dfrac{13}{52} = \dfrac{13 \div 13}{52 \div 13} = \dfrac{1}{4}$

**15.** $\dfrac{21}{36} = \dfrac{21 \div 3}{36 \div 3} = \dfrac{7}{12}$

**16.** $\dfrac{26}{34} = \dfrac{26 \div 2}{34 \div 2}$
$= \dfrac{13}{17}$

**17.** $\dfrac{168}{192} = \dfrac{168 \div 24}{192 \div 24} = \dfrac{7}{8}$

**18.** $\dfrac{51}{105} = \dfrac{51 \div 3}{105 \div 3}$
$= \dfrac{17}{35}$

**19.** $4\dfrac{3}{8} = \dfrac{8 \times 4 + 3}{8} = \dfrac{35}{8}$

**20.** $2\dfrac{19}{23} = \dfrac{23 \times 2 + 19}{23}$
$= \dfrac{65}{23}$

**21.** $8\overline{)45}$ with quotient $5$, $\underline{40}$, remainder $5$     $\dfrac{45}{8} = 5\dfrac{5}{8}$

**22.** $13\overline{)63}$ with quotient $4$, $\underline{52}$, remainder $11$     $\dfrac{63}{13} = 4\dfrac{11}{13}$

**23.** $3\dfrac{15}{55} = 3\dfrac{5 \times 3}{5 \times 11} = 3\dfrac{3}{11}$

**24.** $\dfrac{234}{16} = \dfrac{117 \times 2}{8 \times 2} = \dfrac{117}{8}$

**25.** $240\overline{)385}$ with quotient $1$, $\underline{240}$, remainder $145$     $\dfrac{385}{240} = 1\dfrac{145}{240}$
$= 1\dfrac{29}{48}$

**26.** $\dfrac{4}{7} \times \dfrac{5}{11} = \dfrac{4 \times 5}{7 \times 11}$
$= \dfrac{20}{77}$

**27.** $\dfrac{7}{9} \times \dfrac{21}{35} = \dfrac{1}{3} \times \dfrac{7}{5} = \dfrac{7}{15}$

**28.** $12 \times \dfrac{3}{7} \times 0 = 0$

**29.** $\dfrac{3}{5} \times \dfrac{2}{7} \times \dfrac{10}{27} = \dfrac{1}{1} \times \dfrac{2}{7} \times \dfrac{2}{9} = \dfrac{4}{63}$

**30.** $12 \times 8\dfrac{1}{5} = \dfrac{12}{1} \times \dfrac{41}{5}$
$= \dfrac{492}{5} = 98\dfrac{2}{5}$

**31.** $5\dfrac{1}{4} \times 4\dfrac{6}{7} = \dfrac{21}{4} \times \dfrac{34}{7}$

$\qquad = \dfrac{3}{2} \times \dfrac{17}{1} = \dfrac{51}{2}$

$\qquad = 25\dfrac{1}{2}$

**32.** $5\dfrac{3}{8} \times 3\dfrac{4}{5} = \dfrac{43}{8} \times \dfrac{19}{5}$

$\qquad = \dfrac{817}{40} = 20\dfrac{17}{40}$

**33.** $35 \times \dfrac{7}{10} = \dfrac{35}{1} \times \dfrac{7}{10}$

$\qquad = \dfrac{7}{1} \times \dfrac{7}{2} = \dfrac{49}{2}$

$\qquad = 24\dfrac{1}{2}$

**34.** $37\dfrac{5}{8} \times 18 = \dfrac{301}{8} \times \dfrac{18}{1}$

$\qquad = \dfrac{301}{4} \times \dfrac{9}{1}$

$\qquad = \dfrac{2709}{4} = 677\dfrac{1}{4}$

$\$677\dfrac{1}{4}$

**35.** $18\dfrac{1}{5} \times 26\dfrac{3}{4} = \dfrac{91}{5} \times \dfrac{107}{4} = \dfrac{9737}{20}$

$\qquad = 486\dfrac{17}{20}$ square inches

**36.** $\dfrac{3}{7} \div \dfrac{2}{5} = \dfrac{3}{7} \times \dfrac{5}{2}$

$\qquad = \dfrac{15}{14} = 1\dfrac{1}{14}$

**37.** $\dfrac{9}{17} \div \dfrac{18}{5} = \dfrac{9}{17} \times \dfrac{5}{18} = \dfrac{5}{34}$

**38.** $1200 \div \dfrac{5}{8} = \dfrac{1200}{1} \times \dfrac{8}{5}$

$\qquad = 1920$

**39.** $\dfrac{2\frac{1}{4}}{3\frac{1}{3}} = 2\dfrac{1}{4} \div 3\dfrac{1}{3}$

$\qquad = \dfrac{9}{4} \div \dfrac{10}{3}$

$\qquad = \dfrac{9}{4} \times \dfrac{3}{10}$

$\qquad = \dfrac{27}{40}$

**40.** $\dfrac{20}{4\frac{4}{5}} = 20 \div 4\dfrac{4}{5}$

$\qquad = \dfrac{20}{1} \div \dfrac{24}{5}$

$\qquad = \dfrac{20}{1} \times \dfrac{5}{24}$

$\qquad = \dfrac{5}{1} \times \dfrac{5}{6}$

$\qquad = \dfrac{25}{6} = 4\dfrac{1}{6}$

**41.** $2\dfrac{1}{8} \div 20\dfrac{1}{2} = \dfrac{17}{8} \div \dfrac{41}{2}$

$\qquad = \dfrac{17}{8} \times \dfrac{2}{41}$

$\qquad = \dfrac{17}{164}$

**42.** $0 \div 3\dfrac{7}{5} = 0$

**43.** $4\dfrac{2}{11} \div 3 = \dfrac{46}{11} \div \dfrac{3}{1}$

$\qquad = \dfrac{46}{11} \times \dfrac{1}{3}$

$\qquad = \dfrac{46}{33}$

$\qquad = 1\dfrac{13}{33}$

**44.** $21\dfrac{7}{8} \div 3\dfrac{1}{8} = \dfrac{175}{8} \div \dfrac{25}{8}$

$\qquad = \dfrac{175}{8} \times \dfrac{8}{25}$

$\qquad = 7$ dresses

**45.**   $420 \div 2\dfrac{1}{4} = \dfrac{420}{1} \div \dfrac{9}{4}$

$\qquad\qquad = \dfrac{420}{1} \times \dfrac{4}{9}$

$\qquad\qquad = \dfrac{140}{1} \times \dfrac{4}{3} = \dfrac{560}{3}$

$\qquad\qquad = 186\dfrac{2}{3}$ calories

**46.**   $14 = 2 \times 7$

$\qquad 49 = 7 \times 7$

$\text{LCD} = 2 \times 7 \times 7 = 98$

**47.**   $40 = 2 \times 2 \times 2 \times 5$

$\qquad 30 = 2 \times 3 \times 5$

$\text{LCD} = 2 \times 2 \times 2 \times 3 \times 5 = 120$

**48.**   $18 = 2 \times 3 \times 3$

$\qquad 6 = 2 \times 3$

$\qquad 45 = 3 \times 3 \times 5$

$\text{LCD} = 2 \times 3 \times 3 \times 5 = 90$

**49.**   $\dfrac{3}{7} = \dfrac{3}{7} \times \dfrac{8}{8} = \dfrac{24}{56}$

**50.**   $\dfrac{11}{24} = \dfrac{11}{24} \times \dfrac{3}{3} = \dfrac{33}{72}$

**51.**   $\dfrac{9}{43} = \dfrac{9}{43} \times \dfrac{4}{4} = \dfrac{36}{172}$

**52.**   $\dfrac{17}{18} = \dfrac{17}{18} \times \dfrac{11}{11} = \dfrac{187}{198}$

**53.**   $\dfrac{3}{7} - \dfrac{5}{14} = \dfrac{3}{7} \times \dfrac{2}{2} - \dfrac{5}{14}$

$\qquad\qquad = \dfrac{6}{14} - \dfrac{5}{14}$

$\qquad\qquad = \dfrac{1}{14}$

**54.**   $\dfrac{1}{2} + \dfrac{1}{3} + \dfrac{1}{4} = \dfrac{1}{2} \times \dfrac{6}{6} + \dfrac{1}{3} \times \dfrac{4}{4} + \dfrac{1}{4} \times \dfrac{3}{3}$

$\qquad\qquad\qquad = \dfrac{6}{12} + \dfrac{4}{12} + \dfrac{3}{12}$

$\qquad\qquad\qquad = \dfrac{13}{12} = 1\dfrac{1}{12}$

**55.**   $\dfrac{4}{7} + \dfrac{7}{9} = \dfrac{4}{7} \times \dfrac{9}{9} + \dfrac{7}{9} \times \dfrac{7}{7}$

$\qquad\qquad = \dfrac{36}{63} + \dfrac{49}{63}$

$\qquad\qquad = \dfrac{85}{63} = 1\dfrac{22}{63}$

**56.**   $\dfrac{7}{8} - \dfrac{3}{5} = \dfrac{7}{8} \times \dfrac{5}{5} - \dfrac{3}{5} \times \dfrac{8}{8}$

$\qquad\qquad = \dfrac{35}{40} - \dfrac{24}{40}$

$\qquad\qquad = \dfrac{11}{40}$

**57.**   $\dfrac{7}{30} + \dfrac{2}{21} = \dfrac{7}{30} \times \dfrac{7}{7} + \dfrac{2}{21} \times \dfrac{10}{10}$

$\qquad\qquad = \dfrac{49}{210} + \dfrac{20}{210}$

$\qquad\qquad = \dfrac{69}{210} = \dfrac{23}{70}$

**58.**   $\dfrac{5}{18} + \dfrac{5}{12} = \dfrac{5}{18} \times \dfrac{2}{2} + \dfrac{5}{12} \times \dfrac{3}{3}$

$\qquad\qquad = \dfrac{10 + 15}{36}$

$\qquad\qquad = \dfrac{25}{36}$

**59.**   $\dfrac{15}{16} - \dfrac{13}{24} = \dfrac{15}{16} \times \dfrac{3}{3} - \dfrac{13}{24} \times \dfrac{2}{2}$

$\qquad\qquad = \dfrac{45}{48} - \dfrac{26}{48}$

$\qquad\qquad = \dfrac{19}{48}$

**60.** $\dfrac{14}{15} - \dfrac{3}{25} = \dfrac{14}{15} \times \dfrac{5}{5} - \dfrac{3}{25} \times \dfrac{3}{3}$

$\qquad = \dfrac{70}{75} - \dfrac{9}{75}$

$\qquad = \dfrac{61}{75}$

**61.** $1 - \dfrac{17}{23} = \dfrac{23}{23} - \dfrac{17}{23}$

$\qquad = \dfrac{6}{23}$

**62.** $6 - \dfrac{5}{9} = 5\dfrac{9}{9} - \dfrac{5}{9}$

$\qquad = 5\dfrac{4}{9}$

**63.** $3 + 5\dfrac{2}{3} = 8\dfrac{2}{3}$

**64.** $8 + 12\dfrac{5}{7} = 20\dfrac{5}{7}$

**65.** $3\dfrac{1}{4} + 1\dfrac{5}{8} = 3\dfrac{2}{8} + 1\dfrac{5}{8}$

$\qquad = 4\dfrac{7}{8}$

**66.** $7\dfrac{3}{16} - 2\dfrac{5}{6} = 7\dfrac{9}{48} - 2\dfrac{40}{48}$

$\qquad = 6\dfrac{57}{48} - 2\dfrac{40}{48}$

$\qquad = 4\dfrac{17}{48}$

**67.** $\dfrac{3}{5} \times \dfrac{1}{2} + \dfrac{2}{5} \div \dfrac{2}{3}$

$\qquad = \dfrac{3}{5} \times \dfrac{1}{2} + \dfrac{2}{5} \times \dfrac{3}{2}$

$\qquad = \dfrac{3}{10} + \dfrac{3}{5}$

$\qquad = \dfrac{3}{10} + \dfrac{3}{5} \times \dfrac{3}{2}$

$\qquad = \dfrac{3}{10} + \dfrac{6}{10} = \dfrac{9}{10}$

**68.** $\left(\dfrac{3}{7} - \dfrac{1}{14}\right) \times \dfrac{8}{9} = \left(\dfrac{6}{14} - \dfrac{1}{14}\right) \times \dfrac{8}{9}$

$\qquad = \dfrac{5}{14} \times \dfrac{8}{9}$

$\qquad = \dfrac{40}{126} = \dfrac{20}{63}$

**69.** $1\dfrac{7}{8} + 2\dfrac{3}{4} + 4\dfrac{1}{10}$

$\qquad = 1\dfrac{70}{80} + 2\dfrac{60}{80} + 4\dfrac{8}{80}$

$\qquad = 7\dfrac{138}{80}$

$\qquad = 8\dfrac{58}{80}$

$\qquad = 8\dfrac{29}{40}$ miles

**70.** $\begin{array}{cc} 28\frac{1}{6} & 27\frac{7}{6} \\ -\,1\frac{5}{6} & -\,1\frac{5}{6} \\ \hline & 26\frac{2}{6} = 26\frac{1}{3} \end{array}$

Then, $26\dfrac{1}{3} \times 10\dfrac{3}{4} = \dfrac{79}{3} \times \dfrac{43}{4}$

$\qquad\qquad\qquad = \dfrac{3397}{12}$

$\qquad\qquad\qquad = 283\dfrac{1}{12}$ miles

**71.** $3\dfrac{1}{3} \times \dfrac{1}{2} = \dfrac{10}{3} \times \dfrac{1}{2} = \dfrac{5}{3} = 1\dfrac{2}{3}$ cups sugar

$\quad\ 4\dfrac{1}{4} \times \dfrac{1}{2} = \dfrac{17}{4} \times \dfrac{1}{2} = \dfrac{17}{8} = 2\dfrac{1}{8}$ cups flour

**72.** $24\dfrac{1}{4} \times 8\dfrac{1}{2} = \dfrac{97}{4} \times \dfrac{17}{2}$

$\qquad\qquad = \dfrac{1649}{8}$

$\qquad\qquad = 206\dfrac{1}{8}$ miles

**73.** $48 \div 3\frac{1}{5} = \frac{48}{1} \div \frac{16}{5}$

$\qquad\qquad = \frac{48}{1} \times \frac{5}{16} = \frac{3}{1} \times \frac{5}{1}$

$\qquad\qquad = 15$ lengths

**74.** $15\frac{3}{4} - 6\frac{1}{8} = 15\frac{6}{8} - 6\frac{1}{8}$

$\qquad\qquad\quad = 9\frac{5}{8}$

$9\frac{5}{8}$ liters

**75.** $366 \div 12\frac{1}{5} = \frac{366}{1} \div \frac{61}{5}$

$\qquad\qquad\quad = \frac{366}{1} \times \frac{5}{61} = \frac{6}{1} \times \frac{5}{1}$

$\qquad\qquad\quad = 30$ words per minute

**76.** Regular pay: $\qquad 4\frac{1}{2} \times 8 = \frac{9}{2} \times \frac{8}{1}$

$\qquad\qquad\qquad\qquad\qquad = 36$

Overtime rate: $\quad 1\frac{1}{2} \times 4\frac{1}{2} = \frac{3}{2} \times \frac{9}{2}$

$\qquad\qquad\qquad\qquad\qquad = \frac{27}{4}$

Overtime pay: $\qquad \frac{27}{4} \times 3 = \frac{27}{4} \times \frac{3}{1}$

$\qquad\qquad\qquad\qquad\qquad = \frac{81}{4} = 20\frac{1}{4}$

Total pay: $\quad 36 + 20\frac{1}{4} = 56\frac{1}{4}$

$\qquad\qquad \$56\frac{1}{4}$

**77.** $\qquad 88\frac{3}{8} \qquad\qquad 87\frac{11}{8}$

$\qquad \underline{-\ 79\frac{5}{8}} \qquad\quad \underline{-\ 79\frac{5}{8}}$

$\qquad\qquad\qquad\qquad\qquad 8\frac{6}{8} = 8\frac{3}{4}$

$\$8\frac{3}{4}$ decrease in stocks

**78.** $1\frac{1}{2} + \frac{1}{16} + \frac{1}{8} + \frac{1}{4} = 1\frac{8}{16} + \frac{1}{16} + \frac{2}{16} + \frac{4}{16}$

$\qquad\qquad\qquad\qquad\qquad = 1\frac{15}{16}$

$\qquad 3 - 1\frac{15}{16} = 2\frac{16}{16} - 1\frac{15}{16}$

$\qquad\qquad\qquad\quad = 1\frac{1}{16}$

$1\frac{1}{16}$ inch

**79.** $\qquad \frac{1}{10} \times 880 = \qquad 98$

$\qquad\quad \frac{1}{2} \times 880 = \qquad 440$

$\qquad \underline{+\ \frac{1}{8} \times 880} = \underline{+\ 110}$

$\qquad\qquad\qquad\qquad\qquad 638$

Left over: $\qquad 880$

$\qquad\qquad\qquad \underline{-\ 638}$

$\qquad\qquad\qquad \$242$

**80. a.** $460 \div 18\frac{2}{5} = \frac{460}{1} \div \frac{92}{5}$

$\qquad\qquad\qquad = \frac{460}{1} \times \frac{5}{92} = 25$

$\qquad 25$ miles per gallon

**b.** $\quad 18\frac{2}{5} \times 1\frac{1}{5} = \frac{92}{5} \times \frac{6}{5}$

$\qquad\qquad\qquad\qquad = \frac{552}{25} = 22\frac{2}{25}$

$\quad \$22\frac{2}{25}$

---

## Chapter 2 Test

**1.** $\frac{3}{5}$; 3 of the 5 parts are shaded.

**2.** $\frac{311}{388}$

**3.** $\frac{18}{42} = \frac{18 \div 6}{42 \div 6} = \frac{3}{7}$

**4.** $\frac{15}{70} = \frac{15 \div 5}{70 \div 5} = \frac{3}{14}$

**5.** $\dfrac{225}{50} = \dfrac{225 \div 25}{50 \div 25} = \dfrac{9}{2}$

**6.** $6\dfrac{4}{5} = \dfrac{6 \times 5 + 4}{5} = \dfrac{34}{5}$

**7.** $14\overline{)114}$ with quotient $8$, $\dfrac{112}{2}$

$\dfrac{114}{14} = 8\dfrac{2}{14} = 8\dfrac{1}{7}$

**8.** $42 \times \dfrac{2}{7} = \dfrac{42}{1} \times \dfrac{2}{7} = \dfrac{6 \times 7 \times 2}{1 \times 7} = \dfrac{12}{1} = 12$

**9.** $\dfrac{7}{9} \times \dfrac{2}{5} = \dfrac{7 \times 2}{9 \times 5} = \dfrac{14}{45}$

**10.** $4\dfrac{1}{3} \times 7\dfrac{1}{5} = \dfrac{13}{3} \times \dfrac{36}{5}$

$= \dfrac{13 \times 3 \times 12}{3 \times 5}$

$= \dfrac{156}{5} = 31\dfrac{1}{5}$

**11.** $\dfrac{7}{8} \div \dfrac{5}{11} = \dfrac{7}{8} \times \dfrac{11}{5}$

$= \dfrac{7 \times 11}{8 \times 5} = \dfrac{77}{40}$

$= 1\dfrac{37}{40}$

**12.** $\dfrac{12}{31} \div \dfrac{8}{13} = \dfrac{12}{31} \times \dfrac{13}{8} = \dfrac{3 \times 4 \times 13}{31 \times 2 \times 4} = \dfrac{39}{62}$

**13.** $7\dfrac{1}{5} \div 1\dfrac{1}{25} = \dfrac{36}{5} \div \dfrac{26}{25} = \dfrac{36}{5} \times \dfrac{25}{26}$

$= \dfrac{2 \times 18 \times 5 \times 5}{5 \times 2 \times 13} = \dfrac{18 \times 5}{13}$

$= \dfrac{90}{13} = 6\dfrac{12}{13}$

**14.** $5\dfrac{1}{7} \div 3 = \dfrac{36}{7} \div \dfrac{3}{1} = \dfrac{36}{7} \times \dfrac{1}{3}$

$= \dfrac{3 \times 12 \times 1}{7 \times 3}$

$= \dfrac{12}{7} = 1\dfrac{5}{7}$

**15.** $24 = 2 \times 2 \times 2 \times 3$

$18 = 2 \times 3 \times 3$

$\text{LCD} = 2 \times 2 \times 2 \times 3 \times 3 = 72$

**16.** $16 = 2 \times 2 \times 2 \times 2$

$24 = 2 \times 2 \times 2 \times 3$

$\text{LCD} = 2 \times 2 \times 2 \times 2 \times 3 = 48$

**17.** $4 = 2 \times 2$

$8 = 2 \times 2 \times 2$

$6 = 2 \times 3$

$\text{LCD} = 2 \times 2 \times 2 \times 3 = 24$

**18.** $\dfrac{5}{12} = \dfrac{5}{12} \times \dfrac{6}{6} = \dfrac{30}{72}$

**19.** $\dfrac{7}{9} - \dfrac{5}{12} = \dfrac{28}{36} - \dfrac{15}{36} = \dfrac{13}{36}$

**20.** $\dfrac{2}{15} + \dfrac{5}{12} = \dfrac{8}{60} + \dfrac{25}{60}$

$= \dfrac{33}{60}$

$= \dfrac{11}{20}$

**21.** $\dfrac{1}{4} + \dfrac{3}{7} + \dfrac{3}{14} = \dfrac{7}{28} + \dfrac{12}{28} + \dfrac{6}{28} = \dfrac{25}{28}$

**22.** $8\dfrac{3}{5} + 5\dfrac{4}{7} = 8\dfrac{21}{35} + 5\dfrac{20}{35}$

$= 13\dfrac{41}{35}$

$= 14\dfrac{6}{35}$

**23.** $18\frac{6}{7} - 13\frac{13}{14} = 18\frac{12}{14} - 13\frac{13}{14}$

$\qquad\qquad\quad = 17\frac{26}{14} - 13\frac{13}{14}$

$\qquad\qquad\quad = 4\frac{13}{14}$

**24.** $\frac{6}{7} - \frac{5}{7} \times \frac{1}{4} = \frac{6}{7} - \frac{5}{28}$

$\qquad\qquad\quad = \frac{24}{28} - \frac{5}{28} = \frac{19}{28}$

**25.** $\left(\frac{1}{2} + \frac{1}{3}\right) \times \frac{7}{5} = \left(\frac{3}{6} + \frac{2}{6}\right) \times \frac{7}{5}$

$\qquad\qquad\qquad = \frac{5}{6} \times \frac{7}{5}$

$\qquad\qquad\qquad = \frac{7}{6} = 1\frac{1}{6}$

**26.** $8\frac{1}{6} \times 5\frac{1}{7} = \frac{49}{6} \times \frac{36}{7}$

$\qquad\qquad = \frac{7 \times 7 \times 6 \times 6}{6 \times 7}$

$\qquad\qquad = \frac{6 \times 7}{1} = 42$

The area is 42 square yards.

**27.** $18\frac{2}{3} \div 2\frac{1}{3} = \frac{56}{3} \div \frac{7}{3} = \frac{56}{3} \times \frac{3}{7}$

$\qquad\qquad\quad = \frac{8 \times 7 \times 3}{3 \times 7}$

$\qquad\qquad\quad = 8$

He can make 8 packages.

**28.** $\frac{9}{10} - \frac{1}{5} = \frac{9}{10} - \frac{2}{10} = \frac{7}{10}$

He has $\frac{7}{10}$ of a mile left to walk.

**29.** $4\frac{1}{8} + 3\frac{1}{6} + 6\frac{3}{4} = 4\frac{3}{24} + 3\frac{4}{24} + 6\frac{18}{24}$

$\qquad\qquad\qquad = 13\frac{25}{24}$

$\qquad\qquad\qquad = 14\frac{1}{24}$

She jogged $14\frac{1}{24}$ miles.

**30. a.** $\frac{1}{4} \times 120 = \frac{1}{4} \times \frac{120}{1} = 30$

$\qquad \frac{1}{12} \times 120 = \frac{1}{12} \times \frac{120}{1} = 10$

$\qquad \frac{1}{3} \times 120 = \frac{1}{3} \times \frac{120}{1} = 40$

$120 - 30 - 10 - 40 = 120 - 80 = 40$

They shipped 40 oranges.

**b.** $24 \times 40 = 960$

It cost 960¢ or $9.60.

**31. a.** $48\frac{1}{8} \div \frac{5}{8} = \frac{385}{8} \times \frac{8}{5} = \frac{385}{5}$

$\qquad\qquad\qquad = 77$ candles

**b.** Cost is

$\qquad 48\frac{1}{8} \times 2 \div 77 = \frac{385}{8} \times \frac{2}{1} \times \frac{1}{77}$

$\qquad\qquad\qquad\quad = \frac{385}{308} = 1\frac{77}{308}$

$\qquad\qquad\qquad\quad = 1\frac{1}{4}$

**c.** Profit per candle:

$\qquad 12 - 1\frac{1}{4} = 11\frac{4}{4} - 1\frac{1}{4}$

$\qquad\qquad\qquad = 10\frac{3}{4}$

Profit is

$\qquad 10\frac{3}{4} \times 77 = \frac{43}{4} \times \frac{77}{1}$

$\qquad\qquad\qquad = \frac{3311}{4} = \$827\frac{3}{4}$

# Chapters 1–2 Cumulative Test

**1.** 84,361,208 = Eighty-four million, three hundred sixty-one thousand, two hundred eight.

**2.**
$$
\begin{array}{r}
128 \\
452 \\
178 \\
34 \\
+\ 77 \\
\hline
869
\end{array}
$$

**3.**
$$
\begin{array}{r}
156,200 \\
364,700 \\
+\ 198,320 \\
\hline
719,220
\end{array}
$$

**4.**
$$
\begin{array}{r}
5718 \\
-\ 3643 \\
\hline
2075
\end{array}
$$

**5.**
$$
\begin{array}{r}
1,000,361 \\
-\ 983,145 \\
\hline
17,216
\end{array}
$$

**6.**
$$
\begin{array}{r}
126 \\
\times\ 38 \\
\hline
1008 \\
378 \\
\hline
4788
\end{array}
$$

**7.**
$$
\begin{array}{r}
16,908 \\
\times\ \ \ 12 \\
\hline
33\ 816 \\
169\ 08 \\
\hline
202,896
\end{array}
$$

**8.**
$$
\begin{array}{r}
4\ 658 \\
7\overline{)\ 32,606} \\
\underline{28} \\
4\ 6 \\
\underline{4\ 2} \\
40 \\
\underline{35} \\
56 \\
\underline{56} \\
0
\end{array}
$$

**9.**
$$
\begin{array}{r}
369 \\
18\overline{)\ 6642} \\
\underline{54} \\
124 \\
\underline{108} \\
162 \\
\underline{162} \\
0
\end{array}
$$

**10.** $7^2 = 7 \times 7 = 49$

**11.** 6,037,452 rounds to 6,037,000.

**12.** $4 \times 3^2 \div 12 \div 6 = 4 \times 9 + 12 \div 6 = 36 + 2 = 38$

**13.** $3 \times \$26 + 2 \times \$48 = \$174$

**14.** $516 + 199 + 203 = 918$ for checks
$64 + 1160 - 918 = 1224 - 918 = 306$
Her balance will be \$306.

**15.** $\dfrac{55}{84}$

**16.** $\dfrac{28}{52} = \dfrac{28 \div 4}{52 \div 4} = \dfrac{7}{13}$

**17.** $18\dfrac{3}{4} = \dfrac{4 \times 18 + 3}{4} = \dfrac{75}{4}$

**18.**
$$
\begin{array}{r}
14 \\
7\overline{)\ 100} \\
\underline{7} \\
30 \\
\underline{28} \\
2
\end{array}
\qquad \dfrac{100}{7} = 14\dfrac{2}{7}
$$

**19.** $3\dfrac{7}{8} \times 2\dfrac{5}{6} = \dfrac{31}{8} \times \dfrac{17}{6}$

$\qquad\qquad = \dfrac{527}{48} = 10\dfrac{47}{48}$

**20.** $\dfrac{44}{49} \div 2\dfrac{13}{21} = \dfrac{44}{49} \div \dfrac{55}{21} = \dfrac{44}{49} \times \dfrac{21}{55}$

$\qquad\qquad = \dfrac{4 \times 11 \times 3 \times 7}{7 \times 7 \times 5 \times 11} = \dfrac{12}{35}$

**21.**     $13 = 13$
           $39 = 3 \times 13$
           LCD $= 3 \times 13 = 39$

**22.** $\dfrac{7}{18} + \dfrac{5}{27} = \dfrac{21}{54} + \dfrac{10}{54} = \dfrac{31}{54}$

**23.** $2\dfrac{1}{8} + 6\dfrac{3}{4} = 2\dfrac{1}{8} + 6\dfrac{6}{8}$

$\qquad\qquad = 8\dfrac{7}{8}$

**24.** $12\dfrac{1}{5} - 4\dfrac{2}{3} = 12\dfrac{3}{15} - 4\dfrac{10}{15}$

$\qquad\qquad = 11\dfrac{18}{15} - 4\dfrac{10}{15} = 7\dfrac{8}{15}$

**25.** $\dfrac{1}{3} + \dfrac{5}{8} \div \dfrac{5}{4} = \dfrac{1}{3} + \dfrac{5}{8} \times \dfrac{4}{5}$

$\qquad\qquad = \dfrac{1}{3} + \dfrac{1}{2} = \dfrac{2}{6} + \dfrac{3}{6}$

$\qquad\qquad = \dfrac{5}{6}$

**26.** $9\dfrac{1}{2} + 6\dfrac{3}{8} + 7\dfrac{1}{4} = 9\dfrac{4}{8} + 6\dfrac{3}{8} + 7\dfrac{2}{8}$

$\qquad\qquad\qquad = 22\dfrac{9}{8} = 23\dfrac{1}{8}$

$23\dfrac{1}{8}$ tons were hauled.

**27.** $221\dfrac{2}{5} \div 9 = \dfrac{1107}{5} \div \dfrac{9}{1}$

$\qquad\qquad = \dfrac{1107}{5} \times \dfrac{1}{9} = \dfrac{123}{5}$

$\qquad\qquad = 24\dfrac{3}{5}$ miles per gallon

**28.** $2\dfrac{1}{2} \times 3\dfrac{1}{4} = \dfrac{5}{2} \times \dfrac{13}{4} = \dfrac{65}{8} = 8\dfrac{1}{8}$

$2\dfrac{1}{2} \times 2\dfrac{1}{3} = \dfrac{5}{2} \times \dfrac{7}{3} = \dfrac{35}{6} = 5\dfrac{5}{6}$

She needs $8\dfrac{1}{8}$ cups of sugar and $5\dfrac{5}{6}$ cups
of flour.

**29.**     $\begin{array}{r} 30{,}000 \\ \times \quad 2{,}000 \\ \hline 60{,}000{,}000 \end{array}$ miles

# DECIMALS

## Pretest Chapter 3

**1.** 47.813 = Forty-seven and eight hundred thirteen thousandths

**2.** 0.0567

**3.** $2.11 = 2\frac{11}{100}$

**4.** $0.525 = \frac{525}{1000} = \frac{21}{40}$

**5.** 1.59, 1.6, 1.601, 1.61

**6.** 123.49268 rounds to 123.5

**7.** $1.053458 \approx 1.053$

**8.**
$$
\begin{array}{r}
\overset{1}{5}.12 \\
4.70 \\
8.03 \\
+\ 1.60 \\
\hline
19.45
\end{array}
$$

**9.**
$$
\begin{array}{r}
24.613 \\
0.273 \\
+\ 2.305 \\
\hline
27.191
\end{array}
$$

**10.**
$$
\begin{array}{r}
42.16 \\
-\ 31.57 \\
\hline
10.59
\end{array}
$$

**11.**
$$
\begin{array}{r}
26.000 \\
+\ 18.329 \\
\hline
7.671
\end{array}
$$

**12.**
$$
\begin{array}{r}
11.67 \\
\times\ 0.03 \\
\hline
0.3501
\end{array}
$$

**13.** $4.7805 \times 1000 = 4780.5$

**14.** $0.00037.96 \times 10^5 = 37.96$

**15.**
$$
\begin{array}{r}
0.354 \\
0.09\overline{)0.03186} \\
\underline{27}\phantom{000} \\
48\phantom{00} \\
\underline{45}\phantom{00} \\
36\phantom{0} \\
\underline{36}\phantom{0} \\
0
\end{array}
$$

**16.**
$$
\begin{array}{r}
0.128 \\
2.6\overline{)0.3328} \\
\underline{-26}\phantom{00} \\
72\phantom{0} \\
\underline{52}\phantom{0} \\
208 \\
\underline{208} \\
0
\end{array}
$$

**17.**
$$
\begin{array}{r}
0.6875 \\
16\overline{)11.0000} \\
\underline{9\ 6}\phantom{000} \\
1\ 40\phantom{00} \\
\underline{1\ 28}\phantom{00} \\
120\phantom{0} \\
\underline{112}\phantom{0} \\
80 \\
\underline{80} \\
0
\end{array}
$$

**18.**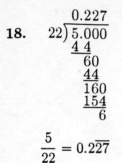

$$\frac{5}{22} = 0.2\overline{27}$$

**19.** $(0.2)^2 + 8.7 \times 0.3 - 1.68$
$$= 0.04 + 8.7 \times 0.3 - 1.68$$
$$= 0.04 + 2.61 - 1.68$$
$$= 2.65 - 1.68$$
$$= 0.97$$

**20.**
```
   57,312.8
 - 57,124.8
   188.0
```

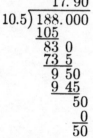

17.9 miles per gallon

**21.**
```
  10.15
 × 3.6
  630
 315
 37.80  square yards
```
```
   37.80
 × 12.95
  18900
  34020
  7560
  3780
 $489.5100
```

$489.51 cost for carpeting.

**22.**
```
     8.53
 35)298.55
    280
     18 5
     17 5
      1 05
      1 05
         0
```

$8.53 per hour

## 3.1  Exercises

**1.** A decimal fraction is a fraction whose denominator is a power of 10.

**3.** Hundred-thousandths

**5.** 0.57 = Fifty-seven hundredths

**7.** 3.8 = Three and eight tenths

**9.** 5.283 = Five and two hundred eighty-three thousandths

**11.** 28.0037 = Twenty-eight and thirty-seven ten thousandths

**13.** $124.20 = One hundred twenty-four and $\frac{20}{100}$ dollars

**15.** $1236.08 = One thousand two hundred thirty-six and $\frac{8}{100}$ dollars

**17.** $10,000.76 = Ten thousand and $\frac{76}{100}$ dollars

**19.** seven tenths = 0.7

**21.** forty-five hundredths = 0.45

**23.** twenty-two thousandths = 0.022

**25.** two hundred eighty-six millionths
= 0.000286

**27.** $\dfrac{7}{10} = 0.7$

**29.** $\dfrac{76}{100} = 0.76$

**31.** $\dfrac{1}{100} = 0.01$

**33.** $\dfrac{53}{1000} = 0.053$

**35.** $\dfrac{2403}{10,000} = 0.2403$

**37.** $8\dfrac{3}{10} = 8.3$

**39.** $84\dfrac{13}{100} = 84.13$

**41.** $1\dfrac{19}{1000} = 1.019$

**43.** $126\dfrac{571}{10,000} = 126.0571$

**45.** $0.02 = \dfrac{2}{100} = \dfrac{1}{50}$

**47.** $3.6 = 3\dfrac{6}{10} = 3\dfrac{3}{5}$

**49.** $8.24 = 8\dfrac{24}{100} = 8\dfrac{6}{25}$

**51.** $12.625 = 12\dfrac{625}{1000} = 12\dfrac{5}{8}$

**53.** $7.0015 = 7\dfrac{15}{10,000} = 7\dfrac{3}{2000}$

**55.** $235.1254 = 235\dfrac{1254}{10,000} = 235\dfrac{627}{5000}$

**57.** $0.0187 = \dfrac{187}{10,000}$

**59.** $8.0108 = 8\dfrac{108}{10,000} = 8\dfrac{27}{2500}$

**61.** $\dfrac{4}{1,000,000} = \dfrac{1}{250,000}$

## Cumulative Review Problems

**63.**
$$\begin{array}{r} 156 \\ 84 \\ 39 \\ 463 \\ +\ 76 \\ \hline 818 \end{array}$$

**65.** 56,800

## 3.2   Exercises

**1.** $1.3 > 1.29$

**3.** $0.68 < 0.681$

**5.** $18.92 < 18.93$

**7.** $0.0006 > 0.0005$

**9.** $1.002 < 1.0021$

**11.** $126.34 > 125.35$

**13.** $\dfrac{72}{1000} = 0.072$

**15.** $\dfrac{8}{10} = 0.8$

$\dfrac{8}{10} > 0.08$

**17.** 12.6, 12.65, 12.8

**19.** 0.007, 0.0071, 0.05

**21.** 5.12, 5.2, 5.23, 5.3

**23.** 26.003, 26.033, 26.034, 26.04

**25.** 18.006, 18.060, 18.065, 18.066, 18.606

**27.** 5.67 rounds to 5.7.

**29.** 29.43 rounds to 29.4.

**31.** 578.064 rounds to 578.1.

**33.** 2176.83 rounds to 2176.8.

**35.** 26.032 rounds to 26.03.

**37.** 5.76582 rounds to 5.77.

**39.** 156.1749 rounds to 156.17.

**41.** 2786.706 rounds to 2786.71.

**43.** 1.06132 rounds to 1.061.

**45.** 0.047357 rounds to 0.0474.

**47.** 5.00761238 rounds to 5.00761.

**49.** 135.564 rounds to 136.

**51.** $2536.85 rounds to $2537.

**53.** $10,098.47 rounds to $10,098.

**55.** $56.9832 rounds to $56.98.

**57.** $5783.716 rounds to $5783.72.

**59.** 0.095 rounds to 0.10 kilogram.
0.066 rounds to 0.07 kilogram.

**61.** 365.24122 rounds to 365.24.

**63.** $\dfrac{6}{100} = 0.06$ and $\dfrac{6}{10} = 0.6$

0.0059, 0.006, 0.0519, $\dfrac{6}{100}$, 0.0601, 0.0612,

0.062, $\dfrac{6}{10}$, 0.61

**65.** You should consider only one digit to the right of the decimal place that you wish to round to. 86.23498 is closer to 86.23 than to 86.24.

---

## Cumulative Review Problems

**67.**   $3\frac{1}{4}$         $3\frac{2}{8}$
          $2\frac{1}{2}$         $2\frac{4}{8}$
        $+ 6\frac{3}{8}$       $+ 6\frac{3}{8}$
        $\overline{\phantom{xxxx}}$   $\overline{11\frac{9}{8} = 12\frac{1}{8}}$

**69.**     47,073
         $-$ 46,381
        $\overline{\phantom{xx}692}$ miles

---

## 3.3   Exercises

**1.**     57.1
        $+ 19.7$
        $\overline{76.8}$

**3.**     718.98
        $+ 496.57$
        $\overline{1215.55}$

**5.**      13.4
             7.6
        + 275.2
           296.2

**7.**      5.60
             9.23
        + 8.17
          23.00

**9.**      4.9637
            28.1200
        + 3.6450
          36.7287

**11.**     12.00
             3.62
        + 51.80
          67.42

**13.**    753.61
            28.75
           162.30
           100.50
        + 67.05
          1112.21

**15.**     6.10
             5.62
        + 8.14
          19.86  m

**17.**     1.75
             2.50
             1.55
        + 2.80
          8.60  or 8.6 pounds

**19.**     4.99
            12.50
            11.85
            28.50
             3.29
        + 16.99
          $78.12

**21.**     18.42
            706.15
            21.03
            45.00
        + 621.37
          $1411.97  total

**23.**     6.8
        − 2.9
          3.9

**25.**    123.00
        − 96.34
          26.66

**27.**    132.20
        − 16.67
          115.53

**29.**    586.513
        − 78.200
          508.313

**31.**     3.00269
        − 0.80368
          2.19901

**33.**    24.0079
        − 19.3614
          4.6465

**35.**     8.000
        − 1.263
          6.737

**37.**    7362.14
        − 6173.07
          1189.07

**39.**     1.5000
        − 0.0365
          1.4635

**41.**     3.264
        − 1.800
        1.464  meters

**43.**     37,026.65
        −       79.49
        $36,947.16

**45.**     47.70
        + 7.00
        54.70

        100.00
        − 54.70
        45.30

        $45.30 change

**47.**     12.62
        − 0.98
        11.64  cm

**49.** 2.45 + 1.35 − 0.85
        = 3.80 − 0.85
        = 2.95 liters

**51.**     0.0150
        − 0.0089
        0.0061  milligrams

**53.**     43.8
        − 37.6
        $6.2  billion

        or $6,200,000,000

**55.**     61.4
        − 48.1
        $13.3  billion

        or $13,300,000,000

**57.**       2.60
            1.50
            1.30
            0.80
        + 2.20
        $8.40

        Exact:     2.63
                 1.47
                 1.26
                 0.79
             + 2.19
             $8.34

        Estimate is close to actual amount.
        Difference is 6¢.

**59.** $x + 7.1 = 15.5$

            15.5
        − 7.1
            8.4

        $x = 8.4$

**61.**   $156.9 + x = 200.6$

            200.6
        − 156.9
            43.7

        $x = 43.7$

**63.**   $4.162 = x + 2.053$

            4.162
        − 2.053
            2.109

        $x = 2.109$

---

## Cumulative Review Problems

**65.**     2536
        ×       8
        20,288

**67.** $\dfrac{22}{7} \times \dfrac{49}{50} = \dfrac{77}{25} = 3\dfrac{2}{25}$

## 3.4    Exercises

**1.**
$$\begin{array}{r} 0.6 \\ \times\ 0.2 \\ \hline 0.12 \end{array}$$

**3.**
$$\begin{array}{r} 0.12 \\ \times\ 0.5 \\ \hline 0.060 = 0.06 \end{array}$$

**5.**
$$\begin{array}{r} 0.0036 \\ \times\ \ \ 0.8 \\ \hline 0.00288 \end{array}$$

**7.**
$$\begin{array}{r} 0.079 \\ \times\ 0.09 \\ \hline 0.00711 \end{array}$$

**9.**
$$\begin{array}{r} 0.043 \\ \times\ 0.012 \\ \hline 0086 \\ 0043 \\ \hline 0.000516 \end{array}$$

**11.**
$$\begin{array}{r} 10.97 \\ \times\ 0.06 \\ \hline 0.6582 \end{array}$$

**13.**
$$\begin{array}{r} 5167 \\ \times\ 0.19 \\ \hline 46503 \\ 5167 \\ \hline 981.73 \end{array}$$

**15.**
$$\begin{array}{r} 2.163 \\ \times\ \ 0.008 \\ \hline 0.017304 \end{array}$$

**17.**
$$\begin{array}{r} 0.7613 \\ \times\ \ \ 1009 \\ \hline 68517 \\ 761300 \\ \hline 768.1517 \end{array}$$

**19.**
$$\begin{array}{r} 2350 \\ \times\ 3.6 \\ \hline 14100 \\ 7050 \\ \hline 8460.0 = 8460 \end{array}$$

**21.**
$$\begin{array}{r} 4.57 \\ \times\ 11.8 \\ \hline 3656 \\ 457 \\ 457 \\ \hline 53.926 \end{array}$$

**23.**
$$\begin{array}{r} 6523.7 \\ \times\ 0.001 \\ \hline 6.5237 \end{array}$$

**25.**
$$\begin{array}{r} 155.40 \\ \times\ \ \ \ \ 60 \\ \hline \$9324.00 = \$9324 \end{array}$$

**27.**
$$\begin{array}{r} 9.55 \\ \times\ 40 \\ \hline \$382.00 = \$382 \end{array}$$

**29.**
$$\begin{array}{r} 19.2 \\ \times\ 15.5 \\ \hline 960 \\ 960 \\ \hline 297.60 = 297.6\ \text{square feet} \end{array}$$

**31.**
$$\begin{array}{r} 36.90 \\ \times\ 18 \\ \hline 29520 \\ 3690 \\ \hline \$664.20 \end{array}$$

**33.**
$$\begin{array}{r} 26.4 \\ \times\ 19.5 \\ \hline 1320 \\ 2376 \\ 264 \\ \hline 514.80 = 514.8\ \text{miles} \end{array}$$

**35.** $2.86 \times 10 = 28.6$

**37.** $43.0 \times 10 = 430$

**39.** $128.65 \times 1000 = 128{,}650$

**41.** $5.60982 \times 10{,}000 = 56{,}098.2$

**43.** $280{,}560.2 \times 10^2 = 28{,}056{,}020$

**45.** $816.32 \times 10^3 = 816{,}320$

**47.** $5.932 \times 100 = 593.2$ centimeters

**49.** $2.98 \times 1000 = 2980$ meters

**51.**

```
    124.00        Amount left is
    110.00
     83.60             820.00
     76.00         −   572.00
     44.60            $248.00
     44.60
     44.60
  +  44.60
   $572.00
```

**53.**

```
     254.2
   ×  19.6
    15252
   22878
   2542
  4982.32  square yards
```

```
    4982.32
  ×  12.50
    000000
  2491160
   996464
   498232
  $62,279.0000
```

**55.** To multiply by numbers such as 0.1, 0.01, 0.001, and 0.0001, count the number of decimal places in the first number. Then, in the other number, move the decimal point to the left from its present position the same number of decimal places as was in the first number.

## Cumulative Review Problems

**57.**

```
        98
   12) 1176
       108
        96
        96
         0
```

**59.**

```
       125 R4
   37) 4629
       37
       92
       74
       189
       185
         4
```

**61.**

```
     $ 0.73  billion
   − $ 0.62  billion
     $ 0.11  billion
```

or $110,000,000

**63.**

```
     $ 2.51   billion
   + $ 0.876  billion
     $ 3.386  billion
```

or $3,386,000,000

## 3.5   Exercises

**1.**

```
        2.1
   6) 12.6
      12
       6
       6
       0
```

**3.**

```
      17.83
   4) 71.32
      4
      31
      28
       3 3
       3 2
        12
        12
         0
```

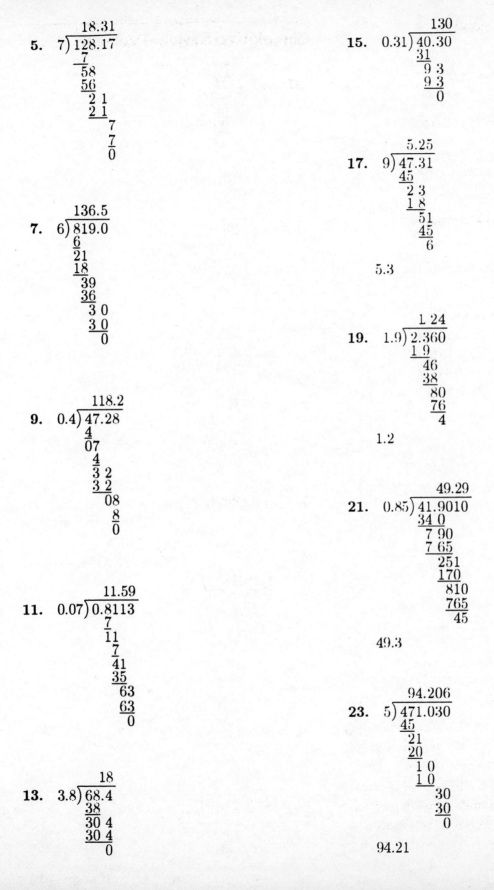

**5.**
```
      18.31
   7)128.17
     7
     58
     56
      2 1
      2 1
        7
        7
        0
```

**7.**
```
     136.5
   6)819.0
     6
     21
     18
      39
      36
       3 0
       3 0
         0
```

**9.**
```
      118.2
   0.4)47.28
      4
      07
       4
       3 2
       3 2
         08
          8
          0
```

**11.**
```
       11.59
   0.07)0.8113
        7
        11
         7
         41
         35
          63
          63
           0
```

**13.**
```
        18
   3.8)68.4
       38
       30 4
       30 4
          0
```

**15.**
```
        130
   0.31)40.30
        31
         9 3
         9 3
           0
```

**17.**
```
      5.25
   9)47.31
     45
      2 3
      1 8
       51
       45
        6
```

5.3

**19.**
```
       1 24
   1.9)2.360
       1 9
        46
        38
         80
         76
          4
```

1.2

**21.**
```
       49.29
   0.85)41.9010
        34 0
         7 90
         7 65
           251
           170
            810
            765
             45
```

49.3

**23.**
```
      94.206
   5)471.030
     45
      21
      20
       1 0
       1 0
         30
         30
          0
```

94.21

**25.**

```
      26.7777
  9) 241.0000
     18
     ──
     61
     54
     ──
      7 0
      6 3
      ───
        70
        63
        ──
        70
        63
        ──
        70
        63
        ──
         7
```

26.778

**27.**

```
           12.2463
  0.69) 8.450000
        6 9
        ───
        1 55
        1 38
        ────
          170
          138
          ───
          320
          276
          ───
          440
          414
          ───
          260
          200
          ───
           53
```

12.246

**29.**

```
         116.3
  12) 1396.0
      12
      ──
      19
      12
      ──
      76
      72
      ──
       4 0
       3 6
       ───
        4
```

116

**31.**

```
         12.7
  0.55) 7.000
        5 5
        ──
        1 50
        1 10
        ────
          400
          385
          ───
           15
```

13

**33.**

```
        6.2
  38) 235.6
      228
      ───
        7 6
        7 6
        ───
          0
```

6.2 ounces in each portion

**35.**  67.6 + 33.6 = 101.2 ounces

**a.**

```
          25.3 ounces
    4) 101.2
       8
       ──
       21
       20
       ──
        12
        1 2
        ───
          0
```

**b.**

```
          20.24 ounces
    5) 101.20
       10
       ──
       01
       00
       ──
        1 2
        1 0
        ───
          20
          20
          ──
           0
```

**37.**

```
         376.50
  9) 3388.50
     27
     ──
     68
     63
     ──
      58
      54
      ──
       4 5
       4 5
       ───
         0
```

$376.50 per band

**39.**

```
                  9
  125.75) 1131.75
          1131 75
          ───────
                0
```

9 payments

**41. a.** $\dfrac{8.1 + 9.2}{2} = \dfrac{17.3}{2} = 8.65$ lb

 **b.** $\dfrac{14.6 - 8.1}{20} = \dfrac{6.5}{20} = 0.325$ lb

$$
\begin{array}{r}
.325 \\
20\overline{)6.500} \\
\underline{6\ 0} \\
50 \\
\underline{40} \\
100 \\
\underline{100} \\
0
\end{array}
$$

**43.** $0.7 \times n = 0.0861$

$$
\begin{array}{r}
0.\ 123 \\
0.7\overline{)0.\ 0861} \\
\underline{7} \\
16 \\
\underline{14} \\
21 \\
\underline{21} \\
0
\end{array}
$$

$n = 0.123$

**45.** $1.6 \times n = 110.4$

$$
\begin{array}{r}
6\ 9 \\
1.6\overline{)110.4} \\
\underline{96} \\
14\ 4 \\
\underline{14\ 4} \\
0
\end{array}
$$

$n = 69$

**47.** $n \times 0.063 = 2.835$

$$
\begin{array}{r}
45 \\
0.063\overline{)2.835} \\
\underline{2\ 52} \\
315 \\
\underline{315} \\
0
\end{array}
$$

$n = 45$

**49.** $\dfrac{3.8702}{0.0523} \times \dfrac{10,000}{10,000} = \dfrac{38,702}{523}$

$$
\begin{array}{r}
74 \\
523\overline{)38,702} \\
\underline{36\ 61} \\
2\ 092 \\
\underline{2\ 092} \\
0
\end{array}
$$

$\dfrac{3.8702}{0.0523} = 74$

## Cumulative Review Problems

**51.** $\dfrac{3}{8} + 1\dfrac{2}{5} = \dfrac{15}{40} + 1\dfrac{16}{40} = 1\dfrac{31}{40}$

**53.** $3\dfrac{1}{2} \times 2\dfrac{1}{6} = \dfrac{7}{2} \times \dfrac{13}{6} = \dfrac{91}{12} = 7\dfrac{7}{12}$

## Putting Your Skills To Work

**1.** $2005 - 1997 = 8$ years

 $8 \times 4 = 32$ inches

 $\dfrac{32}{12} = 2\dfrac{2}{3}$ feet

 $6 - 2\dfrac{2}{3} = 5\dfrac{3}{3} - 2\dfrac{2}{3} = 3\dfrac{1}{3}$ feet

**3.** $\dfrac{15}{100} \times 3.5 = 0.15 \times 3.5 = 0.525$ degrees F

## 3.6 Exercises

**1.** same quantity

**3.** The digits 8942 repeat.

**5.**
$$
\begin{array}{r}
0.25 \\
4\overline{)1.00} \\
\underline{8} \\
20 \\
\underline{20} \\
0
\end{array}
$$

$\dfrac{1}{4} = 0.25$

**7.**
$$
\begin{array}{r}
0.8 \\
5\overline{)4.0} \\
\underline{4\ 0} \\
0
\end{array}
$$

$\dfrac{4}{5} = 0.8$

**9.**

$$
\begin{array}{r}
0.4375 \\
16\overline{)7.0000} \\
\underline{6\ 4} \\
60 \\
\underline{48} \\
120 \\
\underline{112} \\
80 \\
\underline{80} \\
0
\end{array}
$$

$$\frac{7}{16} = 0.4375$$

**11.**

$$
\begin{array}{r}
0.35 \\
20\overline{)7.00} \\
\underline{6\ 0} \\
1\ 00 \\
\underline{1\ 00} \\
0
\end{array}
$$

$$\frac{7}{20} = 0.35$$

**13.**

$$
\begin{array}{r}
0.62 \\
50\overline{)31.00} \\
\underline{30\ 0} \\
1\ 00 \\
\underline{1\ 00} \\
0
\end{array}
$$

$$\frac{31}{50} = 0.62$$

**15.**

$$
\begin{array}{r}
2.25 \\
4\overline{)9.00} \\
\underline{8} \\
1\ 0 \\
\underline{8} \\
20 \\
\underline{20} \\
0
\end{array}
$$

$$\frac{9}{4} = 2.25$$

**17.**

$$
\begin{array}{r}
0.125 \\
8\overline{)1.000} \\
\underline{8} \\
20 \\
\underline{16} \\
40 \\
\underline{40} \\
0
\end{array}
$$

$$2\frac{1}{8} = 2.125$$

**19.**

$$
\begin{array}{r}
0.4375 \\
16\overline{)7.0000} \\
\underline{6\ 4} \\
60 \\
\underline{48} \\
120 \\
\underline{112} \\
80 \\
\underline{80} \\
0
\end{array}
$$

$$1\frac{7}{16} = 1.4375$$

**21.**

$$
\begin{array}{r}
0.933 \\
15\overline{)14.000} \\
\underline{13\ 5} \\
50 \\
\underline{45} \\
50 \\
\underline{45} \\
5
\end{array}
$$

$$\frac{14}{15} = 0.9\overline{3}$$

**23.**

$$
\begin{array}{r}
0.454 \\
11\overline{)5.000} \\
\underline{4\ 4} \\
60 \\
\underline{55} \\
50 \\
\underline{44} \\
6
\end{array}
$$

$$\frac{5}{11} = 0.\overline{45}$$

**25.**

$$
\begin{array}{r}
0.5833 \\
12\overline{)7.0000} \\
\underline{6\ 0} \\
1\ 00 \\
\underline{96} \\
40 \\
\underline{36} \\
40 \\
\underline{36} \\
4
\end{array}
$$

$$3\frac{7}{12} = 3.58\overline{3}$$

**27.**
$$18\overline{)5.000}$$ → $0.277$
$$\begin{array}{r} 3\ 6 \\ \hline 1\ 40 \\ 1\ 26 \\ \hline 140 \\ 126 \\ \hline 14 \end{array}$$

$2\dfrac{5}{18} = 2.2\overline{7}$

**29.**
$$7\overline{)4.0000}$$ → $0.5714$
$$\begin{array}{r} 3\ 5 \\ \hline 50 \\ 49 \\ \hline 10 \\ 7 \\ \hline 30 \\ 28 \\ \hline 2 \end{array}$$

$\dfrac{4}{7}$ rounds to 0.571.

**31.**
$$21\overline{)19.0000}$$ → $0.9047$
$$\begin{array}{r} 18\ 9 \\ \hline 10 \\ 0 \\ \hline 100 \\ 84 \\ \hline 160 \\ 147 \\ \hline 13 \end{array}$$

$\dfrac{19}{21}$ rounds to 0.905

**33.**
$$48\overline{)7.0000}$$ → $0.1458$
$$\begin{array}{r} 4\ 8 \\ \hline 2\ 20 \\ 1\ 92 \\ \hline 280 \\ 240 \\ \hline 400 \\ 384 \\ \hline 16 \end{array}$$

$\dfrac{7}{48}$ rounds to 0.146.

**35.**
$$27\overline{)35.0000}$$ → $1.2962$
$$\begin{array}{r} 27 \\ \hline 8\ 0 \\ 5\ 4 \\ \hline 2\ 60 \\ 2\ 43 \\ \hline 170 \\ 162 \\ \hline 80 \\ 54 \\ \hline 26 \end{array}$$

$\dfrac{35}{27}$ rounds to 1.296

**37.**
$$52\overline{)21.0000}$$ → $0.4038$
$$\begin{array}{r} 20\ 8 \\ \hline 20 \\ 00 \\ \hline 200 \\ 156 \\ \hline 440 \\ 416 \\ \hline 24 \end{array}$$

$\dfrac{21}{52}$ rounds to 0.404.

**39.**
$$18\overline{)17.0}$$ → $0.944$
$$\begin{array}{r} 16\ 2 \\ \hline 80 \\ 72 \\ \hline 80 \\ 72 \\ \hline 8 \end{array}$$

$\dfrac{17}{18} = 0.9\overline{4}$

$\dfrac{17}{18}$ rounds to 0.944.

**41.**
$$7\overline{)22.0000}$$ → $3.1428$
$$\begin{array}{r} 21 \\ \hline 1\ 0 \\ 7 \\ \hline 30 \\ 28 \\ \hline 20 \\ 14 \\ \hline 60 \\ 56 \\ \hline 4 \end{array}$$

$\dfrac{22}{7}$ rounds to 3.143.

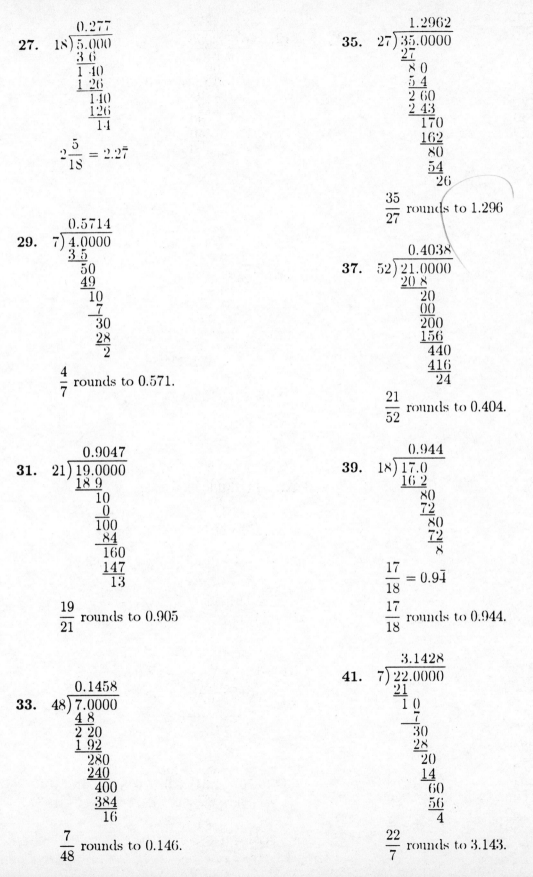

**43.**

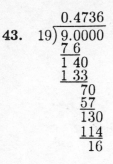

$\dfrac{9}{19}$ rounds to 0.474.

**45.**

$$\begin{array}{r} 0.125 \\ 8\overline{)1.000} \\ \underline{8}\phantom{00} \\ 20 \\ \underline{16} \\ 40 \\ \underline{40} \\ 0 \end{array}$$

$\dfrac{1}{8} = 0.125$ inch

**47.**

$$\begin{array}{r} 0.375 \\ 8\overline{)3.000} \\ \underline{2\,4}\phantom{0} \\ 60 \\ \underline{56} \\ 40 \\ \underline{40} \\ 0 \end{array}$$

$$\begin{array}{r} 0.500 \\ -\ 0.375 \\ \hline 0.125 \end{array}$$

It is too small by 0.125 inch.

**49.**

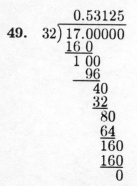

$$\begin{array}{r} 0.53125 \\ -\ 0.53000 \\ \hline 0.00125 \end{array}$$

It is too thick by 0.00125 inch.

**51.** $2.4 + (0.5)^2 - 0.35$
$= 2.4 + 0.25 - 0.35$
$= 2.65 - 0.35$
$= 2.30$ or $2.3$

**53.** $12.2 \times 9.4 - 2.68 + 1.6 \div 0.8$
$= 114.68 - 2.68 + 2$
$= 114$

**55.** $12 \div 0.03 - 50\,(0.5 + 1.5)^3$
$= 12 \div 0.03 - 50\,(2)^3$
$= 12 \div 0.03 - 50\,(8)$
$= 400 - 400$
$= 0$

**57.** $(1.1)^3 + 2.6 \div 0.13 + 0.083$
$= 1.331 + 20 + 0.083$
$= 21.414$

**59.** $(14.73 - 14.61)^2 \div (1.18 + 0.82)$
$= (0.12)^2 \div 2$
$= 0.0144 \div 2$
$= 0.0072$

**61.** $(0.5)^3 + (3 - 2.6) \times 0.5$
$= (0.5)^3 + 0.4 \times 0.5$
$= 0.125 + 0.20$
$= 0.325$

**63.** $(0.76 + 4.24) \div 0.25 + 8.6$
$= 5.00 \div 0.25 + 8.6$
$= 20.0 + 8.6$
$= 28.6$

**65.** $(1.6)^3 + (2.4)^2 + 18.666 \div 3.05 + 4.86$
= 4.096 + 5.76 + 6.12 + 4.86
= 20.836

**67.**
```
            0.5869297
     8921) 5236.0000000
            4460 5
             775 50
             713 68
              61 820
              53 526
               8 2940
               8 0289
                 26510
                 17842
                 86680
                 80289
                  63910
                  62447
                   1463
```

$$\frac{5236}{8921} = 0.586930$$

**69.** **a.**
```
     0.16̄1̄6̄
   − 0.00̄1̄6̄
     0.16
```

**b.**
```
     0.1616̄1̄6̄
   − 0.016666̄
     0.144949
```

**c.** The repeating patterns line up differently.

---

## Cumulative Review Problems

**71.** $12 \times 26 = 312$ square feet

**73.** $56 + 81 + 42 + 198 = \$377$ was deposited.

---

## 3.7   Exercises

**1.**
```
   200,000,000
 + 500,000,000
   700,000,000
```

**3.**
```
   60,000
 − 30,000
   30,000
```

**5.** $400,000 \times 0.8 = 320,000$

**7.** $900,000 \div 6,000 = 150$

**9.** $\$500 \div 50 = \$10$ per week

**11.**
```
      650
    × 7.5
    325 0
   4550
   4875.0
```
4875 kroners

**13.**
```
      48.3
    × 56.9
    43 47
   289 8
   2415
   2748.27
```
2748.27 square feet

**15.**
```
            96
    0.12) 11.52
          10 8
            72
            72
             0
```
96 molds

**17.**
```
    11.68
    10.42
  + 12.67
    34.77
```
```
       11.59
    3) 34.77
       3
        4
        3
        1 7
        1 5
          27
          27
           0
```
11.59 meters of rainfall per year

**19.**  $3.5\overline{)\,42.0}$

$$
\begin{array}{r}
12 \\
3.5\overline{)\,42.0} \\
35\phantom{.0} \\
\hline
7\,0 \\
7\,0 \\
\hline
0
\end{array}
$$

12 days

**21.**
$$
\begin{array}{r}
43.9 \\
11.3 \\
+\;63.4 \\
\hline
118.6
\end{array}
$$

$$
\begin{array}{r}
118.6 \\
\times\;\;10.65 \\
\hline
5930 \\
7116\phantom{0} \\
11860\phantom{00} \\
\hline
1263.090
\end{array}
$$

$1263.09

**23.**  Time $= 40 + 1.5 \times 12$

$\phantom{Time\ } = 40 + 18$

$\phantom{Time\ } = 58$ hours

Salary $= 58 \times 6.20$

$\phantom{Salary\ } = 359.60$

$359.60 for week

**25.**
$$
\begin{array}{r}
420.13 \\
116.32 \\
318.57 \\
+\;\;\;1.86 \\
\hline
\$856.88
\end{array}
$$

$$
\begin{array}{r}
16.50 \\
36.89 \\
+\;376.94 \\
\hline
\$430.33
\end{array}
$$

$$
\begin{array}{r}
856.88 \\
-\;430.33 \\
\hline
\$426.55 \quad \text{balance}
\end{array}
$$

**27.**
$$
\begin{array}{r}
288.65 \\
\times\;\;\;\;60 \\
\hline
17{,}319
\end{array}
$$

$$
\begin{array}{r}
17{,}319 \\
-\;11{,}500 \\
\hline
5819
\end{array}
$$

He will pay $17,319 over 5 years.

He will pay $5819 more than the loan.

**29.**  $7\overline{)\,8.060}$

$$
\begin{array}{r}
1.151 \\
7\overline{)\,8.060} \\
7\phantom{.060} \\
\hline
1\,0 \\
7 \\
\hline
36 \\
35 \\
\hline
10 \\
7 \\
\hline
3
\end{array}
$$

$$
\begin{array}{r}
1.300 \\
-\;1.151 \\
\hline
0.149
\end{array}
$$

Yes, by 0.149 milligram per liter

**31.**  $126.4\overline{)\,17316.\,8}$

$$
\begin{array}{r}
137 \\
126.4\overline{)\,17316.\,8} \\
1264\phantom{0.\,8} \\
\hline
4676\phantom{.\,8} \\
3792\phantom{.\,8} \\
\hline
884\;8 \\
884\;8 \\
\hline
0
\end{array}
$$

137 minutes

**33.**
$$
\begin{array}{r}
89.3 \\
-\;66.4 \\
\hline
22.9 \quad \text{quadrillion Btus}
\end{array}
$$

**35.**  $(33.1 + 43.8 + 66.4) \div 3$

$\phantom{(33.1\ } = 143.3 \div 3$

$\phantom{(33.1\ } \approx 47.8$

Approximately 47.8 quadrillion Btus or
47,800,000,000,000,000 Btus

## Cumulative Review Problems

**37.** $\dfrac{1}{5} + \dfrac{3}{7} \times \dfrac{1}{2} = \dfrac{1}{5} + \dfrac{3}{14}$

$$= \dfrac{1}{5} \times \dfrac{14}{14} + \dfrac{3}{14} \times \dfrac{5}{5}$$

$$= \dfrac{14}{70} + \dfrac{15}{70}$$

$$= \dfrac{29}{70}$$

**39.** $\dfrac{5}{12} \times \dfrac{36}{27} = \dfrac{5 \times 3 \times 12}{12 \times 3 \times 9}$

$$= \dfrac{5}{9}$$

## Chapter 3 Review Problems

**1.** 13.672 = thirteen and six hundred seventy-two thousandths

**2.** Eighty-four hundred thousandths

**3.** $\dfrac{7}{10} = 0.7$

**4.** $\dfrac{81}{100} = 0.81$

**5.** $1\dfrac{523}{1000} = 1.523$

**6.** $\dfrac{79}{10,000} = 0.0079$

**7.** $0.17 = \dfrac{17}{100}$

**8.** $0.365 = \dfrac{365}{1000} = \dfrac{73}{200}$

**9.** $34.24 = 34\dfrac{24}{100}$

$$= 34\dfrac{6}{25}$$

**10.** $1.00025 = 1\dfrac{25}{100,000} = 1\dfrac{1}{4000}$

**11.** $\dfrac{13}{100} = 0.13$

$\dfrac{13}{100} > 0.13$

**12.** $0.716 > 0.706$

**13.** 0.981, 0.918, 0.98, 0.901
0.981, 0.918, 0.980, 0.901
0.901, 0.918, 0.980, 0.981
0.901, 0.918, 0.98, 0.981

**14.** 5.62, 5.2, 5.6, 5.26, 5.59
5.62, 5.20, 5.60, 5.26, 5.59
5.20, 5.26, 5.59, 5.60, 5.62
5.2, 5.26, 5.59, 5.6, 5.62

**15.** 0.6<u>1</u>3 rounds to 0.6.

**16.** 19.20<u>76</u> rounds to 19.21

**17.** 1.099<u>52</u> rounds to 1.100.

**18.** $156.<u>4</u>8 rounds to $156

**19.**
```
     9.6
    11.5
    21.8
 + 34.7
    77.6
```

**20.**
```
    1.800
    2.603
    0.520
 + 1.716
    6.639
```

21.
```
    5.190
 −  1.296
    3.894
```

22.
```
    352.806
 −  195.992
    156.814
```

23.
```
     0.098
 ×   0.032
     0196
    0294
    0.003136
```

24.
```
  126.83
 ×     7
  887.81
```

25.
```
     7.8
 ×   5.2
    15 6
    390
    405.6
```

26.
```
     7053
 ×   0.34
    28212
   21159
   2398.02
```

27. $0.000613 \times 10^3 = 0.613$

28. $1.2354 \times 10^5 = 123,540$

29.
```
     0.35
 ×   3.6
    210
   105
   1.260 or $1.26
```

30.
```
            0.00258
  0.07) 0.0001806
            14
            40
            35
            56
            56
             0
```

31.
```
            36. 8
  5.2) 191. 36
       156
        35 3
        31 2
         4 16
         4 16
            0
```

32.
```
          232.9
  8) 1863.2
     16
     26
     24
     23
     16
      7 2
      7 2
        0
```

33.
```
          574. 42
  1.3) 746. 750
       65
       96
       91
        5 7
        5 2
          55
          52
          30
          26
           4
```

574.42 rounds to 574.4

34.
```
            0. 0589
  0.06) 0.003539
           30
           53
           48
           59
           54
            5
```

0.0589 rounds to 0.059

**35.**

$$
\begin{array}{r}
0.277 \\
18 \overline{)5.000} \\
3\ 6 \\
\hline
1\ 40 \\
1\ 26 \\
\hline
140 \\
126 \\
\hline
14
\end{array}
$$

$$\frac{5}{18} = 0.2\overline{7}$$

**36.**

$$
\begin{array}{r}
0.175 \\
40 \overline{)7.000} \\
4\ 0 \\
\hline
3\ 00 \\
2\ 80 \\
\hline
200 \\
200 \\
\hline
0
\end{array}
$$

$$\frac{7}{40} = 0.175$$

**37.**

$$
\begin{array}{r}
0.833 \\
6 \overline{)5.000} \\
4\ 8 \\
\hline
20 \\
18 \\
\hline
20 \\
18 \\
\hline
2
\end{array}
$$

$$1\frac{5}{6} = 1.8\overline{3}$$

**38.**

$$
\begin{array}{r}
1.1875 \\
16 \overline{)19.0000} \\
16 \\
\hline
3\ 0 \\
1\ 6 \\
\hline
1\ 40 \\
1\ 28 \\
\hline
120 \\
112 \\
\hline
80 \\
80 \\
\hline
0
\end{array}
$$

$$\frac{19}{16} = 1.1875$$

**39.**

$$
\begin{array}{r}
0.7857 \\
14 \overline{)11.0000} \\
9\ 8 \\
\hline
1\ 20 \\
1\ 12 \\
\hline
80 \\
70 \\
\hline
100 \\
98 \\
\hline
2
\end{array}
$$

$$\frac{11}{14} \text{ rounds to } 0.786.$$

**40.**

$$
\begin{array}{r}
0.2941 \\
17 \overline{)5.0000} \\
3\ 4 \\
\hline
1\ 60 \\
1\ 53 \\
\hline
70 \\
68 \\
\hline
20 \\
17 \\
\hline
3
\end{array}
$$

$$2\frac{5}{17} \text{ rounds to } 2.294.$$

**41.** $1.6 \times 2.3 + 0.4 - 0.6 \times 0.8$
$\quad = 3.68 + 0.4 - 0.48$
$\quad = 3.6$

**42.** $0.03 + (1.2)^2 - 5.3 \times 0.06$
$\quad = 0.03 + 1.44 - 5.3 \times 0.06$
$\quad = 0.03 + 1.44 - 0.318$
$\quad = 1.47 - 0.318$
$\quad = 1.152$

**43.** $(1.02)^3 + 5.76 \div 1.2 \times 0.05$
$\quad = 1.061208 + 5.76 \div 1.2 \times 0.05$
$\quad = 1.061208 + 4.8 \times 0.05$
$\quad = 1.061208 + 0.24$
$\quad = 1.301208$

**44.** $6.63 + 8.24 \div (5.76 - 5.68) - 22.5$
$\quad = 6.63 + 8.24 \div 0.08 - 22.5$
$\quad = 6.63 + 103 - 22.5$
$\quad = 109.63 - 22.5$
$\quad = 87.13$

**45.**
$$\begin{array}{r} 2398.26 \\ -\ 1959.07 \\ \hline 439.19 \end{array}$$

**46.**
$$\begin{array}{r} 67.036 \\ \times\ \ 0.006 \\ \hline 0.402216 \end{array}$$

**47.**
$$\begin{array}{r} 0.0610 \\ 0.0023 \\ +\ 0.7770 \\ \hline 0.8403 \end{array}$$

**48.**
$$\begin{array}{r} 69.2 \\ 1.6)\overline{110.72} \\ \underline{96} \\ 14\ 7 \\ \underline{14\ 4} \\ 32 \\ \underline{32} \\ 0 \end{array}$$

**49.** $8 \div 0.4 + 0.1 \times (0.2)^2$
$$= 20 + 0.1 \times 0.04$$
$$= 20 + 0.004$$
$$= 20.004$$

**50.** $(3.8 - 2.8)^3 \div (0.5 + 0.3)$
$$= 1^3 \div 0.8$$
$$= 1 \div 0.8$$
$$= 1.25$$

**51.** Tickets $= 228 + 2.5 \times 388 + 3 \times 430$
$$= 228 + 970 + 1290$$
$$= 2488$$

Not tickets $= 2600 - 2488$
$$= 112 \text{ people}$$

**52.**
$$\begin{array}{r} 26325.8 \\ -\ 26005.8 \\ \hline 320.0 \end{array}$$

$$\begin{array}{r} 24.80 \\ 12.9)\overline{320.000} \\ \underline{258} \\ 62\ 0 \\ \underline{51\ 6} \\ 10\ 40 \\ \underline{10\ 32} \\ 80 \\ \underline{0} \\ 80 \end{array}$$

24.8 miles per gallon

**53.**
$$\begin{array}{r} 189.60 \\ \times\ \ \ \ 48 \\ \hline 151680 \\ 75840 \\ \hline \$9100.80 \end{array}$$

$$\begin{array}{r} 9100.80 \\ -\ 6930.50 \\ \hline \$2170.30 \end{array} \text{ extra}$$

**54.**
$$\begin{array}{r} 8.26 \\ \times\ \ \ 38 \\ \hline 6608 \\ 2478 \\ \hline \$313.88 \end{array}$$

He will earn more at the ABC Company.

**55.**
$$\begin{array}{r} 0.0025 \\ 12)\overline{0.0300} \\ \underline{24} \\ 60 \\ \underline{60} \\ 0 \end{array}$$

$$\begin{array}{r} 0.0025 \\ -\ 0.0020 \\ \hline 0.0005 \end{array}$$

No; by 0.0005 milligram per liter

**56.**
$$\begin{array}{r} 7 \\ 0.046)\overline{0.322} \\ \underline{322} \\ 0 \end{array}$$

7 test tubes

**57.** Fence $= 2 \times 18.3 + 2 \times 9.6$
$= 36.6 + 19.2$
$= 55.8$ feet

**58.**
$$\begin{array}{r} 75.5 \\ \times\ 18.5 \\ \hline 3775 \\ 6040 \\ 755 \\ \hline 1396.75 \end{array}$$

1396.75 square feet

**59.**
$$\begin{array}{r} 16.3 \\ 8.2 \\ 5.7 \\ +\ 18.4 \\ \hline 48.6 \end{array} \text{ miles}$$

**60.**
$$\begin{array}{r} 118.9 \\ 25.6 \\ 18.9 \\ 43.9 \\ 22.6 \\ 13.8 \\ +\ 16.2 \\ \hline 259.9 \end{array}$$

259.9 feet around field.

**61.**
$$\begin{array}{r} 212.50 \\ \times\ \ \ \ 60 \\ \hline \$12,750.00 \end{array}$$

$$\begin{array}{r} 199.50 \\ \times\ \ \ 60 \\ \hline 11,970.00 \\ +\ \ 285.00 \\ \hline \$12,255.00 \end{array}$$

They should change to the new loan.

**62.**
$$\begin{array}{r} 603 \\ -\ 341 \\ \hline 262 \end{array}$$

Increase is $262

**63.**
$$\begin{array}{r} 810 \\ -\ 603 \\ \hline \$207 \end{array}$$

Average is $207

**64.** $479 \div 30 = 15.97$

Average daily benefit is $15.97

**65.** $\dfrac{720}{30} = 24$

Average is $24.00

**66.**
$$\begin{array}{r} 810 \\ -\ 720 \\ \hline 90 \end{array} \qquad \begin{array}{r} 810 \\ +\ 90 \\ \hline 900 \end{array} \qquad \dfrac{900}{30} = 30$$

Average daily benefit is $30.00

**67.**
$$\begin{array}{r} 810 \\ +\ 207 \\ \hline 1017 \end{array}$$

Then $\dfrac{1017}{30} = 33.90$

Average is $33.90

---

## Chapter 3 Test

**1.** One hundred fifty-seven thousandths

**2.** $\dfrac{3977}{10,000} = 0.3977$

**3.** $7.15 = 7\dfrac{15}{100} = 7\dfrac{3}{20}$

**4.** $0.261 = \dfrac{261}{1000}$

**5.** 2.19, 2.91, 2.9, 2.907

    2.190, 2.910, 2,900, 2.907

    2.190, 2.900, 2.907, 2.910

    2.19, 2.9, 2.907, 2.91

**6.** 78.65$\underline{6}$2 rounds to 78.66

**7.** 0.341$\underline{7}$52 rounds to 0.0342

**8.**
$$
\begin{array}{r}
96.200 \\
1.348 \\
+\ 2.150 \\
\hline
99.698
\end{array}
$$

**9.**
$$
\begin{array}{r}
17.00 \\
2.10 \\
16.80 \\
0.04 \\
+\ 1.59 \\
\hline
37.53
\end{array}
$$

**10.**
$$
\begin{array}{r}
1.0075 \\
-\ 0.9096 \\
\hline
0.0979
\end{array}
$$

**11.**
$$
\begin{array}{r}
72.300 \\
-\ 1.145 \\
\hline
71.155
\end{array}
$$

**12.**
$$
\begin{array}{r}
8.31 \\
\times\ 0.07 \\
\hline
0.5817
\end{array}
$$

**13.** $2.189 \times 100 = 218.9$

**14.**
$$
\begin{array}{r}
25.\,7 \phantom{00} \\
0.004\,)\overline{\,0.1028} \\
\underline{8}\phantom{0000} \\
22\phantom{00} \\
\underline{20}\phantom{00} \\
28\phantom{0} \\
\underline{28}\phantom{0} \\
0
\end{array}
$$

**15.**
$$
\begin{array}{r}
47 \\
0.69\,)\overline{\,32.43} \\
\underline{27\,6}\phantom{0} \\
4\,83 \\
\underline{4\,83} \\
0
\end{array}
$$

**16.**
$$
\begin{array}{r}
1.2 \\
9\,)\overline{\,11.0} \\
\underline{9}\phantom{00} \\
2\,0 \\
\underline{1\,8} \\
2
\end{array}
$$

$$\frac{11}{9} = 1.\bar{2}$$

**17.**
$$
\begin{array}{r}
0.5625 \\
16\,)\overline{\,9.0000} \\
\underline{8\,0}\phantom{000} \\
1\,00 \\
\underline{96}\phantom{0} \\
40 \\
\underline{32} \\
80 \\
\underline{80} \\
0
\end{array}
$$

**18.** $(0.3)^3 + 1.02 \div 0.5 - 0.58$

      $= 0.027 + 1.02 \div 0.5 - 0.58$

      $= 0.027 + 2.04 - 0.58$

      $= 2.067 - 0.58$

      $= 1.487$

**19.** $19.36 \div (0.24 + 0.26) \times (0.4)^2$

      $= 19.36 \div 0.5 \times 0.16$

      $= 38.72 \times 0.16$

      $= 6.1952$

**20.**
$$
\begin{array}{r}
4.25 \\
\times\ 7.8 \\
\hline
3400 \\
2975 \\
\hline
33.150
\end{array}
$$

    $33.15

**21.**

$$\begin{array}{r} 42{,}780.5 \\ -\ 42{,}620.5 \\ \hline 160.0 \end{array}$$

$$\begin{array}{r} 18.\,82 \\ 8.5\overline{)160.\,000} \\ \underline{85} \\ 75\ 0 \\ \underline{68\ 0} \\ 7\ 00 \\ \underline{6\ 80} \\ 200 \\ \underline{170} \\ 30 \end{array}$$

18.8 miles per gallon

**22.**

$$\begin{array}{r} 8.01 \\ 5.03 \\ +\ 8.53 \\ \hline 21.57 \end{array}$$

$$\begin{array}{r} 25.00 \\ -\ 21.57 \\ \hline 3.43 \end{array}$$

3.43 centimeters less

**23.** Time $= 40 + 1.5 \times 9$
$$= 40 + 13.5$$
$$= 53.5 \text{ hours}$$

Salary $= \$7.30 \times 53.5$
$$= \$390.55$$

---

# Chapters 1–3 Cumulative Test

**1.** 38,056,954 = Thirty-eight million, fifty-six thousand, nine hundred fifty-four

**2.**
$$\begin{array}{r} 156{,}028 \\ 301{,}579 \\ +\ 21{,}980 \\ \hline 479{,}587 \end{array}$$

**3.**
$$\begin{array}{r} 1{,}091{,}000 \\ -\ 1{,}036{,}520 \\ \hline 54{,}480 \end{array}$$

**4.**
$$\begin{array}{r} 589 \\ \times\ 67 \\ \hline 4123 \\ 3534 \\ \hline 39{,}463 \end{array}$$

**5.**
$$\begin{array}{r} 258 \\ 17\overline{)4386} \\ \underline{34} \\ 98 \\ \underline{85} \\ 136 \\ \underline{136} \\ 0 \end{array}$$

**6.** $20 \div 4 + 2^5 - 7 \times 3$
$$= 20 \div 4 + 32 - 7 \times 3$$
$$= 5 + 32 - 21$$
$$= 37 - 21$$
$$= 16$$

**7.** $\dfrac{33}{88} = \dfrac{33 \div 11}{88 \div 11} = \dfrac{3}{8}$

**8.** $4\dfrac{1}{3} + 3\dfrac{1}{6} = 4\dfrac{2}{6} + 3\dfrac{1}{6}$
$$= 7\dfrac{3}{6}$$
$$= 7\dfrac{1}{2}$$

**9.** $\dfrac{23}{35} - \dfrac{2}{5} = \dfrac{23}{35} - \dfrac{2}{5} \times \dfrac{7}{7}$
$$= \dfrac{23}{35} - \dfrac{14}{35}$$
$$= \dfrac{9}{35}$$

**10.** $\dfrac{7}{10} \times \dfrac{5}{3} - \dfrac{5}{12} \times \dfrac{1}{2} = \dfrac{7}{6} - \dfrac{5}{24}$
$$= \dfrac{7}{6} \times \dfrac{4}{4} - \dfrac{5}{24}$$
$$= \dfrac{28}{24} - \dfrac{5}{24}$$
$$= \dfrac{23}{24}$$

**11.**  $52 \div 3\frac{1}{4} = 52 \div \frac{13}{4}$

$\qquad\quad = 52 \times \frac{4}{13}$

$\qquad\quad = 16$

**12.**  $1\frac{3}{8} \div \frac{5}{12} = \frac{11}{8} \div \frac{5}{12}$

$\qquad\qquad = \frac{11}{8} \times \frac{12}{5}$

$\qquad\qquad = \frac{11 \times 4 \times 3}{4 \times 2 \times 5}$

$\qquad\qquad = \frac{33}{10} = 3\frac{3}{10}$

**13.**  $60{,}000 \times 400{,}000 = 24{,}000{,}000{,}000$

**14.**  $\dfrac{571}{1000} = 0.571$

**15.**  2.01, 2.1, 2.11, 2.12, 20.1

**16.**  26.079$\underline{8}$4 rounds to 26.080

**17.**
```
    1.90
    2.36
   15.20
 +  0.08
   19.54
```

**18.**
```
   28.007
 − 19.368
    8.639
```

**19.**
```
    56.8
 ×  0.02
   1.136
```

**20.**  $365.123 \times 100 = 36{,}512.3$

**21.**
```
            1. 058
  0.06) 0.06348
           6
           3
           0
          34
          30
           48
           48
            0
```

**22.**
```
          0.8125
  16) 13.0000
      12 8
         20
         16
          40
          32
           80
           80
            0
```

$\dfrac{13}{16} = 0.8125$

**23.**  $1.44 \div 0.12 + (0.3)^3 + 1.56$

$\qquad = 1.44 \div 0.12 + 0.027 + 1.57$

$\qquad = 12 + 0.027 + 1.57$

$\qquad = 12.027 + 1.57$

$\qquad = 13.597$

**24.**
```
     28.5
 ×    16
   171 0
   285
   456.0
```

456 miles

**25.**
```
   199.36
     1.03
   166.35
+  93.50
   460.24
```

```
    90.00
    37.49
+  137.18
   264.67
```

```
   460.24
-  264.67
   195.57
```

$195.57

**26.**
```
            60
320.50) 19,230.00
        19 230 0
              0
```

60 months

# RATIO AND PROPORTION

## Pretest Chapter 4

1. $\dfrac{13}{18}$

2. $\dfrac{44}{220} = \dfrac{44 \div 44}{220 \div 44} = \dfrac{1}{5}$

3. $\dfrac{\$72}{\$16} = \dfrac{72 \div 8}{16 \div 8} = \dfrac{9}{2}$

4. $\dfrac{121}{132} = \dfrac{121 \div 11}{132 \div 11} = \dfrac{11}{12}$

5. a. $\dfrac{\$70}{\$240} = \dfrac{70 \div 10}{240 \div 10} = \dfrac{7}{24}$

   b. $\dfrac{\$22}{\$240} = \dfrac{22 \div 2}{240 \div 2} = \dfrac{11}{120}$

6. $\dfrac{9 \text{ attendants}}{300 \text{ passengers}}$

   $= \dfrac{9 \div 3}{300 \div 3} = \dfrac{3 \text{ attendants}}{100 \text{ passengers}}$

7. $\dfrac{620 \text{ gallons}}{840 \text{ square feet}}$

   $= \dfrac{620 \div 20}{840 \div 20} = \dfrac{31 \text{ gallons}}{42 \text{ square feet}}$

8. $\dfrac{122 \text{ miles}}{4 \text{ hours}} = \dfrac{122 \div 4}{4 \div 4}$

   $= \dfrac{30.5 \text{ miles}}{1 \text{ hour}}$ or $30.5$ miles/hour

9. $\dfrac{\$435}{15} = \dfrac{\$435 \div 15}{15 \div 15} = \dfrac{\$29}{1 \text{ CD player}}$

10. $\dfrac{13}{40} = \dfrac{39}{120}$

11. $\dfrac{116}{158} = \dfrac{29}{37}$

12. $\dfrac{14}{31} = \dfrac{42}{93}$

    $14 \times 93 \overset{?}{=} 31 \times 42$

    $1302 = 1302$   True

13. $\dfrac{17}{33} \overset{?}{=} \dfrac{19}{45}$

    $17 \times 45 \overset{?}{=} 33 \times 19$

    $765 \neq 627$   False

14. $9 \times n = 153$

    $\dfrac{9 \times n}{9} = \dfrac{153}{9}$

    $n = 17$

15. $234 = 13 \times n$

    $\dfrac{234}{13} = \dfrac{13 \times n}{13}$

    $18 = n$

16. $\dfrac{36}{16} = \dfrac{9}{n}$

    $36 \times n = 16 \times 9$

    $36 \times n = 144$

    $\dfrac{36 \times n}{36} = \dfrac{144}{36}$

    $n = 4$

**17.**
$$\frac{3}{144} = \frac{n}{336}$$
$$144 \times n = 3 \times 336$$
$$144 \times n = 1008$$
$$\frac{144 \times n}{144} = \frac{1008}{144}$$
$$n = 7$$

**18.**
$$\frac{n}{900} = \frac{15}{22.5}$$
$$n \times 22.5 = 900 \times 15$$
$$n \times 22.5 = 13,500$$
$$\frac{n \times 22.5}{22.5} = \frac{13,500}{22.5}$$
$$n = 600$$

**19.**
$$\frac{\frac{1}{2}}{n} = \frac{\frac{3}{4}}{3}$$
$$\frac{3}{4} \times n = 3 \times \frac{1}{2}$$
$$\frac{3}{4} \times n = \frac{3}{2}$$
$$\frac{\frac{3}{4} \times n}{\frac{3}{4}} = \frac{\frac{3}{2}}{\frac{3}{4}}$$
$$n = \frac{3}{2} \div \frac{3}{4}$$
$$= \frac{3}{2} \times \frac{4}{3} = 2$$

**20.**
$$\frac{1.5 \text{ cups}}{6 \text{ portions}} = \frac{n \text{ cups}}{14 \text{ portions}}$$
$$14 \times 1.5 = 6 \times n$$
$$21 = 6 \times n$$
$$\frac{21}{6} = \frac{6 \times n}{6}$$
$$n = 3.5 \text{ cups}$$

**21.**
$$\frac{81 \text{ miles}}{2 \text{ gallons}} = \frac{n \text{ miles}}{9 \text{ gallons}}$$
$$81 \times 9 = 2 \times n$$
$$729 = 2 \times n$$
$$\frac{729}{2} = \frac{2 \times n}{2}$$
$$364.5 = n$$

364.5 miles

**22.**
$$\frac{5 \text{ inches}}{365 \text{ miles}} = \frac{2 \text{ inches}}{n \text{ miles}}$$
$$5 \times n = 2 \times 365$$
$$5 \times n = 730$$
$$\frac{5 \times n}{5} = \frac{730}{5}$$
$$n = 146 \text{ miles}$$

**23.**
$$\frac{121 \text{ bulbs}}{6 \text{ defective}} = \frac{1089 \text{ bulbs}}{n \text{ defective}}$$
$$121 \times n = 6 \times 1089$$
$$121 \times n = 6534$$
$$\frac{121 \times n}{121} = \frac{6534}{121}$$
$$n = 54$$

54 defective bulbs

**24.**
$$\frac{67 \text{ runs}}{245 \text{ innings}} = \frac{n \text{ runs}}{9 \text{ innings}}$$
$$67 \times 9 = n \times 245$$
$$603 = n \times 245$$
$$\frac{603}{245} = \frac{n \times 245}{245}$$
$$n = 2.5 \text{ runs}$$

**25.**
$$\frac{3 \text{ read}}{10 \text{ people}} = \frac{n \text{ read}}{45,600 \text{ people}}$$
$$3 \times 45,600 = 10 \times n$$
$$136,800 = 10 \times n$$
$$\frac{136,800}{10} = \frac{10 \times n}{10}$$
$$13,680 = n$$

13,680 read Boston Globe

## 4.1   Exercises

**1.** ratio

**3.** 5 to 8

**5.** $6{:}18 = \dfrac{6}{18} = \dfrac{6 \div 6}{18 \div 6} = \dfrac{1}{3}$

**7.** $21{:}18 = \dfrac{21}{18} = \dfrac{21 \div 3}{18 \div 3} = \dfrac{7}{6}$

**9.** $36{:}132 = \dfrac{36}{132} = \dfrac{36 \div 12}{132 \div 12} = \dfrac{3}{11}$

**11.** $150{:}225 = \dfrac{150}{225} = \dfrac{150 \div 75}{225 \div 75} = \dfrac{2}{3}$

**13.** $60 \text{ to } 64 = \dfrac{60}{64} = \dfrac{60 \div 4}{64 \div 4} = \dfrac{15}{16}$

**15.** $28 \text{ to } 42 = \dfrac{28}{42} = \dfrac{28 \div 14}{42 \div 14} = \dfrac{2}{3}$

**17.** $32 \text{ to } 20 = \dfrac{32}{20} = \dfrac{32 \div 4}{20 \div 4} = \dfrac{8}{5}$

**19.** $8 \text{ ounces to } 12 \text{ ounces} = \dfrac{8}{12} = \dfrac{8 \div 4}{12 \div 4} = \dfrac{2}{3}$

**21.** 39 kilograms to 26 kilograms
$$= \dfrac{39}{26} = \dfrac{39 \div 13}{26 \div 13} = \dfrac{3}{2}$$

**23.** $\$54 \text{ to } \$63 = \dfrac{54}{63} = \dfrac{54 \div 9}{63 \div 9} = \dfrac{6}{7}$

**25.** 312 yards to 24 yards
$$= \dfrac{312}{24} = \dfrac{312 \div 24}{24 \div 24} = \dfrac{13}{1}$$

**27.** $2\dfrac{1}{2}$ pounds to $4\dfrac{1}{4}$ pounds
$$= \dfrac{2\frac{1}{2}}{4\frac{1}{4}} = 2\dfrac{1}{2} \div 4\dfrac{1}{4} = \dfrac{5}{2} \div \dfrac{17}{4}$$
$$= \dfrac{5}{2} \times \dfrac{4}{17} = \dfrac{10}{17}$$

**29.** $\dfrac{165}{286} = \dfrac{165 \div 15}{285 \div 15} = \dfrac{11}{19}$

**31.** $\dfrac{35}{165} = \dfrac{35 \div 5}{165 \div 5} = \dfrac{7}{33}$

**33.** $\dfrac{205}{1225} = \dfrac{205 \div 5}{1225 \div 5} = \dfrac{41}{245}$

**35.** $\dfrac{450}{205} = \dfrac{450 \div 5}{205 \div 5} = \dfrac{90}{41}$

**37.** $\dfrac{42}{672} = \dfrac{42 \div 42}{672 \div 42} = \dfrac{1}{16}$

**39.** $\dfrac{\$40}{16 \text{ magazines}} = \dfrac{\$40 \div 8}{16 \text{ magazines} \div 8}$
$$= \dfrac{\$5}{2 \text{ magazines}}$$

**41.** $\dfrac{\$170}{12 \text{ bushes}} = \dfrac{\$170 \div 2}{12 \text{ bushes} \div 2} = \dfrac{\$85}{6 \text{ bushes}}$

**43.** $\dfrac{310 \text{ gallons}}{625 \text{ sq ft}} = \dfrac{310 \text{ gallons} \div 5}{625 \text{ sq ft} \div 5}$
$$= \dfrac{62 \text{ gallons}}{125 \text{ sq ft}}$$

**45.** $\dfrac{6150 \text{ rev}}{15 \text{ miles}} = \dfrac{6150 \text{ rev} \div 15}{15 \text{ miles} \div 15}$
$$= \dfrac{410 \text{ rev}}{1 \text{ mile}} = 410 \text{ rev/mile}$$

**47.** $\dfrac{\$330,000}{12 \text{ employees}} = \dfrac{\$330,000 \div 12}{12 \text{ employees} \div 12}$

$= \dfrac{\$27,500}{1 \text{ employee}}$

$= \$27,500/\text{employee}$

**49.** $\dfrac{\$520}{40 \text{ hours}} = \dfrac{\$520 \div 40}{40 \text{ hours} \div 40}$

$= \dfrac{\$13}{1 \text{ hour}} = \$13/\text{hour}$

**51.** $\dfrac{192 \text{ miles}}{12 \text{ gallons}} = \dfrac{192 \text{ miles} \div 12}{12 \text{ gallons} \div 12}$

$= \dfrac{16 \text{ miles}}{1 \text{ gallon}}$

$= 16 \text{ mi/gal}$

**53.** $\dfrac{2480 \text{ gallons}}{16 \text{ hours}} = \dfrac{2480 \text{ gallons} \div 16}{16 \text{ hours} \div 16}$

$= \dfrac{155 \text{ gallons}}{1 \text{ hour}}$

$= 155 \text{ gal/hr}$

**55.** $\dfrac{2250 \text{ pencils}}{18 \text{ boxes}} = \dfrac{2250 \text{ pencils} \div 18}{18 \text{ boxes} \div 18}$

$= \dfrac{125 \text{ pencils}}{1 \text{ box}}$

$= 125 \text{ pencils/box}$

**57.** $\dfrac{276 \text{ miles}}{4 \text{ hours}} = \dfrac{276 \text{ miles} \div 4}{4 \text{ hours} \div 4}$

$= \dfrac{69 \text{ miles}}{1 \text{ hour}} = 69 \text{ mi/hr}$

**59.** $\dfrac{\$3870}{129 \text{ shares}} = \dfrac{\$3870 \div 129}{129 \text{ shares} \div 129}$

$= \dfrac{\$30}{1 \text{ share}} = \$30/\text{share}$

**61.** Profit $= 1200 - 760 = \$440$

$\dfrac{\$440}{80 \text{ calendars}} = \dfrac{\$440 \div 80}{80 \text{ calendars} \div 80}$

$= \dfrac{\$5.50}{1 \text{ calendar}}$

$= \$5.50/\text{calendar}$

**63. a.** 16-ounce box: $\dfrac{\$1.28}{16 \text{ ounces}} = \$0.08/\text{oz}$

24-ounce box: $\dfrac{\$1.68}{24 \text{ ounces}} = \$0.07/\text{oz}$

**b.** $\begin{array}{r} 0.08 \\ -\ 0.07 \\ \hline \$0.01 \ /\text{ounce} \end{array}$

**c.** $48(0.01) = \$0.48$

**65. a.** $\dfrac{3978 \text{ moose}}{306 \text{ acres}} = 13 \text{ moose/acre}$

**b.** $\dfrac{5520 \text{ moose}}{460 \text{ acres}} = 12 \text{ moose/acre}$

**c.** North Slope

**67.** $\dfrac{\$12,876.50}{525 \text{ shares}} = \$24.53/\text{share}$

**69.** Design:

$\dfrac{750 \text{ meters per second}}{330 \text{ meters per second}} = \text{Mach } 2.3$

Modify:

$\dfrac{810 \text{ meters per second}}{330 \text{ meters per second}} = \text{Mach } 2.5$

$\begin{array}{r} 2.5 \\ -\ 2.3 \\ \hline 0.2 \end{array}$ Increased by Mach 0.2

## Cumulative Review Problems

**71.**  $\begin{aligned} & 2\frac{1}{4} \\ &+\ \frac{3}{8} \end{aligned}$ $\qquad$ $\begin{aligned} & 2\frac{2}{8} \\ &+\ \frac{3}{8} \\ \hline & 2\frac{5}{8} \end{aligned}$

**73.**  $\begin{aligned} \frac{3}{5} \times \frac{5}{8} - \frac{2}{3} \times \frac{1}{4} &= \frac{3}{8} - \frac{2}{12} \\ &= \frac{9}{24} - \frac{4}{24} \\ &= \frac{5}{24} \end{aligned}$

**75.**   $12 \times 5.2 = 62.4 \text{ sq yd}$

$\dfrac{\$764.40}{62.4 \text{ sq yd}} = \$12.25/\text{sq yd}$

## 4.2   Exercises

**1.** equal

**3.**  $\dfrac{18}{9} = \dfrac{2}{1}$

**5.**  $\dfrac{20}{36} = \dfrac{5}{9}$

**7.**  $\dfrac{84}{105} = \dfrac{12}{15}$

**9.**  $\dfrac{5\frac{1}{2}}{16} = \dfrac{7\frac{2}{3}}{23}$

**11.**  $\dfrac{6.5}{14} = \dfrac{13}{28}$

**13.**  $\dfrac{3 \text{ inches}}{40 \text{ miles}} = \dfrac{27 \text{ inches}}{360 \text{ miles}}$

**15.**  $\dfrac{10 \text{ runs}}{45 \text{ games}} = \dfrac{36 \text{ runs}}{162 \text{ games}}$

**17.**  $\dfrac{3 \text{ hours}}{\$525} = \dfrac{7 \text{ hours}}{\$1225}$

**19.**  $\dfrac{3 \text{ teaching assistants}}{40 \text{ children}}$
$= \dfrac{21 \text{ teaching assistants}}{280 \text{ children}}$

**21.**  $\dfrac{4800 \text{ people}}{3 \text{ restaurants}} = \dfrac{11{,}200 \text{ people}}{7 \text{ restaurants}}$

**23.**   $\dfrac{10}{25} \overset{?}{=} \dfrac{6}{15}$
$10 \times 15 \overset{?}{=} 25 \times 6$
$150 = 150 \text{ True}$

**25.**   $\dfrac{11}{7} \overset{?}{=} \dfrac{20}{13}$
$11 \times 13 \overset{?}{=} 7 \times 2$
$143 \neq 140 \text{ False}$

**27.**   $\dfrac{17}{75} \overset{?}{=} \dfrac{22}{100}$
$17 \times 100 \overset{?}{=} 22 \times 75$
$1700 \neq 1650 \text{ False}$

**29.**   $\dfrac{102}{120} \overset{?}{=} \dfrac{85}{100}$
$102 \times 100 \overset{?}{=} 120 \times 85$
$10{,}200 = 10{,}200 \text{ True}$

**31.**   $\dfrac{7}{9} \overset{?}{=} \dfrac{10.5}{13.5}$
$7 \times 13.5 \overset{?}{=} 9 \times 10.5$
$94.5 = 94.5 \text{ True}$

**33.**   $\dfrac{3}{17} \overset{?}{=} \dfrac{4.5}{24.5}$
$3 \times 24.5 \overset{?}{=} 17 \times 4.5$
$73.5 \neq 76.5 \text{ False}$

**35.**
$$\frac{2\frac{1}{3}}{3} \stackrel{?}{=} \frac{7}{15}$$
$$2\frac{1}{3} \times 15 \stackrel{?}{=} 3 \times 7$$
$$35 \neq 21 \quad \text{False}$$

**37.**
$$\frac{9}{22} \stackrel{?}{=} \frac{3}{7\frac{1}{3}}$$
$$9 \times 7\frac{1}{3} \stackrel{?}{=} 22 \times 3$$
$$66 = 66 \quad \text{True}$$

**39.**
$$\frac{135 \text{ miles}}{3 \text{ hours}} \stackrel{?}{=} \frac{225 \text{ miles}}{5 \text{ hours}}$$
$$135 \times 5 \stackrel{?}{=} 3 \times 225$$
$$675 = 675 \quad \text{True}$$

**41.**
$$\frac{166 \text{ gallons}}{14 \text{ acres}} \stackrel{?}{=} \frac{249 \text{ gallons}}{21 \text{ acres}}$$
$$166 \times 21 \stackrel{?}{=} 14 \times 249$$
$$3486 = 3486 \quad \text{True}$$

**43.**
$$\frac{27 \text{ points}}{45 \text{ games}} \stackrel{?}{=} \frac{22 \text{ points}}{40 \text{ games}}$$
$$27 \times 40 \stackrel{?}{=} 45 \times 22$$
$$1080 \neq 990 \quad \text{False}$$

**45.**
$$\frac{96 \text{ male}}{54 \text{ female}} \stackrel{?}{=} \frac{144 \text{ male}}{81 \text{ female}}$$
$$96 \times 81 \stackrel{?}{=} 54 \times 144$$
$$7776 = 7776$$

Yes, the ratio is the same.

**47.**
$$\frac{550 \text{ miles}}{15 \text{ hours}} \stackrel{?}{=} \frac{230 \text{ miles}}{6 \text{ hours}}$$
$$550 \times 6 \stackrel{?}{=} 15 \times 230$$
$$3300 \neq 3450 \quad \text{No}$$

**49.**
$$\frac{75 \text{ feet}}{20 \text{ feet}} \stackrel{?}{=} \frac{105 \text{ feet}}{28 \text{ feet}}$$
$$75 \times 28 \stackrel{?}{=} 20 \times 105$$
$$2100 = 2100 \quad \text{Yes}$$

**51. a.**
$$\frac{169}{221} = \frac{169 \div 13}{221 \div 13} = \frac{13}{17}$$
$$\frac{247}{323} = \frac{247 \div 19}{323 \div 19} = \frac{13}{17}$$
True

**b.**
$$\frac{169}{221} \stackrel{?}{=} \frac{247}{323}$$
$$169 \times 323 \stackrel{?}{=} 221 \times 247$$
$$54{,}587 = 54{,}587 \quad \text{True}$$

**c.** For most students it is faster to multiply than to reduce fractions.

## Cumulative Review Problems

**53.**
$$\begin{array}{r} 2.83 \\ \times \ 5.002 \\ \hline 566 \\ 14\ 1500 \\ \hline 14.15566 \end{array}$$

**55.**
$$\begin{array}{r} 25.8 \\ 7.03 \overline{)\ 181.374} \\ \underline{140\ 6} \\ 40\ 77 \\ \underline{35\ 15} \\ 5\ 624 \\ \underline{5\ 624} \\ 0 \end{array}$$

## 4.3 Exercises

**1.** Divide each side of the equation by the number $a$. Calculate $\frac{b}{a}$. The value of $n$ is $\frac{b}{a}$.

**3.** $12 \times n = 132$
$$\frac{12 \times n}{12} = \frac{132}{12}$$
$$n = 11$$

**5.** $n \times 3.8 = 16.8$
$$\frac{3 \times n}{3} = \frac{16.8}{3}$$
$$n = 5.6$$

**7.** $n \times 3.8 = 95$

$$\frac{n \times 3.8}{3.8} = \frac{95}{3.8}$$

$$n = 25$$

Check:

$$\frac{6}{16} \overset{?}{=} \frac{3}{8}$$

$$6 \times 8 \overset{?}{=} 16 \times 3$$

$$48 = 48$$

**9.** $40.6 = 5.8 \times n$

$$\frac{40.6}{5.8} = \frac{5.8 \times n}{5.8}$$

$$7 = n$$

**11.** $\dfrac{3}{4} \times n = 26$

$$\frac{\frac{3}{4} \times n}{\frac{3}{4}} = \frac{26}{\frac{3}{4}}$$

$$n = 26 \div \frac{3}{4}$$

$$= 26 \times \frac{4}{3}$$

$$= \frac{104}{3} = 34\frac{2}{3}$$

**13.** $\dfrac{n}{20} = \dfrac{3}{4}$

$$n \times 4 = 20 \times 3$$

$$\frac{n \times 4}{4} = \frac{60}{4}$$

$$n = 15$$

Check:

$$\frac{15}{20} \overset{?}{=} \frac{3}{4}$$

$$4 \times 15 \overset{?}{=} 20 \times 3$$

$$60 = 60$$

**15.** $\dfrac{6}{n} = \dfrac{3}{8}$

$$6 \times 8 = n \times 3$$

$$\frac{48}{3} = \frac{n \times 3}{3}$$

$$16 = n$$

**17.** $\dfrac{12}{40} = \dfrac{n}{25}$

$$12 \times 25 = 40 \times n$$

$$\frac{300}{40} = \frac{40 \times n}{40}$$

$$7.5 = n$$

Check:

$$\frac{12}{40} \overset{?}{=} \frac{7.5}{25}$$

$$12 \times 25 \overset{?}{=} 40 \times 7.5$$

$$300 = 300$$

**19.** $\dfrac{25}{100} = \dfrac{8}{n}$

$$25 \times n = 100 \times 8$$

$$\frac{25 \times n}{25} = \frac{800}{25}$$

$$n = 32$$

Check:

$$\frac{25}{100} \overset{?}{=} \frac{8}{32}$$

$$25 \times 32 \overset{?}{=} 100 \times 8$$

$$800 = 800$$

**21.** $\dfrac{n}{6} = \dfrac{150}{12}$

$$n \times 12 = 6 \times 150$$

$$\frac{n \times 12}{12} = \frac{900}{12}$$

$$n = 75$$

Check:

$$\frac{75}{6} \overset{?}{=} \frac{150}{12}$$

$$75 \times 12 \overset{?}{=} 6 \times 150$$

$$900 = 900$$

**23.**  $\dfrac{15}{4} = \dfrac{n}{6}$

$15 \times 6 = 4 \times n$

$\dfrac{90}{4} = \dfrac{4 \times n}{4}$

$22.5 = n$

Check:

$\dfrac{15}{4} \overset{?}{=} \dfrac{22.5}{6}$

$15 \times 6 \overset{?}{=} 4 \times 22.5$

$90 = 90$

**25.**  $\dfrac{18}{n} = \dfrac{3}{11}$

$18 \times 11 = 3 \times n$

$\dfrac{198}{3} = \dfrac{3 \times n}{3}$

$66 = n$

Check:

$\dfrac{18}{66} \overset{?}{=} \dfrac{3}{11}$

$18 \times 11 \overset{?}{=} 3 \times 66$

$198 = 198$

**27.**  $\dfrac{21}{n} = \dfrac{2}{3}$

$21 \times 3 = n \times 2$

$\dfrac{63}{2} = \dfrac{n \times 2}{2}$

$31.5 = n$

**29.**  $\dfrac{9}{26} = \dfrac{n}{52}$

$9 \times 52 = 26 \times n$

$\dfrac{468}{26} = \dfrac{26 \times n}{26}$

$18 = n$

**31.**  $\dfrac{15}{12} = \dfrac{10}{n}$

$15 \times n = 12 \times 10$

$\dfrac{15 \times n}{15} = \dfrac{120}{15}$

$n = 8$

**33.**  $\dfrac{n}{36} = \dfrac{4.5}{1}$

$n \times 1 = 36 \times 4.5$

$n = 162$

**35.**  $\dfrac{1.5}{n} = \dfrac{0.3}{8}$

$1.5 \times 8 = n \times 0.3$

$\dfrac{12}{0.3} = \dfrac{n \times 0.3}{0.3}$

$40 = n$

**37.**  $\dfrac{7}{8} = \dfrac{n}{4.2}$

$7 \times 4.2 = 8 \times n$

$\dfrac{29.4}{8} = \dfrac{8 \times n}{8}$

$3.7 \approx n$

**39.**  $\dfrac{13.8}{15} = \dfrac{n}{6}$

$13.8 \times 6 = 15 \times n$

$82.8 = 15 \times n$

$\dfrac{82.8}{15} = \dfrac{15 \times n}{15}$

$5.5 \approx n$

**41.**  $\dfrac{22\frac{2}{9}}{100} = \dfrac{2}{n}$

$22\dfrac{2}{9} \times n = 100 \times 2$

$\dfrac{\frac{200}{9} \times n}{\frac{200}{9}} = \dfrac{200}{\frac{200}{9}}$

$n = \dfrac{200}{1} \times \dfrac{9}{200}$

$n = 9$

**43.** $\dfrac{n \text{ grams}}{14 \text{ liters}} = \dfrac{6 \text{ grams}}{22 \text{ liters}}$

$n \times 22 = 14 \times 6$

$\dfrac{n \times 22}{22} = \dfrac{84}{22}$

$n \approx 3.82$

3.82 grams

**45.** $\dfrac{128 \text{ miles}}{4 \text{ hours}} = \dfrac{80 \text{ miles}}{n \text{ hours}}$

$128 \times n = 4 \times 80$

$\dfrac{128 \times n}{128} = \dfrac{320}{128}$

$n = 2.5$

2.5 hours

**47.** $\dfrac{80 \text{ gallons}}{26 \text{ acres}} = \dfrac{42 \text{ gallons}}{n \text{ acres}}$

$80 \times n = 26 \times 42$

$\dfrac{80 \times n}{80} = \dfrac{1092}{80}$

$n = 13.65$

13.65 acres

**49.** $\dfrac{3 \text{ km}}{1.86 \text{ mi}} = \dfrac{n \text{ km}}{4 \text{ mi}}$

$3 \times 4 = 1.86 \times n$

$\dfrac{12}{1.86} = \dfrac{1.86 \times n}{1.86}$

$6.45 \approx n$

6.45 km

**51.** $\dfrac{35 \text{ dimes}}{3.5 \text{ dollars}} = \dfrac{n \text{ dimes}}{8 \text{ dollars}}$

$35 \times 8 \quad = 3.5 \times n$

$\dfrac{280}{3.5} \quad = \dfrac{3.5 \times n}{3.5}$

$80 \quad\quad = n$

80 dimes

**53.** $\dfrac{3\frac{1}{4} \text{ feet}}{8 \text{ pounds}} = \dfrac{n \text{ feet}}{12 \text{ pounds}}$

$3\dfrac{1}{4} \times 12 = 8 \times n$

$\dfrac{39}{8} = \dfrac{8 \times n}{8}$

$4\dfrac{7}{8} = n$

$4\dfrac{7}{8}$ feet

**55.** $\dfrac{5 \text{ inches}}{3 \text{ inches}} = \dfrac{n \text{ inches}}{5 \text{ inches}}$

$5 \times 5 = 3 \times n$

$\dfrac{25}{3} = \dfrac{3 \times n}{3}$

$8\dfrac{1}{3} = n$

$8\dfrac{1}{3}$ inches wide

**57.** $\dfrac{n}{7\frac{1}{4}} = \dfrac{2\frac{1}{5}}{4\frac{1}{8}}$

$n \times 4\dfrac{1}{8} = 7\dfrac{1}{4} \times 2\dfrac{1}{5}$

$n \times \dfrac{33}{8} = \dfrac{29}{4} \times \dfrac{11}{5}$

$n \times \dfrac{33}{8} = \dfrac{319}{20}$

$\dfrac{n \times \frac{33}{8}}{\frac{33}{8}} = \dfrac{\frac{319}{20}}{\frac{33}{8}}$

$n = \dfrac{319}{20} \times \dfrac{8}{33}$

$n = \dfrac{58}{15} = 3\dfrac{13}{15}$

**59.**

$$\frac{9\frac{3}{4}}{n} = \frac{8\frac{1}{2}}{4\frac{1}{3}}$$

$$9\frac{3}{4} \times 4\frac{1}{3} = n \times 8\frac{1}{2}$$

$$\frac{39}{4} \times \frac{13}{3} = n \times \frac{17}{2}$$

$$\frac{169}{4} = n \times \frac{17}{2}$$

$$\frac{\frac{169}{4}}{\frac{17}{2}} = \frac{n \times \frac{17}{2}}{\frac{17}{2}}$$

$$\frac{169}{4} \times \frac{2}{17} = n$$

$$\frac{169}{34} = n \text{ or } n = 4\frac{33}{34}$$

## Cumulative Review Problems

**61.** $(1.6)^2 - 0.12 \times 3.5 + 36.8 \div 2.5$

$$= 2.56 - 0.12 \times 3.5 + 36.8 \div 2.5$$

$$= 2.56 - 0.42 + 14.72$$

$$= 2.14 + 14.72$$

$$= 16.86$$

**63.** 0.0034

## 4.4    Exercises

**1.** $\dfrac{19 \text{ desserts}}{16 \text{ people}} = \dfrac{n \text{ desserts}}{320 \text{ people}}$

$$16 \times n = 19 \times 320$$

$$\frac{16 \times n}{16} = \frac{6080}{16}$$

$$n = 380$$

380 desserts

**3.** $\dfrac{\frac{3}{4} \text{ cup}}{1 \text{ gallon}} = \dfrac{n \text{ cups}}{4 \text{ gallons}}$

$$\frac{3}{4} \times 4 = 1 \times n$$

$$3 = n$$

3 cups

**5.** $\dfrac{\$10 \text{ U.S.}}{15 \text{ Swiss}} = \dfrac{\$430 \text{ U.S.}}{n \text{ Swiss}}$

$$10 \times n = 15 \times 430$$

$$\frac{10 \times n}{10} = \frac{6450}{10}$$

$$n = 645$$

645 Swiss francs

**7.** $\dfrac{6.5 \text{ ft}}{5 \text{ ft}} = \dfrac{n \text{ ft}}{152 \text{ ft}}$

$$6.5 \times 152 = 5 \times n$$

$$\frac{988}{5} = \frac{5 \times n}{5}$$

$$197.6 = n$$

197.6 feet

**9.** $\dfrac{4 \text{ inches}}{250 \text{ miles}} = \dfrac{5.7 \text{ inches}}{n \text{ miles}}$

$$4 \times n = 250 \times 5.7$$

$$\frac{4 \times n}{4} = \frac{1425}{4}$$

$$n = 356.25$$

356 miles

**11.** $\dfrac{5 \text{ cups}}{12 \text{ people}} = \dfrac{n \text{ cups}}{28 \text{ people}}$

$$5 \times 28 = 12 \times n$$

$$\frac{140}{12} = \frac{12 \times n}{12}$$

$$11.\overline{6} = n$$

$11\frac{2}{3}$ cups

**13.** $\dfrac{17 \text{ made}}{25 \text{ throws}} = \dfrac{n \text{ made}}{150 \text{ throws}}$

$$17 \times 150 = 25 \times n$$

$$\frac{2250}{25} = \frac{25 \times n}{25}$$

$$102 = n$$

102 made free throws

**15.** $\dfrac{7 \text{ flights}}{79 \text{ people}} = \dfrac{434 \text{ flights}}{n \text{ people}}$

$7 \times n = 79 \times 434$

$\dfrac{7 \times n}{7} = \dfrac{34{,}286}{7}$

$n = 4898$

4898 people

**17.** $\dfrac{26 \text{ tagged}}{n \text{ total}} = \dfrac{6 \text{ tagged}}{18 \text{ total}}$

$26 \times 18 = 6 \times n$

$\dfrac{468}{6} = \dfrac{6 \times n}{6}$

$78 = n$

78 giraffes

**19.** $\dfrac{425 \text{ pounds}}{3 \text{ acres}} = \dfrac{n \text{ pounds}}{14 \text{ acres}}$

$425 \times 14 = 3 \times n$

$\dfrac{5950}{3} = \dfrac{3 \times n}{3}$

$1983.\overline{3} = n$

$1983\dfrac{1}{3}$ pounds

$1983\dfrac{1}{3} \times 1.8 = 3570$

\$3570

**21.** $\dfrac{5 \text{ defective}}{100 \text{ made}} = \dfrac{n \text{ defective}}{5400 \text{ made}}$

$5 \times 5400 = 100 \times n$

$\dfrac{27{,}000}{100} = \dfrac{100 \times n}{n}$

$270 = n$

270 defective chips

**23.** Water: $\dfrac{n \text{ cups}}{3 \text{ servings}} = \dfrac{2 \text{ cups}}{6 \text{ servings}}$

$n \times 6 = 3 \times 2$

$\dfrac{n \times 6}{6} = \dfrac{6}{6}$

$n = 1$

1 cup of water

Milk: $\dfrac{n \text{ cups}}{3 \text{ servings}} = \dfrac{\frac{3}{4} \text{ cups}}{6 \text{ servings}}$

$n \times 6 = 3 \times \dfrac{3}{4}$

$n \times 6 = \dfrac{9}{4}$

$\dfrac{n \times 6}{6} = \dfrac{\frac{9}{4}}{6}$

$n = \dfrac{9}{4} \div 6$

$= \dfrac{9}{4} \times \dfrac{1}{6} = \dfrac{3}{8}$

$\dfrac{3}{8}$ cup of milk

**25.** Water: $\dfrac{n \text{ cups}}{8 \text{ servings}} = \dfrac{2 \text{ cups}}{6 \text{ servings}}$

$n \times 6 = 8 \times 2$

$\dfrac{n \times 6}{6} = \dfrac{16}{6}$

$n = \dfrac{16}{6} = \dfrac{8}{3}$

High altitude: $\dfrac{8}{3} \times \dfrac{3}{4} = 2$

2 cups of water

Milk: $\dfrac{n \text{ cups}}{8 \text{ servings}} = \dfrac{\frac{3}{4} \text{ cups}}{6 \text{ servings}}$

$n \times 6 = 8 \times \dfrac{3}{4}$

$n \times 6 = 6$

$\dfrac{n \times 6}{6} = \dfrac{6}{6}$

$n = 1$

1 cup of milk

**27.** Griffey:
$$\frac{\$8,510,532}{48} \approx \$177,303 \text{ per home run}$$

Bell:
$$\frac{\$6,000,000}{38} \approx \$157,895 \text{ per home run}$$

**29.** Clemens:
$$\frac{\$8,250,000}{30} = \$275,000 \text{ per game}$$

Martinez:
$$\frac{\$11,000,000}{31} \approx \$354,839 \text{ per game}$$

## Cumulative Review Problems

**31.** 56,179 rounds to 56,200.

**33.** 56.148 rounds to 56.1

**35. a.** $1\frac{3}{16} \times \frac{4}{5} = \frac{19}{16} \times \frac{4}{5} = \frac{76}{80}$
$$= \frac{19}{20} \text{ of a square foot}$$

**b.** $\frac{19}{20} \times 1500 = 19 \times 75$
$$= 1452 \text{ square feet}$$

## Putting Your Skills To Work

**1.**
$$23 + 27 = 50 \text{ total}$$
$$\frac{27 \text{ males}}{50 \text{ total}} = \frac{h \text{ males}}{163,400}$$
$$27 \times 163,400 = 50 \times n$$
$$\frac{4,411,800}{50} = \frac{50 \times n}{50}$$
$$88,236 = n$$

88,236 males

**3.**
$$10 \times 3000 = 30,000$$
$$163,400 + 30,000 = 193,400 \text{ moose in 2010}$$
$$\frac{23 \text{ females}}{50 \text{ total}} = \frac{n \text{ females}}{193,400 \text{ total}}$$
$$23 \times 193,400 = 50 \times n$$
$$\frac{4,448,200}{50} = \frac{50 \times n}{50}$$
$$88,964 = n$$

88,964 females

## Chapter 4 Review Problems

**1.** $88{:}40 = \frac{88}{40} = \frac{88 \div 8}{40 \div 8} = \frac{11}{5}$

**2.** $65{:}39 = \frac{65}{39} = \frac{65 \div 13}{39 \div 13} = \frac{5}{3}$

**3.** $28{:}35 = \frac{28}{35} = \frac{28 \div 7}{35 \div 7} = \frac{4}{5}$

**4.** $250{:}475 = \frac{250}{475} = \frac{250 \div 25}{475 \div 25} = \frac{10}{19}$

**5.** $2\frac{1}{3}$ to $4\frac{1}{4} = \frac{2\frac{1}{3}}{4\frac{1}{4}} = 2\frac{1}{3} \div 4\frac{1}{4}$
$$= \frac{7}{3} \div \frac{17}{4} = \frac{7}{3} \times \frac{4}{17}$$
$$= \frac{28}{51}$$

**6.** $\frac{27}{81} = \frac{27 \div 27}{81 \div 27} = \frac{1}{3}$

**7.** $\frac{280}{651} = \frac{280 \div 7}{651 \div 7} = \frac{40}{93}$

**8.** $\frac{156}{441} = \frac{156 \div 3}{441 \div 3} = \frac{52}{147}$

**9.** $\frac{26}{65} = \frac{26 \div 13}{65 \div 13} = \frac{2}{5}$

10. $\dfrac{35}{215} = \dfrac{35 \div 5}{215 \div 5} = \dfrac{7}{43}$

11. $\dfrac{20}{215} = \dfrac{20 \div 5}{215 \div 5} = \dfrac{4}{43}$

12. $\dfrac{10 \text{ gallons}}{18 \text{ people}} = \dfrac{5 \text{ gallons}}{9 \text{ people}}$

13. $\dfrac{44 \text{ revolutions}}{121 \text{ minutes}} = \dfrac{4 \text{ revolutions}}{11 \text{ minutes}}$

14. $\dfrac{188 \text{ vibrations}}{16 \text{ seconds}} = \dfrac{47 \text{ vibrations}}{4 \text{ seconds}}$

15. $\dfrac{12 \text{ cups}}{38 \text{ people}} = \dfrac{6 \text{ cups}}{19 \text{ people}}$

16. $\dfrac{\$2125}{125 \text{ shares}} = \dfrac{\$17}{1 \text{ share}} = \$17/\text{share}$

17. $\dfrac{\$2244}{132 \text{ chairs}} = \dfrac{\$17}{1 \text{ chair}} = \$17/\text{chair}$

18. $\dfrac{\$742.50}{55 \text{ sq yd}} = \dfrac{\$13.50}{1 \text{ yd}} = \$13.50/\text{sq yd}$

19. $\dfrac{\$768.0}{62 \text{ tickets}} = \dfrac{\$12.40}{1 \text{ ticket}} = \$12.40/\text{ticket}$

20. a. $\dfrac{\$2.75}{12.5 \text{ oz}} = \dfrac{\$2.75 \div 12.5}{12.5 \text{ oz} \div 12.5} = \$0.22/\text{oz}$

   b. $\dfrac{\$1.75}{7.0 \text{ oz}} = \dfrac{\$1.75 \div 7.0}{7.0 \text{ oz} \div 7.0} = \$0.25/\text{oz}$

   c. $0.25 - 0.22 = 0.03$
   $\$0.03/\text{oz}$

21. a. $\dfrac{\$2.96}{4 \text{ oz}} = \dfrac{\$296 \div 4}{4 \text{ oz} \div 4} = \$0.74/\text{oz}$

   b. $\dfrac{\$5.22}{9 \text{ oz}} = \dfrac{\$5.22 \div 9}{9 \text{ oz} \div 9} = \$0.58/\text{oz}$

c. $\begin{array}{r} \$0.74 \\ -\ 0.58 \\ \hline \$0.16/\text{oz} \end{array}$

22. $\dfrac{12}{48} = \dfrac{7}{28}$

23. $\dfrac{1\frac{1}{2}}{5} = \dfrac{4}{13\frac{1}{3}}$

24. $\dfrac{7.5}{45} = \dfrac{22.5}{135}$

25. $\dfrac{15 \text{ pounds}}{\$4.50} = \dfrac{27 \text{ pounds}}{\$8.10}$

26. $\dfrac{138 \text{ passengers}}{3 \text{ buses}} = \dfrac{230 \text{ passengers}}{5 \text{ buses}}$

27. $\dfrac{16}{48} \stackrel{?}{=} \dfrac{2}{12}$
   $16 \times 12 \stackrel{?}{=} 48 \times 2$
   $192 \neq 96 \quad \text{False}$

28. $\dfrac{20}{25} \stackrel{?}{=} \dfrac{8}{10}$
   $25 \times 10 \stackrel{?}{=} 25 \times 8$
   $200 = 200 \quad \text{True}$

29. $\dfrac{24}{20} \stackrel{?}{=} \dfrac{18}{15}$
   $24 \times 15 \stackrel{?}{=} 20 \times 18$
   $360 = 360 \quad \text{True}$

30. $\dfrac{84}{48} \stackrel{?}{=} \dfrac{14}{8}$
   $84 \times 8 \stackrel{?}{=} 48 \times 14$
   $672 = 672 \quad \text{True}$

31. $\dfrac{37}{33} \stackrel{?}{=} \dfrac{22}{19}$
   $37 \times 19 \stackrel{?}{=} 33 \times 22$
   $703 \neq 726 \quad \text{False}$

**32.** $\dfrac{15}{18} \overset{?}{=} \dfrac{18}{22}$

$15 \times 22 \overset{?}{=} 18 \times 18$

$330 \neq 324$ False

**33.** $\dfrac{84 \text{ miles}}{7 \text{ gallons}} \overset{?}{=} \dfrac{108 \text{ miles}}{9 \text{ gallons}}$

$84 \times 9 \overset{?}{=} 7 \times 108$

$756 = 756$ True

**34.** $\dfrac{156 \text{ rev}}{6 \text{ min}} \overset{?}{=} \dfrac{181 \text{ rev}}{7 \text{ min}}$

$156 \times 7 \overset{?}{=} 6 \times 181$

$1092 \neq 1086$ False

**35.** $7 \times n = 161$

$\dfrac{7 \times n}{7} = \dfrac{161}{7}$

$n = 23$

**36.** $8 \times n = 42$

$\dfrac{8 \times n}{8} = \dfrac{42}{8}$

$n = 5\dfrac{1}{4} = 5.25$

**37.** $558 = 18 \times n$

$\dfrac{558}{18} = \dfrac{18 \times n}{18}$

$31 = n$

**38.** $663 = 39 \times n$

$\dfrac{663}{39} = \dfrac{39 \times n}{39}$

$17 = n$

**39.** $\dfrac{3}{11} = \dfrac{9}{n}$

$3 \times n = 11 \times 9$

$\dfrac{3 \times n}{3} = \dfrac{99}{3}$

$n = 33$

**40.** $\dfrac{2}{7} = \dfrac{12}{n}$

$2 \times n = 7 \times 12$

$\dfrac{2 \times n}{2} = \dfrac{84}{2}$

$n = 42$

**41.** $\dfrac{n}{28} = \dfrac{6}{24}$

$24 \times n = 28 \times 6$

$\dfrac{24 \times n}{24} = \dfrac{168}{24}$

$n = 7$

**42.** $\dfrac{n}{32} = \dfrac{15}{20}$

$n \times 20 = 32 \times 15$

$\dfrac{n \times 20}{20} = \dfrac{480}{20}$

$n = 24$

**43.** $\dfrac{2\frac{1}{4}}{9} = \dfrac{4\frac{3}{4}}{n}$

$2\dfrac{1}{4} \times n = 9 \times 4\dfrac{3}{4}$

$\dfrac{9}{4} \times n = 9 \times \dfrac{19}{4}$

$\dfrac{\frac{9}{4} \times n}{\frac{9}{4}} = \dfrac{\frac{171}{4}}{\frac{9}{4}}$

$n = \dfrac{171}{4} \div \dfrac{9}{4}$

$= \dfrac{171}{4} \times \dfrac{4}{9}$

$= \dfrac{171}{9}$

$= 19$

**44.**
$$\frac{3\frac{1}{3}}{2\frac{2}{3}} = \frac{7}{n}$$

$$3\frac{1}{3} \times n = 2\frac{2}{3} \times 7$$

$$\frac{10}{3} \times n = \frac{8}{3} \times 7$$

$$\frac{\frac{10}{3} \times n}{\frac{10}{3}} = \frac{\frac{56}{3}}{\frac{10}{3}}$$

$$n = \frac{56}{3} \div \frac{10}{3}$$

$$= \frac{56}{3} \times \frac{3}{10}$$

$$= \frac{56}{10} = \frac{28}{5}$$

$$= 5\frac{3}{5} = 5.6$$

**45.**
$$\frac{54}{72} = \frac{n}{4}$$

$$54 \times 4 = 72 \times n$$

$$\frac{216}{72} = \frac{72 \times n}{72}$$

$$3 = n$$

**46.**
$$\frac{45}{135} = \frac{n}{3}$$

$$45 \times 3 = 135 \times n$$

$$\frac{135}{135} = \frac{135 \times n}{135}$$

$$1 = n$$

**47.**
$$\frac{6}{n} = \frac{2}{29}$$

$$6 \times 29 = 2 \times n$$

$$\frac{174}{2} = \frac{2 \times n}{2}$$

$$87 = n$$

**48.**
$$\frac{8}{n} = \frac{2}{81}$$

$$8 \times 81 = n \times 2$$

$$\frac{648}{2} = \frac{n \times 2}{2}$$

$$324 = n$$

**49.**
$$\frac{25}{7} = \frac{60}{n}$$

$$25 \times n = 7 \times 60$$

$$\frac{25 \times n}{25} = \frac{420}{25}$$

$$n = 16.8$$

**50.**
$$\frac{60}{9} = \frac{31}{n}$$

$$60 \times n = 9 \times 31$$

$$60 \times n = 279$$

$$\frac{60 \times n}{60} = \frac{279}{60}$$

$$n \approx 4.7$$

**51.**
$$\frac{35 \text{ miles}}{28 \text{ gallons}} = \frac{15 \text{ miles}}{n \text{ gallons}}$$

$$35 \times n = 28 \times 15$$

$$\frac{35 \times n}{35} = \frac{420}{35}$$

$$n = 12$$

**52.**
$$\frac{8 \text{ defective}}{100 \text{ perfect}} = \frac{44 \text{ defective}}{n \text{ perfect}}$$

$$8 \times n = 100 \times 44$$

$$\frac{8 \times n}{8} = \frac{4400}{8}$$

$$n = 550$$

550 perfect parts

**53.** $\dfrac{3 \text{ gallons}}{2 \text{ rooms}} = \dfrac{n \text{ gallons}}{10 \text{ rooms}}$

$3 \times 10 = 2 \times n$

$\dfrac{30}{2} = \dfrac{2 \times n}{2}$

$15 = n$

15 gallons

**54.** $\dfrac{49 \text{ coffee}}{100 \text{ adults}} = \dfrac{n \text{ coffee}}{3450 \text{ adults}}$

$49 \times 3450 = 100 \times n$

$\dfrac{169{,}050}{100} = \dfrac{100 \times n}{100}$

$1691 \approx n$

1691 drink coffee

**55.** $\dfrac{24 \text{ francs}}{5 \text{ dollars}} = \dfrac{n \text{ francs}}{420 \text{ dollars}}$

$24 \times 420 = 5 \times n$

$\dfrac{10{,}080}{5} = \dfrac{5 \times n}{5}$

$2016 = n$

2016 francs

**56.** $\dfrac{7.5 \text{ lb}}{18 \text{ people}} = \dfrac{n \text{ lb}}{120 \text{ people}}$

$7.5 \times 120 = 18 \times n$

$900 = 18 \times n$

$\dfrac{900}{18} = \dfrac{18 \times n}{18}$

$50 = n$

50 pounds

**57.** $\dfrac{225 \text{ miles}}{3 \text{ inches}} = \dfrac{n \text{ miles}}{8 \text{ inches}}$

$8 \times 225 = 3 \times n$

$\dfrac{1800}{3} = \dfrac{3 \times n}{3}$

$600 = n$

600 miles

**58.** $\dfrac{84 \text{ rev/min}}{14 \text{ miles per hour}} = \dfrac{96 \text{ rev/min}}{n \text{ miles per hour}}$

$84 \times n = 14 \times 96$

$\dfrac{84 \times n}{84} = \dfrac{1344}{84}$

$n = 16$

16 miles per hour

**59.** $\dfrac{6 \text{ feet}}{16 \text{ feet}} = \dfrac{n \text{ feet}}{320 \text{ feet}}$

$6 \times 320 = 16 \times n$

$\dfrac{1920}{16} = \dfrac{16 \times n}{16}$

$120 = n$

**60. a.** $\dfrac{680 \text{ miles}}{26 \text{ gallons}} = \dfrac{200 \text{ miles}}{n \text{ gallons}}$

$680 \times n = 26 \times 200$

$\dfrac{680 \times n}{680} = \dfrac{5200}{680}$

$n \approx 7.65$

7.65 gallons

**b.**
$$
\begin{array}{r}
7.65 \\
\times\ 1.15 \\
\hline
3825 \\
765\phantom{0} \\
7\,65\phantom{00} \\
\hline
8.7975
\end{array}
$$

$8.80

**61.** $\dfrac{3.5 \text{ cm}}{2.5 \text{ cm}} = \dfrac{8 \text{ cm}}{n \text{ cm}}$

$3.5 \times n = 2.5 \times 8$

$\dfrac{3.5 \times n}{3.5} = \dfrac{20}{3.5}$

$n \approx 5.71$

5.71 centimeters tall

**62.** $\dfrac{3 \text{ grams}}{50 \text{ pounds}} = \dfrac{n \text{ grams}}{125 \text{ pounds}}$

$3 \times 125 = 50 \times n$

$\dfrac{375}{50} = \dfrac{50 \times n}{50}$

$7.5 = n$

7.5 grams

**63.** $\dfrac{33 \text{ bricks}}{2 \text{ feet}} = \dfrac{n \text{ bricks}}{11 \text{ feet}}$

$33 \times 11 = 2 \times n$

$\dfrac{363}{2} = \dfrac{2 \times n}{2}$

$181.5 = n$

She needs 181.5 bricks which means she needs 182 bricks.

**64.** $\dfrac{8 \text{ vote}}{13 \text{ people}} = \dfrac{n \text{ vote}}{15,600 \text{ people}}$

$8 \times 15,600 = 13 \times n$

$\dfrac{124,800}{13} = \dfrac{13 \times n}{13}$

$9600 = n$

9600 people will vote

**65.** $\dfrac{3 \text{ gallons}}{500 \text{ sq ft}} = \dfrac{n \text{ gallons}}{1400 \text{ sq ft}}$

$3 \times 1400 = 500 \times n$

$\dfrac{4200}{500} = \dfrac{500 \times n}{500}$

$8.4 = n$

He needs 8.4 gallons which means he really needs 5 gallons.

**66.** $\dfrac{5 \text{ split}}{3 \text{ owned}} = \dfrac{n \text{ split}}{51 \text{ owned}}$

$5 \times 51 = 3 \times n$

$\dfrac{255}{3} = \dfrac{3 \times n}{3}$

$85 = n$

85 shares

**67.** $\dfrac{14 \text{ centimeters}}{145 \text{ feet}} = \dfrac{11 \text{ centimeters}}{n \text{ feet}}$

$14 \times n = 145 \times 11$

$\dfrac{14 \times n}{14} = \dfrac{1595}{14}$

$n \approx 113.93$

Width is 113.93 feet

**68.** $\dfrac{40 \text{ days}}{3 \text{ minutes}} = \dfrac{365 \text{ days}}{n \text{ minutes}}$

$40 \times n = 3 \times 365$

$\dfrac{40 \times n}{40} = \dfrac{1095}{40}$

$n \approx 27.38$

27.38 minutes

**69.** $\dfrac{68 \text{ goals}}{27 \text{ games}} = \dfrac{n \text{ goals}}{34 \text{ games}}$

$68 \times 34 = 27 \times n$

$\dfrac{2312}{27} = \dfrac{27 \times n}{27}$

$86 \approx n$

86 goals

**70.** $\dfrac{345 \text{ calories}}{10 \text{ ounces}} = \dfrac{n \text{ calories}}{16 \text{ ounces}}$

$345 \times 16 = 10 \times n$

$\dfrac{5520}{10} = \dfrac{10 \times n}{10}$

$552 = n$

552 calories

## Chapter 4 Test

**1.** $\dfrac{18}{52} = \dfrac{18 \div 2}{52 \div 2} = \dfrac{9}{26}$

**2.** $\dfrac{70}{185} = \dfrac{70 \div 5}{185 \div 5} = \dfrac{14}{37}$

**3.** $\dfrac{784 \text{ miles}}{24 \text{ gal}} = \dfrac{784 \text{ miles} \div 8}{24 \text{ gal} \div 8} = \dfrac{98 \text{ miles}}{3 \text{ gal}}$

**4.** $\dfrac{2100 \text{ sq ft}}{45 \text{ lb}} = \dfrac{2100 \text{ sq ft} \div 15}{45 \text{ lb} \div 15} = \dfrac{140 \text{ sq ft}}{3 \text{ lb}}$

**5.** $\dfrac{19 \text{ tons}}{5 \text{ days}} = \dfrac{19 \text{ tons} \div 5}{5 \text{ days} \div 5} = 3.8 \text{ tons/day}$

**6.** $\dfrac{\$57.96}{7 \text{ hr}} = \dfrac{\$57.96 \div 7}{7 \text{ hr} \div 7} = \$8.28/\text{hr}$

**7.** $\dfrac{5400 \text{ ft}}{22 \text{ poles}} = \dfrac{5400 \text{ ft} \div 22}{22 \text{ poles} \div 22} = 245.45 \text{ ft/pole}$

**8.** $\dfrac{\$9373}{110 \text{ shares}} = \dfrac{\$9373 \div 110}{110 \text{ shares} \div 110}$
$= \$85.21/\text{share}$

**9.** $\dfrac{17}{29} = \dfrac{51}{87}$

**10.** $\dfrac{3\frac{1}{2}}{5} = \dfrac{18}{25\frac{5}{7}}$

**11.** $\dfrac{490 \text{ miles}}{21 \text{ gallons}} = \dfrac{280 \text{ miles}}{12 \text{ gallons}}$

**12.** $\dfrac{5 \text{ tablespoons}}{18 \text{ people}} = \dfrac{15 \text{ tablespoons}}{54 \text{ people}}$

**13.** $\dfrac{50}{24} \overset{?}{=} \dfrac{35}{16}$
$50 \times 16 \overset{?}{=} 24 \times 34$
$800 \neq 816 \quad \text{False}$

**14.** $\dfrac{18.4}{20} \overset{?}{=} \dfrac{46}{50}$
$18.4 \times 50 \overset{?}{=} 20 \times 46$
$920 = 920 \quad \text{True}$

**15.** $\dfrac{32 \text{ smokers}}{46 \text{ nonsmokers}} \overset{?}{=} \dfrac{160 \text{ smokers}}{230 \text{ nonsmokers}}$
$32 \times 230 \overset{?}{=} 46 \times 160$
$7360 = 7360 \quad \text{True}$

**16.** $\dfrac{\$0.74}{16 \text{ oz}} \overset{?}{=} \dfrac{\$1.84}{40 \text{ oz}}$
$0.74 \times 40 \overset{?}{=} 16 \times 1.84$
$29.6 \neq 29.44 \quad \text{False}$

**17.** $\dfrac{n}{20} = \dfrac{4}{5}$
$n \times 5 = 20 \times 4$
$\dfrac{n \times 5}{5} = \dfrac{80}{5}$
$n = 16$

**18.** $\dfrac{9}{2} = \dfrac{63}{n}$
$9 \times n = 2 \times 63$
$\dfrac{9 \times n}{9} = \dfrac{126}{9}$
$n = 14$

**19.** $\dfrac{2\frac{2}{3}}{8} = \dfrac{6\frac{1}{3}}{n}$
$2\frac{2}{3} \times n = 8 \times 6\frac{1}{3}$
$\dfrac{8}{3} \times n = 8 \times \dfrac{19}{3}$
$\dfrac{\frac{8}{3} \times n}{\frac{8}{3}} = \dfrac{\frac{152}{3}}{\frac{8}{3}}$
$n = \dfrac{152}{3} \div \dfrac{8}{3} = \dfrac{152}{3} \times \dfrac{3}{8}$
$= 19$

**20.** $\dfrac{4.2}{11} = \dfrac{n}{77}$
$4.2 \times 77 = 11 \times n$
$\dfrac{323.4}{11} = \dfrac{11 \times n}{11}$
$29.4 = n$

**21.** $\dfrac{45 \text{ women}}{15 \text{ men}} = \dfrac{n \text{ women}}{40 \text{ men}}$

$45 \times 40 = 15 \times n$

$\dfrac{1800}{15} = \dfrac{15 \times n}{15}$

$120 = n$

120 women

**22.** $\dfrac{3.5 \text{ oz}}{4.2 \text{ grams}} = \dfrac{7 \text{ oz}}{n \text{ grams}}$

$3.5 \times n = 4.2 \times 7$

$\dfrac{3.5 \times n}{3.5} = \dfrac{29.4}{3.5}$

$n = 8.4$

8.4 grams

**23.** $\dfrac{n \text{ inches of snow}}{14 \text{ inches of rain}} = \dfrac{12 \text{ inches of snow}}{1.4 \text{ inches of rain}}$

$n \times 1.4 = 14 \times 12$

$\dfrac{n \times 1.4}{1.4} = \dfrac{168}{1.4}$

$n = 120$

120 inches of snow

**24.** $\dfrac{28 \text{ lb}}{\$n} = \dfrac{3 \text{ lb}}{\$0.55}$

$28 \times 0.55 = n \times 3$

$\dfrac{15.4}{3} = \dfrac{n \times 3}{3}$

$5.13 \approx n$

$5.13

**25.** $\dfrac{3 \text{ eggs}}{11 \text{ people}} = \dfrac{n \text{ eggs}}{22 \text{ people}}$

$3 \times 22 = 11 \times n$

$\dfrac{66}{11} = \dfrac{11 \times n}{11}$

$6 = n$

6 eggs

**26.** $\dfrac{42 \text{ ft}}{170 \text{ lb}} = \dfrac{20 \text{ ft}}{n \text{ lb}}$

$42 \times n = 170 \times 20$

$\dfrac{42 \times n}{42} = \dfrac{3400}{42}$

$n \approx 80.95$

80.95 pounds

**27.** $\dfrac{9 \text{ inches}}{57 \text{ miles}} = \dfrac{3 \text{ inches}}{n \text{ miles}}$

$9 \times n = 57 \times 3$

$\dfrac{9 \times n}{9} = \dfrac{171}{9}$

$n = 19$

19 miles

**28.** $\dfrac{\$240}{4000 \text{ sq ft}} = \dfrac{\$n}{6000 \text{ sq ft}}$

$240 \times 6000 = 4000 \times n$

$\dfrac{1,440,000}{4000} = \dfrac{4000 \times n}{4000}$

$360 = n$

$360

**29.** $\dfrac{1.5 \text{ quarts}}{3000 \text{ miles}} = \dfrac{n \text{ quarts}}{8000 \text{ miles}}$

$1.5 \times 8000 = 3000 \times n$

$\dfrac{12,000}{3000} = \dfrac{3000 \times n}{3000}$

$4 = n$

4 quarts

**30.** $\dfrac{570 \text{ km}}{9 \text{ hr}} = \dfrac{n \text{ km}}{11 \text{ hr}}$

$570 \times 11 = 9 \times n$

$6270 = 9 \times n$

$\dfrac{6270}{9} = \dfrac{9 \times n}{9}$

$696.67 \approx n$

696.67 km

**31.** $\dfrac{0.6 \text{ gallons}}{2 \text{ hours}} = \dfrac{n \text{ gallons}}{46 \text{ hours}}$

$0.6 \times 46 = 2 \times n$

$\dfrac{27.6}{2} = \dfrac{2 \times n}{2}$

$13.8 = n$

13.8 gallons

**32.** $\dfrac{7 \text{ hits}}{34 \text{ bats}} = \dfrac{n \text{ hits}}{155 \text{ bats}}$

$7 \times 155 = 34 \times n$

$\dfrac{1085}{34} = \dfrac{34 \times n}{34}$

$32 \approx n$

32 hits

---

# Chapters 1–4 Cumulative Test

**1.** 26,597,089 = Twenty-six million, five hundred ninety-seven thousand, eighty nine

**2.** $23 \overline{)1564}$ with quotient 68

$\begin{array}{r} 68 \\ 23{\overline{)1564}} \\ \underline{138} \\ 184 \\ \underline{184} \\ 0 \end{array}$

**3.** $\dfrac{1}{4} + \dfrac{1}{8} \times \dfrac{1}{4} = \dfrac{1}{4} + \dfrac{3}{24}$

$\qquad = \dfrac{1}{4} \times \dfrac{8}{8} + \dfrac{3}{32}$

$\qquad = \dfrac{8}{32} + \dfrac{3}{32} = \dfrac{11}{32}$

**4.** $2\dfrac{1}{2} - 1\dfrac{3}{7} = 2\dfrac{7}{35} - 1\dfrac{15}{35}$

$\qquad = 1\dfrac{42}{35} - 1\dfrac{15}{35}$

$\qquad = \dfrac{27}{35}$

**5.** $4\dfrac{1}{2} \times 3\dfrac{1}{4} = \dfrac{9}{2} \times \dfrac{13}{4} = \dfrac{117}{8} = 14\dfrac{5}{8}$

**6.** $\begin{array}{r} 12.1000 \\ -\ .8416 \\ \hline 8.2584 \end{array}$

**7.** $\begin{array}{r} 0.8163 \\ \times\ 0.22 \\ \hline 16326 \\ 16326 \\ \hline 0.179586 \end{array}$

**8.** $\dfrac{\$1.68}{12 \text{ bananas}} = \dfrac{\$1.68 \div 12}{12 \text{ bananas} \div 12}$

$\qquad\qquad = \dfrac{\$0.14}{1 \text{ banana}}$

$\qquad\qquad = \$0.14/\text{banana}$

**9.** $\dfrac{12 \text{ yen}}{3 \text{ pesos}} = \dfrac{4 \text{ yen}}{1 \text{ peso}}$

**10.** $\dfrac{12}{17} \overset{?}{=} \dfrac{30}{42.5}$

$12 \times 42.5 \overset{?}{=} 17 \times 30$

$510 = 510 \quad \text{True}$

**11.** $\dfrac{4\frac{1}{3}}{13} \overset{?}{=} \dfrac{2\frac{2}{3}}{8}$

$4\dfrac{1}{3} \times 8 \overset{?}{=} 13 \times 2\dfrac{2}{3}$

$\dfrac{13}{3} \times 8 \overset{?}{=} 13 \times \dfrac{8}{3}$

$\dfrac{104}{3} = \dfrac{104}{3} \quad \text{True}$

**12.** $\dfrac{9}{2.1} = \dfrac{n}{0.7}$

$9 \times 0.7 = 2.1 \times n$

$\dfrac{6.3}{2.1} = \dfrac{2.1 \times n}{2.1}$

$3 = n$

**13.**
$$\frac{50}{20} = \frac{5}{n}$$
$$50 \times n = 20 \times 5$$
$$\frac{50 \times n}{50} = \frac{100}{50}$$
$$n = 2$$

**14.**
$$\frac{n}{56} = \frac{16}{7}$$
$$n \times 7 = 56 \times 16$$
$$\frac{n \times 7}{7} = \frac{896}{7}$$
$$n = 128$$

**15.**
$$\frac{7}{n} = \frac{28}{36}$$
$$7 \times 36 = n \times 28$$
$$\frac{252}{28} = \frac{n \times 28}{28}$$
$$n = 9$$

**16.**
$$\frac{n}{11} = \frac{5}{16}$$
$$n \times 16 = 11 \times 5$$
$$\frac{n \times 16}{16} = \frac{55}{16}$$
$$n \approx 3.4$$

**17.**
$$\frac{3\frac{1}{3}}{7} = \frac{10}{n}$$
$$3\frac{1}{3} \times n = 7 \times 10$$
$$\frac{\frac{10}{3} \times n}{\frac{10}{3}} = \frac{70}{\frac{10}{3}}$$
$$n = 70 \times \frac{3}{10}$$
$$n = 21$$

**18.**
$$\frac{300 \text{ miles}}{4 \text{ inches}} = \frac{625 \text{ miles}}{n \text{ inches}}$$
$$300 \times n = 4 \times 625$$
$$\frac{300 \times n}{300} = \frac{2500}{300}$$
$$n \approx 8.33$$

8.33 inches

**19.**
$$\frac{\$7}{\$84} = \frac{n}{\$9000}$$
$$7 \times 9000 = 84 \times n$$
$$\frac{63,000}{84} = \frac{84 \times n}{84}$$
$$n = \$750$$

**20.**
$$\frac{14 \text{ people}}{3.5 \text{ lb}} = \frac{20 \text{ people}}{n \text{ lb}}$$
$$14 \times n = 3.5 \times 20$$
$$\frac{14 \times n}{14} = \frac{70}{14}$$
$$n = 5$$

5 pounds

**21.**
$$\frac{39 \text{ gallons sap}}{2 \text{ gallons syrup}} = \frac{n \text{ gallons sap}}{11 \text{ gallons syrup}}$$
$$39 \times 11 = 2 \times n$$
$$\frac{429}{2} = \frac{2 \times n}{2}$$
$$214.5 = n$$

214.5 gallons of maple sap

# PERCENT

**Pretest Chapter 5**

1. $0.17 = 17\%$

2. $0.387 = 38.7\%$

3. $1.34 = 134\%$

4. $8.94 = 894\%$

5. $0.006 = 0.6\%$

6. $0.004 = 0.4\%$

7. $\dfrac{17}{100} = 17\%$

8. $\dfrac{27}{100} = 27\%$

9. $\dfrac{13.4}{100} = 13.4\%$

10. $\dfrac{19.8}{100} = 19.8\%$

11. $\dfrac{6\frac{1}{2}}{100} = 6\frac{1}{2}\%$

12. $\dfrac{1\frac{3}{8}}{100} = 1\frac{3}{8}\%$

13. $\begin{array}{r} 0.80 \\ 10\overline{)8.00} \\ \underline{8\ 0} \\ 0 \end{array}$

$\dfrac{8}{10} = 80\%$

14. $\begin{array}{r} 0.025 \\ 40\overline{)1.000} \\ \underline{80} \\ 200 \\ \underline{200} \\ 0 \end{array}$

$\dfrac{1}{40} = 0.025 = 2.5\%$

15. $\begin{array}{r} 2.60 \\ 20\overline{)52.00} \\ \underline{40} \\ 12\ 0 \\ \underline{12\ 0} \\ 0 \end{array}$

$\dfrac{52}{20} = 260\%$

16. $\begin{array}{r} 1.0625 \\ 16\overline{)17.0000} \\ \underline{16} \\ 1\ 0 \\ \underline{0} \\ 1\ 00 \\ \underline{96} \\ 40 \\ \underline{32} \\ 80 \\ \underline{80} \\ 0 \end{array}$

$\dfrac{17}{16} = 1.0625 = 106.25\%$

**17.**

$$
\begin{array}{r}
0.71428 \\
7\overline{)5.00000} \\
\underline{4\ 9} \\
10 \\
\underline{7} \\
30 \\
\underline{28} \\
20 \\
\underline{14} \\
60 \\
\underline{56} \\
4
\end{array}
$$

$$\frac{5}{7} = 71.43\%$$

**20.**

$$
\begin{array}{r}
0.68421 \\
19\overline{)13.00000} \\
\underline{11\ 4} \\
1\ 60 \\
\underline{1\ 52} \\
80 \\
\underline{76} \\
40 \\
\underline{38} \\
20 \\
\underline{19} \\
1
\end{array}
$$

$$\frac{13}{19} = 0.6842 = 68.42\%$$

**18.**

$$
\begin{array}{r}
0.28571 \\
7\overline{)2.00000} \\
\underline{1\ 4} \\
60 \\
\underline{56} \\
40 \\
\underline{35} \\
50 \\
\underline{49} \\
10 \\
\underline{7} \\
3
\end{array}
$$

$$\frac{2}{7} = 0.2857 = 28.57\%$$

**21.**

$$
\begin{array}{r}
0.40 \\
5\overline{)2.00} \\
\underline{2\ 0} \\
0 \\
\underline{0} \\
0
\end{array}
$$

$$4\frac{2}{5} = 4.40 = 440\%$$

**22.**

$$
\begin{array}{r}
0.75 \\
4\overline{)3.00} \\
\underline{2\ 8} \\
20 \\
\underline{20} \\
0
\end{array}
$$

$$2\frac{3}{4} = 2.75 = 275\%$$

**19.**

$$
\begin{array}{r}
0.95652 \\
23\overline{)22.00000} \\
\underline{20\ 7} \\
1\ 30 \\
\underline{1\ 15} \\
150 \\
\underline{138} \\
120 \\
\underline{115} \\
50 \\
\underline{46} \\
4
\end{array}
$$

$$\frac{22}{23} = 95.65\%$$

**23.**

$$
\begin{array}{r}
0.00333 \\
300\overline{)1.00000} \\
\underline{900} \\
1000 \\
\underline{900} \\
1000 \\
\underline{900} \\
100
\end{array}
$$

$$\frac{1}{300} = 0.33\%$$

24.

$$400 \overline{)\begin{array}{c} 0.0025 \\ 1.0000 \\ \underline{800} \\ 2000 \\ \underline{2000} \\ 0 \end{array}}$$

$$\frac{1}{400} = 0.0025 = 0.25\%$$

25. $22\% = \dfrac{22}{100} = \dfrac{11}{50}$

26. $53\% = \dfrac{53}{100}$

27. $150\% = \dfrac{150}{100} = \dfrac{3}{2} = 1\dfrac{1}{2}$

28. $160\% = \dfrac{160}{100} = \dfrac{8}{5} = 1\dfrac{3}{5}$

29. $6\dfrac{1}{3}\% = \dfrac{6\frac{1}{3}}{100} = 6\dfrac{1}{3} \div 100$

$$= \dfrac{19}{3} \times \dfrac{1}{100} = \dfrac{19}{300}$$

30. $4\dfrac{2}{3}\% = \dfrac{4\frac{2}{3}}{100}$

$$= 4\dfrac{2}{3} \div 100$$

$$= \dfrac{14}{3} \times \dfrac{1}{100}$$

$$= \dfrac{7}{150}$$

31. $51\dfrac{1}{4}\% = \dfrac{\frac{205}{4}}{100} = \dfrac{205}{4} \div 100$

$$= \dfrac{205}{4} \times \dfrac{1}{100} = \dfrac{41}{80}$$

32. $43\dfrac{3}{4}\% = \dfrac{43\frac{3}{4}}{100}$

$$= 43\dfrac{3}{4} \div 100$$

$$= \dfrac{175}{4} \times \dfrac{1}{100}$$

$$= \dfrac{7}{16}$$

33. Find 24% of $230 = 0.24\,(230) = 55.2$

34. What is 78% of 62?

$$n = 78\% \times 62$$
$$n = 0.78 \times 62$$
$$= 48.36$$

35. 68 is what percent of 72?

$$68 = n \times 72$$

$$n = \dfrac{68}{72} = 0.9\overline{4} = 94.44\%$$

$$\dfrac{68}{72} = \dfrac{n \times 72}{72}$$

$$n = 0.9\overline{4}$$

$$\approx 94.44\%$$

36. What percent of 76 is 34?

$$n \times 76 = 34$$

$$\dfrac{n \times 76}{76} = \dfrac{34}{76}$$

$$n \approx 0.4474 = 44.74\%$$

37. 8% of what number is 240?

$$8\% \times n = 240$$

$$\dfrac{0.08 \times n}{0.08} = \dfrac{240}{0.08}$$

$$n = 3000$$

**38.** 354 is 40% of what number?

$$354 = 40\% \times n$$

$$\frac{354}{0.40} = \frac{0.40 \times n}{0.40}$$

$$885 = n$$

**39.** $\dfrac{24}{38} = 0.6316$

They won 63.16% of the games.

**40.** 5% of what number is \$0.72

$$5\% \times n = 0.72$$

$$0.05 \times n = 0.72$$

$$\frac{0.05 \times n}{0.05} = \frac{0.72}{0.05}$$

$$n = \$14.40$$

**41.** $480 - 336 - 144$

$$144 = n \times 480$$

$$\frac{144}{480} = \frac{n \times 480}{480}$$

$$n = 0.3 = 30\%$$

**42.** $24\% \times 22{,}500 = n$

$$0.24 \times 22{,}500 = n$$

$$5400 = n$$

\$5400

**43.** Interest $= \$1700 \times 0.12 \times 2$

$$= \$408$$

## 5.1   Exercises

**1.** hundred

**3.** Move the decimal point *two* places to the *left*. *Drop* the % symbol.

**5.** $\dfrac{48}{100} = 48\%$

**7.** $\dfrac{9}{100} = 9\%$

**9.** $\dfrac{80}{100} = 80\%$

**11.** $\dfrac{245}{100} = 245\%$

**13.** $\dfrac{5.3}{100} = 5.3\%$

**15.** $\dfrac{0.07}{100} = 0.07\%$

**17.** $\dfrac{13}{100} = 13\%$

**19.** $\dfrac{28}{100} = 28\%$

**21.** $51\% = 0.51$

**23.** $7\% = 0.07$

**25.** $20\% = 0.2$

**27.** $43.6\% = 0.436$

**29.** $0.03\% = 0.003$

**31.** $0.72\% = 0.0072$

**33.** $126\% = 1.26$

**35.** $366\% = 3.66$

**37.** $0.74 = 74\%$

**39.** $0.50 = 50\%$

**41.** $0.08 = 8\%$

**43.** $0.563 = 56.3\%$

**45.** $0.002 = 0.2\%$

**47.** $0.0057 = 0.57\%$

**49.** $1.35 = 135\%$

**51.** $2.72 = 272\%$

**53.** $0.27 = 27\%$

**55.** $0.09 = 9\%$

**57.** $0.94 = 94\%$

**59.** $2.31 = 231\%$

**61.** $\dfrac{10}{100} = 10\%$

**63.** $0.009 = 0.9\%$

**65.** $62\% = \dfrac{62}{100} = 0.62$

**67.** $\dfrac{138}{100} = 1.38$

**69.** $\dfrac{0.3}{100} = 0.003$

**71.** $\dfrac{80}{100} = 0.8$

**73.** $\dfrac{59}{100} = 59\%$

**75.** $36\% = 36$ percent

$\quad\quad = 36$ "per one hundred"

$\quad\quad = 36 \times \dfrac{1}{100} = \dfrac{36}{100}$

$\quad\quad = 0.36$

The rule is using the fact that $36\%$ means 36 per one hundred.

**77. a.** $55{,}562\% = 555.62$

$\quad$ **b.** $55{,}562\% = \dfrac{55{,}562}{100}$

$\quad$ **c.** $55{,}562\% = \dfrac{55{,}562}{100} = \dfrac{27{,}781}{50}$

## Cumulative Review Problems

**79.** $0.56 = \dfrac{56}{100} = \dfrac{14}{25}$

**81.**

$$
\begin{array}{r}
0.6875 \\
16\overline{)11.0000} \\
9\;6\phantom{000} \\
\hline
1\;40\phantom{00} \\
1\;28\phantom{00} \\
\hline
120\phantom{0} \\
112\phantom{0} \\
\hline
80
\end{array}
$$

$\dfrac{11}{16} = 0.6875$

**83.** $3 \times 246 + 7 \times 380 + 5 \times 168 + 9 \times 122$

$\quad\quad = 738 + 2660 + 840 + 1098$

$\quad\quad = 5336$ vases

## 5.2   Exercises

**1.** Write the number in front of the percent symbol as the numerator of the fraction. Write the number 100 as the denominator of the fraction. Reduce the fraction if possible.

**3.** $10\% = \dfrac{10}{100} = \dfrac{1}{10}$

**5.** $33\% = \dfrac{33}{100}$

**7.** $55\% = \dfrac{55}{100} = \dfrac{11}{20}$

**9.** $75\% = \dfrac{75}{100} = \dfrac{3}{4}$

**11.** $20\% = \dfrac{20}{100} = \dfrac{1}{5}$

**13.** $9.5\% = 0.095 = \dfrac{95}{1000} = \dfrac{19}{200}$

**15.** $17.5\% = 0.175 = \dfrac{175}{1000} = \dfrac{7}{40}$

**17.** $64.8\% = 0.648 = \dfrac{648}{1000} = \dfrac{81}{125}$

**19.** $71.25\% = 0.7125 = \dfrac{7125}{10,000} = \dfrac{57}{80}$

**21.** $176\% = \dfrac{176}{100} = \dfrac{44}{25} = 1\dfrac{19}{25}$

**23.** $340\% = \dfrac{340}{100} = \dfrac{17}{5} = 3\dfrac{2}{5}$

**25.** $1200\% = \dfrac{1200}{200} = 12$

**27.** $2\dfrac{1}{6}\% = \dfrac{\frac{13}{6}}{100} = \dfrac{13}{600}$

**29.** $12\dfrac{1}{2}\% = \dfrac{\frac{25}{2}}{100} = \dfrac{25}{200} = \dfrac{1}{8}$

**31.** $8\dfrac{4}{5}\% = \dfrac{\frac{44}{5}}{100} = \dfrac{44}{500} = \dfrac{11}{125}$

**33.** $15\dfrac{5}{9}\% = \dfrac{\frac{140}{9}}{100} = \dfrac{140}{900} = \dfrac{7}{45}$

**35.** $2\dfrac{4}{11}\% = \dfrac{\frac{26}{11}}{100} = \dfrac{26}{1100} = \dfrac{13}{550}$

**37.** $\dfrac{3}{4} = 0.75 = 75\%$

**39.** $\dfrac{1}{3} \approx 0.3333 = 33.33\%$

**41.** $\dfrac{7}{20} = 0.35 = 35\%$

**43.** $\dfrac{7}{25} = 0.28 = 28\%$

**45.** $\dfrac{11}{40} = 0.275 = 27.5\%$

**47.** $\dfrac{5}{12} \approx 0.4167 = 41.67\%$

**49.** $\dfrac{18}{5} = 3.6 = 360\%$

**51.** $2\dfrac{5}{6} = \dfrac{17}{6} \approx 2.8333 = 283.33\%$

**53.** $4\dfrac{1}{8} = \dfrac{33}{8} = 4.125 = 412.5\%$

**55.** $\dfrac{3}{7} \approx 0.4286 = 42.86\%$

**57.** $\dfrac{31}{16} = 1.9375 = 193.75\%$

**59.** $\dfrac{26}{50} = 0.52 = 52\%$

**61.**

$$
\begin{array}{r}
0.025 \\
40\overline{)1.000} \\
\underline{80}\phantom{00} \\
200 \\
\underline{200} \\
0
\end{array}
$$

$$\dfrac{1}{40} = 0.025 = 2.5\%$$

**63.**

$$3000 \overline{)\begin{array}{l} 0.01166 \\ 35.00000 \end{array}}$$

$$\begin{array}{r} 30\ 00 \\ \hline 5\ 000 \\ 3\ 000 \\ \hline 2\ 0000 \\ 1\ 8000 \\ \hline 20000 \\ 18000 \\ \hline 2000 \end{array}$$

$$\frac{35}{3000} \approx 0.0117 = 1.17\%$$

**65.**

$$6 \overline{)\begin{array}{l} 0.83 \\ 5.00 \end{array}}$$

$$\begin{array}{r} 4\ 8 \\ \hline 20 \\ 18 \\ \hline 2 \end{array}$$

$$\frac{5}{6} = 0.83\frac{2}{6} = 0.83\frac{1}{3} = 83\frac{1}{3}\%$$

**67.**

$$14 \overline{)\begin{array}{l} 0.35 \\ 5.00 \end{array}}$$

$$\begin{array}{r} 4\ 2 \\ \hline 80 \\ 70 \\ \hline 10 \end{array}$$

$$\frac{5}{14} = 0.35\frac{10}{14} = 0.35\frac{5}{7} = 35\frac{5}{7}\%$$

**69.**

$$8 \overline{)\begin{array}{l} 0.37 \\ 3.00 \end{array}}$$

$$\begin{array}{r} 2\ 4 \\ \hline 60 \\ 56 \\ \hline 4 \end{array}$$

$$\frac{3}{8} = 0.37\frac{4}{8} = 0.37\frac{1}{2} = 37\frac{1}{2}\%$$

**71.**

$$40 \overline{)\begin{array}{l} 0.07 \\ 3.00 \end{array}}$$

$$\begin{array}{r} 2\ 80 \\ \hline 20 \end{array}$$

$$\frac{3}{40} = 0.07\frac{20}{40} = 0.07\frac{1}{2} = 7\frac{1}{2}\%$$

**73.**

$$12 \overline{)\begin{array}{l} 0.91666 \\ 11.00000 \end{array}}$$

$$\begin{array}{r} 10\ 8 \\ \hline 20 \\ 12 \\ \hline 80 \\ 72 \\ \hline 80 \\ 72 \\ \hline 80 \\ 72 \\ \hline 8 \end{array}$$

$$\frac{11}{12} \approx 0.9167 = 91.67\%$$

$$\frac{11}{12}; \ 0.9167; \ 91.67\%$$

**75.**  $0.56 = 56\%$

$$0.56 = \frac{56}{100} = \frac{14}{25}$$

$$\frac{14}{25}; \ 0.56; \ 56\%$$

**77.**

$$200 \overline{)\begin{array}{l} 0.015 \\ 3.000 \end{array}}$$

$$\begin{array}{r} 2\ 00 \\ \hline 1\ 000 \\ 1\ 000 \\ \hline 0 \end{array}$$

$$\frac{3}{200}; \ 0.015; \ 1.5\%$$

**79.**

$$9 \overline{)\begin{array}{l} 0.555 \\ 5.0 \end{array}}$$

$$\begin{array}{r} 4\ 5 \\ \hline 50 \\ 45 \\ \hline 50 \\ 45 \\ \hline 5 \end{array}$$

$$\frac{5}{9} = 0.\overline{5}$$

$$\frac{5}{9}; \ 0.5556; \ 55.56\%$$

**81.**      $\frac{1}{8} = 0.125$

$3\frac{1}{8}\% = 0.03125 \approx 0.0313$

$0.03125 = \frac{3125}{100,000} = \frac{1}{32}$

$\frac{1}{32}; \ 0.0313; \ 3\frac{1}{8}\%$

**83.** $28\frac{15}{16}\% = \frac{28\frac{15}{16}}{100}$

$= 28\frac{15}{16} \div 100$

$= \frac{463}{16} \times \frac{1}{100}$

$= \frac{463}{1600}$

**85.**    $\frac{123}{800} = \frac{n}{100}$

$800 \times n = 123 \times 100$

$\frac{800 \times n}{800} = \frac{12,300}{800}$

$n = 15.375$

$\frac{123}{800} = 15.375\%$

**87.** It will have at least two zeros to the right of the decimal point.

---

## Cumulative Review Problems

**89.**    $\frac{15}{n} = \frac{8}{3}$

$15 \times 3 = n \times 8$

$\frac{45}{8} = \frac{n \times 8}{8}$

$n = 5.625$

**91.**     10,041
            986
          4,283
     + 533,855
     549,165  documents

---

## 5.3 A   Exercises

**1.** What is 5% of 90?

$n = 5\% \times 90$

**3.** 70% of what is 2?

$70\% \times n = 2$

**5.** 17 is what percent of 85?

$17 = n \times 85$

**7.** What is 20% of 140?

$n = 20\% \times 140$

$n = 0.20 \times 40$

$n = 28$

**9.** Find 152% of 600.

$n = 152\% \times 600$

$n = 1.52 \times 600$

$n = 912$

**11.** 6% of $106.50 is what?

$6\% \times 106.50 = n$

$0.06 \times 106.50 = n$

$\$6.39 = n$

**13.** 25% of what is 14?

$25\% \times n = 14$

$0.25 \times n = 14$

$\frac{0.25 \times n}{0.25} = \frac{14}{0.25}$

$n = 56$

**15.** 52 is 4% of what?

$$52 = 4\% \times n$$
$$52 = 0.04 \times n$$
$$\frac{52}{0.04} = \frac{0.04 \times n}{0.04}$$
$$1300 = n$$

**17.** 22% of what is $33?

$$22\% \times n = 33$$
$$0.22 \times n = 33$$
$$\frac{0.22 \times n}{0.22} = \frac{33}{0.22}$$
$$n = 150$$

$150

**19.** What percent of 64 is 32?

$$n \times 64 = 32$$
$$\frac{n \times 64}{64} = \frac{32}{64}$$
$$n = 0.5$$
$$n = 50\%$$

**21.** 56 is what percent of 200?

$$56 = n \times 200$$
$$\frac{56}{200} = \frac{n \times 200}{200}$$
$$0.28 = n$$
$$28\% = n$$

**23.** 78 is what percent of 120?

$$78 = n \times 120$$
$$\frac{78}{120} = \frac{n \times 120}{120}$$
$$0.65 = n$$

65% of the points

**25.** 20% of 155 is what?

$$20\% \times 155 = n$$
$$0.20 \times 155 = n$$
$$31 = n$$

**27.** 170% of what is 144.5?

$$170\% \times n = 144.5$$
$$1.7 \times n = 144.5$$
$$\frac{1.7 \times n}{1.7} = \frac{144.5}{1.7}$$
$$n = 85$$

**29.** 84 is what percent of 700?

$$84 = n \times 700$$
$$\frac{84}{700} = \frac{n \times 700}{700}$$
$$0.12 = n$$
$$12\% = n$$

**31.** Find 0.4% of 820.

$$n = 0.4\% \times 820$$
$$n = 0.004 \times 820$$
$$n = 3.28$$

**33.** What percent of 35 is 22.4?

$$n \times 35 = 22.4$$
$$\frac{n \times 35}{35} = \frac{22.4}{35}$$
$$n = 0.64$$
$$n = 64\%$$

**35.** 89 is 20% of what?

$$89 = 20\% \times n$$
$$89 = 0.2 \times n$$
$$\frac{89}{0.2} = \frac{0.2 \times n}{0.2}$$
$$445 = n$$

**37.** 42 is what percent of 120?

$$42 = n \times 120$$
$$\frac{42}{120} = \frac{n \times 120}{120}$$
$$0.35 = n$$
$$35\% = n$$

**39.** What is 16.5% of 240?

$$n = 16.5\% \times 240$$
$$n = 0.165 \times 240$$
$$n = 39.6$$

**41.** 44 is what percent of 55?

$$44 = n \times 55$$
$$\frac{44}{55} = \frac{n \times 55}{55}$$
$$0.8 = n$$
$$n = 80\%$$

**43.** 68 is what percent of 80?

$$68 = n \times 80$$
$$\frac{68}{80} = \frac{n \times 80}{80}$$
$$0.85 = n$$
$$85\% = n$$

85% were acceptable

**45.** 44% of 1260 is what?

$$44\% \times 1260 = n$$
$$0.44 \times 1260 = n$$
$$554.4 = n$$

554 students

**47.** 60% of what is 24?

$$60\% \times n = 24$$
$$0.60 \times n = 34$$
$$\frac{0.60 \times n}{0.60} = \frac{24}{0.60}$$
$$n = 40$$

40 years

**49.** Find 12% of 30% of $1600.

$$n = 12\% \times 30\% \times 1600$$
$$n = 0.12 \times 0.30 \times 1600$$
$$n = 57.60$$

**51.**
$$L + 30\% \times L = 520 \text{ (semiperimeter)}$$
$$100\% \times L + 30\%L = 520$$
$$130\% \times L = 520$$
$$1.3 \times L = 520$$
$$\frac{1.3 \times L}{1.3} = \frac{520}{1.3}$$
$$L = 400 \text{ feet}$$
$$\text{and } W = 30\% \times 400$$
$$= 120 \text{ feet}$$

## Cumulative Review Problems

**53.**
$$\begin{array}{r} 1.36 \\ \times\ 1.8 \\ \hline 1088 \\ 136\phantom{8} \\ \hline 2.448 \end{array}$$

**55.**
$$\begin{array}{r} 2834. \\ 0.06\overline{)170.04} \\ \underline{12}\phantom{0000} \\ 50\phantom{000} \\ \underline{48}\phantom{000} \\ 20\phantom{00} \\ \underline{18}\phantom{00} \\ 24\phantom{0} \\ \underline{24}\phantom{0} \\ 0 \end{array}$$

## 5.3B Exercises

|                                   | p  | b   | a   |
|-----------------------------------|----|-----|-----|
| **1.** 75% of 660 is 495.         | 75 | 660 | 495 |
| **3.** What is 22% of 60?         | 22 | 60  | a   |
| **5.** 49% of what is 2450?       | 49 | b   | 2450 |
| **7.** 30 is what percent of 50?  | p  | 50  | 30  |

**9.** 30% of 120 is what?

$$\frac{a}{120} = \frac{30}{100}$$
$$100a = 120 \times 30$$
$$\frac{100a}{100} = \frac{3600}{100}$$
$$a = 36$$

**11.** Find 280% of 70.

$$\frac{a}{70} = \frac{280}{100}$$
$$100a = 70 \times 280$$
$$\frac{100a}{100} = \frac{19,600}{100}$$
$$a = 196$$

**13.** 0.7% of 8000 is what?

$$\frac{a}{8000} = \frac{0.7}{100}$$
$$100a = 8000 \times 0.7$$
$$\frac{100a}{100} = \frac{5600}{100}$$
$$a = 56$$

**15.** 20 is 25% of what?

$$\frac{20}{b} = \frac{25}{100}$$
$$20 \times 100 = 25b$$
$$\frac{2000}{25} = \frac{25b}{25}$$
$$80 = b$$

**17.** 250% of what is 200?

$$\frac{200}{b} = \frac{250}{100}$$
$$200 \times 100 = 250b$$
$$\frac{20,000}{250} = \frac{250b}{250}$$
$$80 = b$$

**19.** 3000 is 0.5% of what?

$$\frac{3000}{b} = \frac{0.5}{100}$$
$$3000 \times 100 = 0.5b$$
$$\frac{300,000}{0.5} = \frac{0.5b}{0.5}$$
$$600,000 = b$$

**21.** 56 is what percent of 280?

$$\frac{p}{100} = \frac{56}{280}$$
$$280p = 56 \times 100$$
$$\frac{280p}{280} = \frac{5600}{280}$$
$$p = 20$$

**23.** What percent of 260 is 10.4?

$$\frac{p}{100} = \frac{10.4}{260}$$
$$260p = 100 \times 10.4$$
$$\frac{260p}{260} = \frac{1040}{260}$$
$$p = 4$$

**25.** 18% of 150 is what?

$$\frac{a}{150} = \frac{18}{100}$$
$$100a = 150 \times 18$$
$$\frac{100a}{100} = \frac{2700}{100}$$
$$a = 27$$

**27.** 150% of what is 120?

$$\frac{150}{100} = \frac{120}{b}$$
$$150b = 100 \times 120$$
$$\frac{150b}{150} = \frac{12,000}{150}$$
$$b = 80$$

**29.** 82 is what percent of 500?

$$\frac{p}{100} = \frac{82}{500}$$
$$500p = 100 \times 82$$
$$\frac{500p}{500} = \frac{8200}{500}$$
$$p = 16.4$$

**31.** Find 0.7% of 520.

$$\frac{a}{520} = \frac{0.7}{100}$$
$$100a = 520 \times 0.7$$
$$\frac{100a}{100} = \frac{364}{100}$$
$$a = 3.64$$

**33.** What percent of 66 is 16.5?

$$\frac{p}{100} = \frac{16.5}{66}$$
$$66p = 100 \times 16.5$$
$$\frac{66p}{66} = \frac{1650}{66}$$
$$p = 25$$

25%

**35.** 68 is 40% of what?

$$\frac{40}{100} = \frac{68}{b}$$
$$40b = 100 \times 68$$
$$\frac{40b}{40} = \frac{6800}{40}$$
$$b = 170$$

**37.** What is 24% of 9500?

$$n = 24\% \times 9500$$
$$n = 0.24 \times 9500$$
$$n = 2280$$

$2280

**39.** 3.90 is what percent of 26.00?

$$\frac{p}{100} = \frac{3.90}{26.00}$$
$$26.00p = 100 \times 3.90$$
$$\frac{26.00p}{26.00} = \frac{390}{26.00}$$
$$p = 15$$

15%

**41.** 8 is what percent of 40?

$$\frac{p}{100} = \frac{8}{40}$$
$$40p = 100 \times 8$$
$$\frac{40p}{40} = \frac{800}{40}$$
$$p = 20$$

20%

**43.** 18% of what is 720?

$$\frac{720}{b} = \frac{18}{100}$$
$$720 \times 100 = 18b$$
$$\frac{72,000}{18} = \frac{18b}{18}$$
$$4000 = b$$

$4000

**45.** Total is

$$
\begin{array}{r}
572 \\
67 \\
843 \\
1239 \\
+\ 2253 \\
\hline
4974
\end{array}
$$

2253 is what percent of 4974?

$$\frac{p}{100} = \frac{2253}{4974}$$

$$4974p = 100 \times 2253$$

$$\frac{4974p}{4974} = \frac{225,300}{4974}$$

$$p \approx 45.3$$

45.3%

**47.** Corrections:

$$2253 + 25\% \times 2253 = 2253 + 563.25$$
$$= 2816.25$$

Government/Administration:

$$1239 + 30\% \times 1239 = 1239 + 371.7$$
$$= 1610.7$$

New total is

$$
\begin{array}{r}
572 \\
67 \\
843 \\
1610.7 \\
+\ 2816.25 \\
\hline
5908.95
\end{array}
$$

572 is what percent of 5908.95?

$$\frac{p}{100} = \frac{572}{5908.95}$$

$$5908.95p = 100 \times 572$$

$$\frac{5908.95p}{5908.95} = \frac{57,200}{5908.95}$$

$$p \approx 9.7$$

9.7%

## Cumulative Review Problems

**49.**
$$\frac{4}{5} + \frac{8}{9} = \frac{4}{5} \times \frac{9}{9} + \frac{8}{9} \times \frac{5}{5}$$
$$= \frac{36}{45} + \frac{40}{45}$$
$$= \frac{76}{45}$$
$$= 1\frac{31}{45}$$

**51.**
$$\left(2\frac{4}{5}\right)\left(1\frac{1}{2}\right) = \frac{14}{5} \times \frac{3}{2}$$
$$= \frac{42}{10} = \frac{21}{5}$$
$$= 4\frac{1}{5}$$

## 5.4  Exercises

**1.** $n \times 2.5\% = 4500$

$$\frac{n \times 0.025}{0.025} = \frac{4500}{0.025}$$

$$n = 180,000$$

180,000 pencils

**3.** $80.50 = 115\% \times n$

$$80.50 = 1.15n$$

$$\frac{80.50}{1.15} = \frac{1.15n}{1.15}$$

$$\$70 = n$$

**5.** $40 \times n = 36$

$$\frac{400 \times n}{400} = \frac{36}{400}$$

$$n = 0.09$$

9%

**7.** $n = 450 \times 74\%$

$n = 450 \times 0.74$

$n = 333$

333 people

**9.** $n \times 4\% = 9.60$

$0.04n = 9.6$

$\dfrac{0.04n}{0.04} = \dfrac{9.6}{0.04}$

$n = 240$

$240

**11.** $n \times 5060 = 1265$

$\dfrac{n \times 5060}{5060} = \dfrac{1265}{5060}$

$n = 0.25$

25%

**13.** $n \times 75\% = 7{,}200{,}000$

$\dfrac{0.75n}{0.75} = \dfrac{7{,}200{,}000}{0.75}$

$n = 9{,}600{,}000$

$9,600,000

**15.** $n \times 206 = 26$

$\dfrac{n \times 206}{206} = \dfrac{26}{206}$

$n \approx 0.1262$

12.62% bones in foot

**17.** $n = 0.9\% \times 24{,}000$

$n = 0.009 \times 24{,}000$

$n = 216$

216 babies

**19.** $100\% \times n + 20\% \times n = 66$

$120\% \times n = 66$

$1.2n = 66$

$\dfrac{1.2n}{1.2} = \dfrac{66}{1.2}$

$n = 55$

$55

**21.** $100\% \times n + 9\% \times n = 163{,}500$

$109\% \times n = 163{,}500$

$1.09n = 163{,}500$

$\dfrac{109n}{1.09} = \dfrac{163{,}500}{1.09}$

$n = 150{,}000$

$150,000

**23.** $3\% + 8\% = 11\%$

$n = 11\% \times 20{,}000$

$n = 0.11 \times 20{,}000$

$n = 2200$

2200 pounds

**25.**

$(15\% + 12\% + 10\%) \times 33{,}000{,}000 = n$

$(37\%) \times 33{,}000{,}000 = n$

$0.37 \times 33{,}000{,}000 = n$

$12{,}210{,}000 = n$

$33{,}000{,}000 - 12{,}210{,}000 = 20{,}790{,}000$

$12,210,000 for personnel, food, and decorations.

$20,790,000 for security, facility rental and all other expenses.

**27.** $n = 35\% \times 190$

$n = 0.35 \times 190$

$n = 66.6$

Discount is $65.50

Sales price is $190 - 65.50 = \$123.50$

**29.** $n = 15\% \times 400$

$n = 0.15 \times 400$

$n = 60$

Save $60 with coupon book.

Total savings are $60 - $30 = $30.

**31. a.** $n = 8\% \times 16,000$

$n = 0.08 \times 16,000$

$n = 1280$

Discount is $1280.

**b.** Cost of motorcycle is

$16,000 - $1280 = $14,720

**33. a.** $n = 33\% \times 1110$

$n = 0.33 \times 1110$

$n = 333$

Discount is $333.

**b.** Cost of clock is

$1110 - $333 = $777

---

## Cumulative Review Problems

**35.** 1,698,481 rounds to 1,698,000.

**37.** 1.63474 rounds to 1.63.

**39.** 0.055613 rounds to 0.0556.

---

## 5.5    Exercises

**1.** Commission $= 2\% \times 170,000$

$= 0.02 \times 170,000$

$= 3400$

$3400

**3.** Total income $= 300 + 4\% \times 96,000$

$= 300 + 0.04 \times 96,000$

$= 300 + 3840$

$= 4140$

$4140

**5.** Increase $= 7600 - 5000$

$= 2600$

Percent $= \dfrac{2600}{5000}$

$= 0.52$

$= 52\%$

52%

**7.** Decrease $= 985 - 394$

$= 591$

Percent $= \dfrac{591}{985}$

$= 0.60$

$= 60\%$

60%

**9.** $I = P \times R \times T$

$= 2000 \times 7\% \times 1$

$= 2000 \times 0.07 \times 1$

$= 140$

$140

**11.** $I = P \times R \times T$

$= 500 \times 1.5\% \times 1$

$= 500 \times 0.015 \times 1$

$= 7.5$

$7.50

**13.** 3 months $= \frac{3}{12} = \frac{1}{4}$ year

$$I = P \times R \times T$$
$$= 12{,}000 \times 16\% \times \frac{1}{4}$$
$$= 12{,}000 \times 0.16 \times \frac{1}{4}$$
$$= 1920 \times \frac{1}{4}$$
$$= 480$$

$480

**15.** Rate $= \dfrac{72{,}000}{12{,}000{,}000}$
$$= 0.006$$
$$= 0.6\%$$
$$= 0.6\%$$

0.6%

**17.** Sales total $= \dfrac{48{,}000}{3\%}$
$$= \dfrac{48{,}000}{0.03}$$
$$= 1{,}600{,}000$$

$1,600,000

**19.** Spending $= 15\% \times 265$
$$= 0.15 \times 265$$
$$= 39.75$$

$39.75

**21.** Thin mints $= 25\% \times 156$
$$= 0.25 \times 156$$
$$= 39$$

39 boxes

**23.** Inspected parts $= \dfrac{32}{4\%}$
$$= \dfrac{32}{0.04}$$
$$= 800$$

800 parts

**25. a.** $I = P \times R \times T$
$$= 3700 \times 3.2\% \times 1$$
$$= 3700 \times 0.032 \times 1$$
$$= 118.4$$
$118.40

   **b.** Total money $= \$3700 + \$118.40$
$$= \$3818.40$$

**27. a.** Purchases $= 52 + 38 + 26$
$$= 116$$

   Tax $= 6\% \times 116$
$$= 0.06 \times 116$$
$$= 6.96$$

$6.96

   **b.** Total cost $= \$116 + \$6.96$
$$= \$122.96$$

**29.** Tax $= 4.6\% \times 18{,}456.82$
$$= 0.046 \times 18{,}456.82$$
$$\approx 849.01$$
$849.01

## Cumulative Review Problems

**31.** $3(12 - 6) - 4(12 \div 3) = 3(6) - 4(4)$
$$= 18 - 16$$
$$= 2$$

**33.** $\left(\dfrac{5}{3}\right)\left(\dfrac{3}{8}\right) - \left(\dfrac{1}{2} - \dfrac{1}{3}\right)$

$$= \left(\dfrac{5}{3}\right)\left(\dfrac{3}{8}\right) - \left(\dfrac{3}{6} - \dfrac{2}{6}\right)$$

$$= \left(\dfrac{5}{3}\right)\left(\dfrac{3}{8}\right) - \dfrac{1}{6}$$

$$= \dfrac{5}{8} - \dfrac{1}{6}$$

$$= \dfrac{5}{8} \times \dfrac{3}{3} - \dfrac{1}{6} \times \dfrac{4}{4}$$

$$= \dfrac{15}{24} - \dfrac{4}{24}$$

$$= \dfrac{11}{24}$$

## Putting Your Skills To Work

1. A 2000 calorie diet requires 3500 mg of potassium.

$$\text{milk} = 200 \text{ mg}$$
$$\text{cereal} = \frac{3}{4} \times 360 = 270 \text{ mg}$$
$$\text{total} = 200 + 270 = 470 \text{ mg}$$
$$\text{percent} = \frac{470}{3500} \approx 0.134 = 13.4\%$$

This diet also requires 300 g of carbohydrates.

$$\text{milk} = 6 \text{ g}$$
$$\text{cereal} = \frac{3}{4} \times 45 = 33.75$$
$$\text{total} = 6 + 33.75 = 39.75$$
$$\text{percent} = \frac{39.75}{300} = 0.1325 = 13.25\%$$

## Chapter 5 Review Problems

1. $0.62 = 62\%$

2. $0.43 = 43\%$

3. $0.372 = 37.2\%$

4. $0.529 = 52.9\%$

5. $0.0828 = 8.28\%$

6. $0.0719 = 7.19\%$

7. $2.52 = 252\%$

8. $4.37 = 437\%$

9. $1.036 = 103.6\%$

10. $1.052 = 105.2\%$

11. $0.006 = 0.6\%$

12. $0.002 = 0.2\%$

13. $\frac{0.52}{100} = 0.52\%$

14. $\frac{0.67}{100} = 0.67\%$

15. $\frac{4\frac{1}{12}}{100} = 4\frac{1}{12}\%$

16. $\frac{3\frac{5}{12}}{100} = 3\frac{5}{12}$

17. $\frac{317}{100} = 317\%$

18. $\frac{225}{100} = 225\%$

19. $\frac{19}{25} = 0.76 = 76\%$

20. $\frac{13}{25} = 0.52 = 52\%$

21. $\frac{11}{20} = 0.55 = 55\%$

22. $\frac{9}{40} = 0.225 = 22.5\%$

23. $\frac{5}{11} \approx 0.4545 = 45.45\%$

24. $\frac{4}{9} \approx 0.4444 = 44.44\%$

25. $2\frac{1}{4} = 2.25 = 225\%$

**26.** $3\dfrac{3}{4} = 3.75 = 375\%$

**27.** $4\dfrac{3}{7} \approx 4.4286 = 442.86\%$

**28.** $5\dfrac{5}{9} \approx 5.5556 = 555.56\%$

**29.** $\dfrac{152}{80} = 1.9 = 190\%$

**30.** $\dfrac{165}{90} \approx 1.8333 = 183.33\%$

**31.** $\dfrac{3}{800} \approx 0.0038 = 0.38\%$

**32.** $\dfrac{5}{800} \approx 0.0063 = 0.63\%$

**33.** $0.8\% = 0.008$

**34.** $0.6\% = 0.006$

**35.** $82.7\% = 0.827$

**36.** $59.6\% = 0.596$

**37.** $236\% = 2.36$

**38.** $177\% = 1.77$

**39.** $32\dfrac{1}{8} = 32.125\%$
$\qquad = 0.32125$

**40.** $26\dfrac{3}{8}\% = 26.375\%$
$\qquad\quad = 0.26375$

**41.** $72\% = \dfrac{72}{100} = \dfrac{72 \div 4}{100 \div 4}$
$\qquad = \dfrac{18}{25}$

**42.** $92\% = \dfrac{92}{100} = \dfrac{92 \div 4}{100 \div 4}$
$\qquad = \dfrac{23}{25}$

**43.** $185\% = \dfrac{185}{100} = \dfrac{185 \div 5}{100 \div 5}$
$\qquad\quad = \dfrac{37}{20}$

**44.** $225\% = \dfrac{225}{100} = \dfrac{225 \div 25}{100 \div 25}$
$\qquad\quad = \dfrac{9}{4}$

**45.** $16.4\% = 0.164$
$\qquad = \dfrac{164}{1000} = \dfrac{164 \div 4}{1000 \div 4}$
$\qquad = \dfrac{41}{250}$

**46.** $30.5\% = 0.305 = \dfrac{305}{1000}$
$\qquad\quad = \dfrac{305 \div 5}{1000 \div 5}$
$\qquad\quad = \dfrac{61}{200}$

**47.** $31\dfrac{1}{4}\% = \dfrac{31\frac{1}{4}}{100} = 31\dfrac{1}{4} \div 100$
$\qquad\quad = \dfrac{125}{4} \times \dfrac{1}{100} = \dfrac{125}{400}$
$\qquad\quad = \dfrac{125 \div 25}{400 \div 25} = \dfrac{5}{16}$

**48.** $43\frac{3}{4}\% = \dfrac{43\frac{3}{4}}{100}$

$= 43\frac{3}{4} \div 100$

$= \dfrac{175}{4} \times \dfrac{1}{100}$

$= \dfrac{175}{400} = \dfrac{175 \div 25}{400 \div 25}$

$= \dfrac{7}{16}$

**49.** $0.05\% = 0.0005$

$= \dfrac{5}{10,000}$

$= \dfrac{5 \div 5}{10,000 \div 5}$

$= \dfrac{1}{2000}$

**50.** $0.06\% = 0.0006$

$= \dfrac{6}{10,000} = \dfrac{6 \div 2}{10,000 \div 2}$

$= \dfrac{3}{5000}$

**51.** $5\overline{)3.0}$ with quotient $0.6$

$\dfrac{3}{5}$; 0.6; 60%

**52.** $8\overline{)7.000}$ with quotient $0.875$

$\dfrac{7}{8}$; 0.875; 87.5%

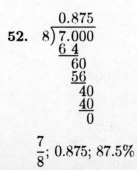

**53.** $37.5\% = 0.375$

$= \dfrac{375}{1000} = \dfrac{3}{8}$

$\dfrac{3}{8}$; 0.375; 37.5%

**54.** $56.25\% = 0.5625$

$= \dfrac{5625}{10,000} = \dfrac{9}{16}$

$\dfrac{9}{16}$; 0.5625; 56.25%

**55.** $0.008 = 0.8\%$

$0.008 = \dfrac{8}{1000} = \dfrac{1}{125}$

$\dfrac{1}{125}$; 0.008; 0.8%

**56.** $0.45 = 45\%$

$0.45 = \dfrac{45}{100} = \dfrac{9}{20}$

$\dfrac{9}{20}$; 0.45; 45%

**57.** What is 83% of 400?

$n = 83\% \times 400$

$n = 0.83 \times 400$

$n = 332$

**58.** What is 45% of 900?

$n = 45\% \times 900$

$n = 0.45 \times 900$

$n = 405$

**59.** 18 is 20% of what number?

$20\% \times n = 18$

$0.2n = 18$

$\dfrac{0.2n}{0.2} = \dfrac{18}{0.2}$

$n = 90$

**60.** 70 is 40% of what number?

$$40\% \times n = 70$$
$$0.4n = 70$$
$$\frac{0.4n}{0.4} = \frac{70}{0.4}$$
$$n = 175$$

**61.** 50 is what percent of 125?

$$50 = 125 \times n$$
$$\frac{50}{125} = \frac{125 \times n}{125}$$
$$0.4 = n$$
$$40\% = n$$

**62.** 70 is what percent of 175?

$$70 = 175 \times n$$
$$\frac{70}{175} = \frac{175 \times n}{175}$$
$$0.4 = n$$
$$40\% = n$$

**63.** Find 162% of 60.

$$n = 162\% \times 60$$
$$n = 1.62 \times 60$$
$$n = 97.2$$

**64.** Find 124% of 80.

$$n = 124\% \times 80$$
$$n = 1.24 \times 80$$
$$n = 99.2$$

**65.** 92% of what number is 147.2?

$$92\% \times n = 147.2$$
$$0.92n = 147.2$$
$$\frac{0.92n}{0.92} = \frac{147.2}{0.92}$$
$$n = 160$$

**66.** 68% of what number is 95.2?

$$68\% \times n = 95.2$$
$$0.68n = 95.2$$
$$\frac{0.68n}{0.68} = \frac{95.2}{0.68}$$
$$n = 140$$

**67.** What percent of 70 is 14?

$$70 \times n = 14$$
$$\frac{70 \times n}{70} = \frac{14}{70}$$
$$n \qquad = 0.2$$
20%

**68.** What percent of 200 is 116?

$$200 \times n = 116$$
$$\frac{200 \times n}{200} = \frac{116}{200}$$
$$n = 0.58$$
58%

**69.** $n = 34\% \times 150$
$n = 0.34 \times 150$
$n = 51$
51 students

**70.** $n = 64\% \times 150$
$n = 0.64 \times 150$
$n = 96$
96 trucks

**71.** $n \times 61\% = 6832$
$$0.61n = 6832$$
$$\frac{0.61n}{0.61} = \frac{6832}{0.61}$$
$$n = 11,200$$
$11,200

**72.** $n \times 12\% = 9624$

$0.12n = 9624$

$$\frac{0.12n}{0.12} = \frac{9624}{0.12}$$

$n = 80,200$

$80,200

**73.** Days $= 29 + 31 + 30 = 90$

Rain days $= 20 + 18 + 16 = 54$

$90 \times n = 54$

$$\frac{90 \times n}{90} = \frac{54}{90}$$

$n = 0.6$

60%

**74.** $600 \times n = 45$

$$\frac{600 \times n}{600} = \frac{45}{600}$$

$n = 0.075$

7.5%

**75.** Difference:

$45.8 - 44 = 1.8$

Percent increase:

$$\frac{1.8}{44} \approx 0.0409$$

$= 4.09\%$

**76.** $n = 3\% \times 18,600$

$n = 0.03 \times 18,600$

$n = 558$

$558

**77.** $38\% \times n = 684$

$0.38n = 684$

$$\frac{0.38n}{0.38} = \frac{684}{0.38}$$

$n = 1800$

$1800

**78.** Rate $= \dfrac{26,000}{650,500}$

$= 0.04 = 4\%$

**79.** Rate $= \dfrac{5010}{83,500}$

$= 0.06 = 6\%$

**80.** Commission $= 7.5\% \times 16,000$

$= 0.075 \times 16,000$

$= 1200$

$1200

**81.** **a.** Discount $= 20\% \times 1595$

$= 0.2 \times 1595$

$= 319$

$319

**b.** Total pay $= 1595 - 319$

$= 1276$

$1276

**82.** **a.** Rebate $= 12\% \times 2125$

$= 0.12 \times 2125$

$= 255$

$255

**b.** Computer costs $= \$2125 - \$255$

$= \$1870$

**83.** **a.** Discount $= 14\% \times 24,000$

$= 0.14 \times 2400$

$= 3360$

$3360

**b.** Sale price $= \$24,000 - \$3360$

$= \$20,640$

**84. a.** 6 months $= \dfrac{6}{12} = \dfrac{1}{2}$ year

$I = P \times R \times T$

$\phantom{I} = 6000 \times 11\% \times \dfrac{1}{2}$

$\phantom{I} = 6000 \times 0.11 \times \dfrac{1}{2}$

$\phantom{I} = 330$

$330

**b.** $I = P \times R \times T$

$\phantom{I} = 6000 \times 11\% \times 2$

$\phantom{I} = 6000 \times 0.11 \times 2$

$\phantom{I} = 1320$

$1320

**85. a.** 3 months $= \dfrac{3}{12} = \dfrac{1}{4}$ year

$I = P \times R \times T$

$\phantom{I} = 3000 \times 8\% \times \dfrac{1}{4}$

$\phantom{I} = 3000 \times 0.08 \times \dfrac{1}{4}$

$\phantom{I} = 240 \times \dfrac{1}{4}$

$\phantom{I} = 60$

$60

**b.** $I = P \times R \times T$

$\phantom{I} = 3000 \times 8\% \times 3$

$\phantom{I} = 3000 \times 0.08 \times 3$

$\phantom{I} = 240 \times 3$

$\phantom{I} = 720$

$720

---

# Chapter 5 Test

**1.** $0.57 = 57\%$

**2.** $0.01 = 1\%$

**3.** $0.008 = 0.8\%$

**4.** $0.139 = 13.9\%$

**5.** $3.56 = 356\%$

**6.** $\dfrac{71}{100} = 71\%$

**7.** $\dfrac{1.8}{100} = 1.8\%$

**8.** $\dfrac{3\frac{1}{7}}{100} = 3\dfrac{1}{7}\%$

**9.** $40\overline{)19.000}$

$$\begin{array}{r} 0.475 \\ \underline{16\ 0} \\ 3\ 00 \\ \underline{2\ 80} \\ 200 \\ \underline{200} \\ 0 \end{array}$$

$\dfrac{19}{40} = 0.475 = 47.5\%$

**10.** $450\overline{)180.0}$

$$\begin{array}{r} 0.4 \\ \underline{180\ 0} \\ 0 \end{array}$$

$\dfrac{180}{450} = 0.4 = 40\%$

**11.** $75\overline{)225}$

$$\begin{array}{r} 3 \\ \underline{225} \\ 0 \end{array}$$

$\dfrac{225}{75} = 3 = 300\%$

**12.** $4\overline{)3.00}$

$$\begin{array}{r} 0.75 \\ \underline{2\ 8} \\ 20 \\ \underline{20} \\ 0 \end{array}$$

$1\dfrac{3}{4} = 1.75 = 175\%$

**13.** $0.1713 = 17.13\%$

**14.** $3.024 = 302.4\%$

**15.** $152\% = \dfrac{152}{100} = \dfrac{38}{25} = 1\dfrac{13}{25}$

**16.** $7\dfrac{3}{4}\% = \dfrac{7\frac{3}{4}}{100}$

$\quad = 7\dfrac{3}{4} \div 100$

$\quad = \dfrac{31}{4} \times \dfrac{1}{100}$

$\quad = \dfrac{31}{100}$

**17.** $n = 17\% \times 157$
$n = 0.17 \times 157$
$n = 26.69$

**18.** $33.8 = 26\% \times n$
$33.8 = 0.26n$
$\dfrac{33.8}{0.26} = \dfrac{0.26n}{0.26}$
$130 = n$

**19.** $n \times 72 = 40$
$\dfrac{n \times 72}{72} = \dfrac{40}{72}$
$n \approx 0.5556 = 55.56\%$

**20.** $n = 0.8\% \times 25{,}000$
$n = 0.008 \times 25{,}000$
$n = 200$

**21.** $16\% \times n = 800$
$0.16n = 800$
$\dfrac{0.16n}{0.16} = \dfrac{800}{0.16}$
$n = 5000$

**22.** $92 = n \times 200$
$\dfrac{92}{200} = \dfrac{n \times 200}{200}$
$n = 0.46$
$n = 46\%$

**23.** $132\% \times 530 = n$
$1.32 \times 530 = n$
$699.6 = n$

**24.** $8 = n \times 350$
$\dfrac{8}{350} = \dfrac{n \times 350}{350}$
$n \approx 0.0229$
$n = 2.29\%$

**25.** $n = 4\% \times 152{,}300$
$n = 0.04 \times 152{,}300$
$n = 6092$
$\$6092$

**26. a.** $33\% \times 457 = n$
$0.33 \times 457 = n$
$150.81 = n$
$\$150.81$

**b.** $457 - 150.81 = 306.19$
$\$306.19$

**27.** $75 = n \times 84$
$\dfrac{75}{84} = \dfrac{n \times 84}{84}$
$n \approx 0.8929 = 89.29\%$

**28.** $56 - 51 = 5$
$5 = n \times 56$
$\dfrac{5}{56} = \dfrac{n \times 56}{56}$
$n \approx 0.0893 = 8.93\%$
$8.93\%$

**29.**  
$5160 = 43\% \times n$  
$5160 = 0.43n$  
$\dfrac{5160}{0.43} = \dfrac{0.43n}{0.43}$  
$12{,}000 = n$  
12,000 registered voters

**30. a.** $I = P \times R \times T$  
$\phantom{I} = 3000 \times 0.16 \times \dfrac{6}{12}$  
$\phantom{I} = 240$  

$240

**b.** $I = P \times R \times T$  
$\phantom{I} = 3000 \times 0.16 \times 2$  
$\phantom{I} = 960$  

$960

# Chapters 1–5 Cumulative Test

**1.**
```
      38
     196
 +  2007
    2241
```

**2.**
```
   23,007
 − 14,563
    8 444
```

**3.**
```
    126
 ×   42
    252
   504
   5292
```

**4.**
```
        89
 36) 3204
     288
     324
     324
       0
```

**5.**
$$\begin{array}{r} 2\frac{1}{4} \\ +\ 3\frac{1}{3} \end{array} \qquad \begin{array}{r} 2\frac{3}{12} \\ +\ 3\frac{4}{12} \\ \hline 5\frac{7}{12} \end{array}$$

**6.** $\dfrac{11}{12} - \dfrac{5}{6} = \dfrac{11}{12} - \dfrac{10}{12} = \dfrac{1}{12}$

**7.** $3\dfrac{17}{36} \times \dfrac{21}{25} = \dfrac{125}{36} \times \dfrac{21}{25}$  
$\phantom{3\dfrac{17}{36} \times \dfrac{21}{25}} = \dfrac{2625}{900}$  
$\phantom{3\dfrac{17}{36} \times \dfrac{21}{25}} = \dfrac{35}{12} = 2\dfrac{11}{12}$

**8.** $\dfrac{5}{12} \div 1\dfrac{3}{4} = \dfrac{5}{12} \div \dfrac{7}{4}$  
$\phantom{\dfrac{5}{12} \div 1\dfrac{3}{4}} = \dfrac{5}{12} \times \dfrac{4}{7}$  
$\phantom{\dfrac{5}{12} \div 1\dfrac{3}{4}} = \dfrac{5}{21}$

**9.** 5731.652 rounds to 5731.7.

**10.**
```
    5.600
    3.210
   18.300
 +  7.008
   34.118
```

**11.**
```
    5.62
 ×  0.3
   1.686
```

**12.**
```
          0.368
 1.4) 0.5152
       42
       95
       84
      112
      112
        0
```

**13.** $\dfrac{78 \div 26}{130 \div 26} = \dfrac{3 \text{ pounds}}{5 \text{ square feet}}$

**14.** $\dfrac{20}{25} \overset{?}{=} \dfrac{300}{375}$

$20 \times 375 \overset{?}{=} 25 \times 300$

$7500 = 7500$ True

**15.** $\dfrac{8}{2.5} = \dfrac{n}{7.5}$

$8 \times 7.5 = 2.5 \times n$

$\dfrac{60}{2.5} = \dfrac{2.5 \times n}{2.5}$

$n = 24$

**16.** $\dfrac{3 \text{ faculty}}{19 \text{ students}} = \dfrac{n \text{ faculty}}{4263 \text{ students}}$

$3 \times 4263 = 19 \times n$

$12{,}789 = 19 \times n$

$\dfrac{12{,}789}{19} = \dfrac{19 \times n}{19}$

$673 \approx n$

$673 = \text{faculty}$

**17.** $0.023 = 2.3\%$

**18.** $\dfrac{46.8}{100} = 46.8\%$

**19.** $1.98 = 198\%$

**20.**
$$\begin{array}{r}
0.0375 \\
80\overline{)3.0000} \\
\underline{2\ 40}\phantom{00} \\
600\phantom{0} \\
\underline{560}\phantom{0} \\
400 \\
\underline{400} \\
0
\end{array}$$

$\dfrac{3}{80} = 0.0375 = 3.75\%$

**21.** $243\% = 2.43$

**22.**
$$\begin{array}{r}
0.75 \\
4\overline{)3.00} \\
\underline{2\ 0}\phantom{0} \\
20 \\
\underline{20} \\
0
\end{array}$$

$6\dfrac{3}{4}\% = 6.75\% = 0.0675$

**23.** What percent of 214 is 38?

$n \times 214 = 38$

$\dfrac{n \times 214}{214} = \dfrac{38}{214}$

$n \approx 0.1776$

$n = 17.76\%$

**24.** $n = 1.7\% \times 6740$

$n = 0.017 \times 6740$

$n = 114.58$

**25.** 219 is 73% of what number?

$219 = 73\% \times n$

$\dfrac{219}{0.73} = \dfrac{0.73 \times n}{0.73}$

$n = 300$

**26.** $114\% \times 630 = n$

$1.14 \times 630 = n$

$718.2 = n$

**27.** $9000 \times 0.07 = 630$

$900 - 630 = \$8370$

**28.** $28\% \times n = 896$

$0.28n = 896$

$\dfrac{0.28n}{0.28} = \dfrac{896}{0.28}$

$n = 3200$

3200 students

**29.** $8.86 - 7.96 = 0.9$

$\qquad 7.96 \times n = 0.9$

$\qquad \dfrac{7.96 \times n}{7.96} = \dfrac{0.9}{7.96}$

$\qquad\qquad n \approx 0.1131$

$\qquad\qquad\quad = 11.31\%$ increase

**30.** $I = P \times R \times T$

$\qquad = 1600 \times 0.11 \times 2$

$\qquad = 352$

$\quad$ \$352

# MEASUREMENT

## Pretest Chapter 6

**1.** $19 \text{ ft} \times \dfrac{12 \text{ in.}}{1 \text{ ft}} = 228 \text{ in.}$

**2.** $5 \text{ gal} \times \dfrac{4 \text{ qt}}{1 \text{ gal}} \times \dfrac{2 \text{ pt}}{1 \text{ qt}} = 40 \text{ pt}$

**3.** $3 \text{ mi} \times \dfrac{1760 \text{ yd}}{1 \text{ mi}} = 5280 \text{ yd}$

**4.** $3.2 \text{ tons} \times \dfrac{2000 \text{ pounds}}{1 \text{ ton}} = 6400 \text{ pounds}$

**5.** $22 \text{ min} \times \dfrac{60 \text{ sec}}{1 \text{ min}} = 1320 \text{ sec}$

**6.** $6 \text{ gal} \times \dfrac{4 \text{ qt}}{1 \text{ gal}} = 24 \text{ qt}$

**7.** $6.75 \text{ km} = 6750 \text{ m}$
(move 3 places right)

**8.** $73.9 \text{ m} = 7390 \text{ cm}$
(move 2 places right)

**9.** $986 \text{ mm} = 98.6 \text{ cm}$
(move 1 place left)

**10.** $27 \text{ mm} = 0.027 \text{ m}$
(move 3 places left)

**11.** $5296 \text{ mm} = 529.6 \text{ cm}$
(move 1 place left)

**12.** $482 \text{ m} = 0.482 \text{ km}$
(move 3 places left)

**13.** $1.2 \text{ km} + 192 \text{ m} + 984 \text{ m}$
$= 1200 \text{ m} + 192 \text{ m} + 984 \text{ m}$
$= 2376 \text{ m}$

**14.** $3862 \text{ cm} + 9342 \text{ mm} + 46.3 \text{ m}$
$= 38.62 \text{ m} + 9.342 \text{ m} + 46.3 \text{ m}$
$= 94.262 \text{ m}$

**15.** $5.66 \text{ L} = 5660 \text{ mL}$ (move 3 places left)

**16.** $7835 \text{ g} = 7.835 \text{ kg}$ (move 3 places left)

**17.** $56.3 \text{ kg} = 0.0563 \text{ t}$ (move 3 places left)

**18.** $4.8 \text{ kL} = 4800 \text{ L}$ (move 3 places right)

**19.** $568 \text{ mg} = 0.568 \text{ g}$ (move 3 places left)

**20.** $8.9 \text{ L} = 8900 \text{ mL} = 8900 \text{ cm}^3$

**21.** $14 \text{ cm} \times \dfrac{0.394 \text{ in.}}{1 \text{ cm}} = 5.52 \text{ in.}$

**22.** $4.2 \text{ ft} \times \dfrac{0.305 \text{ m}}{1 \text{ ft}} = 1.28 \text{ m}$

**23.** $96 \text{ km} \times \dfrac{0.62 \text{ mi}}{1 \text{ km}} = 59.52 \text{ mi}$

**24.** $482 \text{ gal} \times \dfrac{3.79 \text{ L}}{1 \text{ gal}} = 1826.78 \text{ L}$

**25.** $1.4 \text{ oz} \times \dfrac{28.35 \text{ g}}{1 \text{ oz}} = 39.69 \text{ g}$

**26.** $54 \text{ kg} \times \dfrac{2.2 \text{ lb}}{1 \text{ kg}} = 118.8 \text{ lb}$

**27.** $7\dfrac{4}{5} \text{ yd} + 4\dfrac{1}{5} \text{ yd} + 6\dfrac{3}{5} \text{ yd}$

$= 18\dfrac{3}{5} \text{ yd}$

$= \dfrac{93}{5} \text{ yd} \times \dfrac{3 \text{ feet}}{1 \text{ yd}}$

$= \dfrac{279}{5} \text{ feet}$

$= 55.8 \text{ feet}$

**28. a.** $F = 1.8 \times C + 32$
$= 1.8 \times 35 + 32$
$= 95$
$95°\text{F}$

**b.** No

**29.**
$2 \times 95 \text{ km} = 190 \text{ km}$
$190 \text{ km} \times \dfrac{0.62 \text{ mi}}{1 \text{ km}} = 117.8 \text{ mi}$
$130 - 117.8 = 12.2 \text{ miles farther}$

**30.** $\dfrac{1.5 \text{ qt}}{1 \text{ min}} \times \dfrac{1 \text{ gal}}{4 \text{ qt}} \times \dfrac{60 \text{ min}}{1 \text{ hr}} = 22.5 \text{ gal/hr}$

---

## 6.1   Exercises

**1.** We know that each mile is 5280 feet. Each foot is 12 inches. So we know that one mile is $5280 \times 12 = 63{,}360$ inches. The unit fraction is $\frac{63{,}360 \text{ inches}}{1 \text{ mile}}$. So we multiply $23 \text{ miles} \times \frac{63{,}360 \text{ inches}}{1 \text{ mile}}$. The mile unit divides out. We obtain $1{,}457{,}280$ inches. Thus, $23 \text{ miles} = 1{,}457{,}280$ inches. (Whew!!)

**3.** $1760 \text{ yards} = 1 \text{ mile}$

**5.** $1 \text{ ton} = 2000 \text{ pounds}$

**7.** $4 \text{ quarts} = 1 \text{ gallon}$

**9.** $1 \text{ quart} = 2 \text{ pints}$

**11.** $21 \text{ feet} = 21 \text{ feet} \times \dfrac{1 \text{ yard}}{3 \text{ feet}}$
$= 7 \text{ yards}$

**13.** $108 \text{ inches} = 108 \text{ inches} \times \dfrac{1 \text{ foot}}{12 \text{ inches}}$
$= 9 \text{ feet}$

**15.** $10{,}560 \text{ feet}$
$= 10{,}560 \text{ feet} \times \dfrac{1 \text{ mile}}{5280 \text{ feet}}$
$= 2 \text{ miles}$

**17.** $7 \text{ miles} = 7 \text{ miles} \times \dfrac{1760 \text{ yards}}{1 \text{ mile}}$
$= 12{,}320 \text{ yards}$

**19.** $12 \text{ feet} = 12 \text{ feet} \times \dfrac{12 \text{ inches}}{1 \text{ foot}}$
$= 144 \text{ inches}$

**21.** $16 \text{ cups}$
$= 16 \text{ cups} \times \dfrac{8 \text{ fluid ounces}}{1 \text{ cup}}$
$= 128 \text{ fluid ounces}$

**23.** $75 \text{ inches} = 75 \text{ inches} \times \dfrac{1 \text{ foot}}{12 \text{ inches}}$
$= 6.25 \text{ feet}$

**25.** $192 \text{ ounces} = 192 \text{ ounces} \times \dfrac{1 \text{ pound}}{16 \text{ ounces}}$
$= 12 \text{ pounds}$

**27.** $12{,}000 \text{ pounds} \times \dfrac{1 \text{ ton}}{2000 \text{ pounds}} = 6 \text{ tons}$

**29.** $2.25 \text{ pounds} = 2.25 \text{ pounds} \times \dfrac{16 \text{ ounces}}{1 \text{ pound}}$

$= 36 \text{ ounces}$

**31.** $7 \text{ gallons} = 7 \text{ gallons} \times \dfrac{4 \text{ quarts}}{1 \text{ gallon}}$

$= 28 \text{ quarts}$

**33.** $48 \text{ quarts} = 48 \text{ quarts} \times \dfrac{1 \text{ gallon}}{4 \text{ quarts}}$

$= 12 \text{ gallons}$

**35.** $31 \text{ pints} = 31 \text{ pints} \times \dfrac{2 \text{ cups}}{1 \text{ pint}}$

$= 62 \text{ cups}$

**37.** $8 \text{ gallons}$

$= 8 \text{ gallons} \times \dfrac{4 \text{ quarts}}{1 \text{ gallon}} \times \dfrac{2 \text{ pints}}{1 \text{ quart}}$

$= 64 \text{ pints}$

**39.** $12 \text{ weeks} = 12 \text{ weeks} \times \dfrac{7 \text{ days}}{1 \text{ week}}$

$= 84 \text{ days}$

**41.** $18 \text{ hours}$

$= 18 \text{ hours} \times \dfrac{60 \text{ minutes}}{1 \text{ hour}} \times \dfrac{60 \text{ seconds}}{1 \text{ minute}}$

$= 64{,}800 \text{ seconds}$

**43.** $26 \text{ ounces} \times \dfrac{1 \text{ pound}}{16 \text{ ounces}} \times \dfrac{\$6.00}{1 \text{ pound}} = \$9.75$

**45.** $26.2 \text{ miles} \times \dfrac{5280 \text{ feet}}{1 \text{ mile}} = 138{,}336 \text{ feet}$

**47.** $44 \text{ minutes} \times \dfrac{60 \text{ seconds}}{1 \text{ minute}} = 2640 \text{ seconds}$

$44 \text{ minutes } 25 \text{ seconds} = 2640 + 25$

$= 2665 \text{ seconds}$

**49. a.** $2 \text{ feet} \times \dfrac{12 \text{ inches}}{1 \text{ foot}} = 24 \text{ inches}$

$2 \text{ feet } 3 \text{ inches} = 24 + 3$

$= 27 \text{ inches}$

$3 \text{ feet} \times \dfrac{12 \text{ inches}}{1 \text{ foot}} = 36 \text{ inches}$

$3 \text{ feet } 9 \text{ inches} = 36 + 9$

$= 45 \text{ inches}$

$\text{perimeter} = 2\,(27) + 2\,(45)$

$= 54 + 90$

$= 144 \text{ inches}$

**b.** $\text{Cost} = \$0.85 \times 144 = \$122.40$

**51.** $7200 \text{ quarts} \times \dfrac{4 \text{ cups}}{1 \text{ quart}} = 28{,}800 \text{ cups}$

**53.** $12{,}800 \text{ nautical miles} \times \dfrac{38 \text{ land miles}}{33 \text{ nautical miles}}$

$\approx 14{,}739 \text{ land miles}$

## Cumulative Review Problems

**55.** $560 - 515 = \$45 \text{ per month}$

$\text{Savings} = 45 \times 20 \times 12$

$= \$10{,}800$

**57.** $\dfrac{n}{115} = \dfrac{7}{5}$

$n \times 5 = 7 \times 115$

$\dfrac{n \times 5}{5} = \dfrac{805}{5}$

$n = 161$

$161 \text{ miles}$

## 6.2 Exercises

**1.** hecto-

**3.** deci-

**5.** kilo-

**7.** 46 centimeters = 460 millimeters

**9.** 5.2 kilometers = 5200 meters

**11.** 1670 millimeters = 1.67 meter

**13.** 7.32 centimeters = 0.0732 meter

**15.** 2 kilometers = 200,000 centimeters

**17.** 78,000 millimeters = 0.078 kilometer

**19.** 35 mm = 3.5 cm = 0.035 m

**21.** 3582 mm = 3.582 m = 0.003582 km

**23.** (a)

**25.** (c)

**27.** (b)

**29.** 390 decimeters = 39 meters

**31.** 198 millimeters = 1.98 decimeters

**33.** 48.2 meters = 0.482 hectometers

**35.** 243 m + 2.7 km + 312 m
$$= 243 \text{ m} + 2700 \text{ m} + 312 \text{ m}$$
$$= 3255 \text{ m}$$

**37.** 5.2 cm + 361 cm + 968 mm
$$= 5.2 \text{ cm} + 361 \text{ cm} + 96.8 \text{ cm}$$
$$= 463 \text{ cm}$$

**39.** 15 mm + 2 dm + 42 cm
$$= 1.5 \text{ cm} + 20 \text{ cm} + 42 \text{ cm}$$
$$= 63.5 \text{ cm}$$

**41.** 0.95 cm + 1.35 cm + 2.464 mm
$$= 0.95 \text{ cm} + 1.35 \text{ cm} + 0.2464 \text{ cm}$$
$$= 2.5464 \text{ cm or } 25.464 \text{ mm}$$

**43.** 4.8 cm + 607 cm + 834 mm
$$= 4.8 \text{ cm} + 607 \text{ cm} + 83.4 \text{ cm}$$
$$= 695.2 \text{ cm}$$

**45.** 46 m + 986 cm + 0.884 km
$$= 46 \text{ m} + 9.86 \text{ m} + 884 \text{ m}$$
$$= 939.86 \text{ m}$$

**47.** 96.4 centimeters = 0.964 meters

**49.** false

**51.** true

**53.** true

**55.** true

**57.** **a.** 4818 meters = 481,800 centimeters

  **b.** 4818 meters = 4.818 kilometers

**59.** 0.000000254 centimeters
$\qquad$ = 0.00000000254 meters

**61.**  $\qquad$ 36,980,000
$\quad -\ 31,940,000$
$\quad\overline{\qquad 5,040,000}$  metric tons

**63.** 18,560,000 × 1000
$\qquad$ = 18,560,000,000 kilograms

**65.**  $\qquad$ 36,980,000
$\quad -\ 12,100,000$
$\quad\overline{\qquad 24,880,000}$

$\qquad$ Percent $= \dfrac{24,880,000}{12,100,000}$
$\qquad\qquad \approx 2.0562$
$\qquad\qquad = 205.62\%$

Year 2010 $= 305.62 \times 36,980,000$
$\qquad\qquad \approx 113,020,000$ metric tons

## Cumulative Review Problems

**67.** $n = 0.03\%$ of 5900
$\qquad n = 0.0003 \times 5900$
$\qquad n = 1.77$

## 6.3   Exercises

**1.** 1 kL

**3.** 1 mg

**5.** 1 g

**7.** 58 kL = 58,000 L

**9.** 2.6 L = 2600 mL

**11.** 18.9 mL = 0.0189 L

**13.** 752 L = 0.752 kL

**15.** 2.43 kL = 2,430,000 mL

**17.** 82 mL = 82 cm$^3$

**19.** 5261 mL = 0.005261 kL

**21.** 74 L = 74,000 cm$^3$

**23.** 1620 g = 1.62 kg

**25.** 35 mg = 0.035 g

**27.** 6328 mg = 6.328 g

**29.** 2.92 kg = 2920 g

**31.** 17 t = 17,000 kg

**33.** 7 mL = 0.007 L = 0.000007 kL

**35.** 128 cm$^3$ = 0.128 L = 0.000128 kL

**37.** 0.033 kg = 33 g = 33,000 mg

**39.** 2.58 metric tons = 2580 kg
$\qquad\qquad\qquad\qquad = 2,580,000$ g

**41.** (b)

**43.** (a)

**45.** $83 \text{ L} + 822 \text{ mL} + 30.1 \text{ L}$
$$= 83 \text{ L} + 0.822 \text{ L} + 30.1 \text{ L}$$
$$= 113.922 \text{ L}$$

**47.** $20 \text{ g} + 52 \text{ mg} + 1.5 \text{ kg}$
$$= 20 \text{ g} + 0.052 \text{ g} + 1500 \text{ g}$$
$$= 1520.052 \text{ g}$$

**49.** true

**51.** false

**53.** false

**55.** true

**57.**      $75 \text{ kg} = 75,000 \text{ g}$
$$\frac{n}{7.25} = \frac{75,000}{5000}$$
$$n \times 5000 = 7.25 \times 75,000$$
$$\frac{n \times 5000}{5000} = \frac{543,750}{5000}$$
$$n = \$108.75$$

**59.**      $0.4 \text{ L} = 400 \text{ mL}$
$$850 \times 400 = 340,000$$
$$\$340,000$$

**61.**              $0.45 \text{ t} = 450 \text{ kg}$
$$450 \text{ kg} \times \frac{\$22,450}{1 \text{ kg}} = \$10,102,500$$

**63.** 5632 picograms $= 0.005632$ micrograms

**65.**    $\begin{array}{r} 3860 \\ - \ 1810 \\ \hline 2050 \text{ kg} \end{array}$

**67.** $\text{Value} = \dfrac{24,000,000}{2610}$
$$\approx \$9195.40 \text{ per kilogram}$$
or $\$9.20$ per gram.

**69.** 1997

---

## Cumulative Review Problems

**71.** $\text{percent} = \dfrac{14}{70}$
$$= 0.20$$
$$= 20\%$$

**73.** $n = 1.7\%$ of $18,900$
$$n = 0.017 \times 18,900$$
$$n = 321.30$$
$$\$321.30$$

---

## 6.4   Exercises

**1.** $7 \text{ ft} \times \dfrac{0.305 \text{ m}}{1 \text{ ft}} \approx 2.14 \text{ m}$

**3.** $9 \text{ in.} \times \dfrac{2.54 \text{ cm}}{1 \text{ in.}} \approx 22.86 \text{ cm}$

**5.** $14 \text{ m} \times \dfrac{1.09 \text{ yards}}{1 \text{ m}} \approx 15.26 \text{ yd}$

**7.** $30.8 \text{ yd} \times \dfrac{0.914 \text{ m}}{1 \text{ yd}} \approx 28.15 \text{ m}$

**9.** $82 \text{ mi} \times \dfrac{1.61 \text{ km}}{1 \text{ mi}} \approx 132.02 \text{ km}$

**11.** $16 \text{ ft} \times \dfrac{0.305 \text{ m}}{1 \text{ ft}} \approx 4.88 \text{ m}$

**13.** $17.5 \text{ cm} \times \dfrac{0.394 \text{ in.}}{1 \text{ cm}} \approx 6.90 \text{ in.}$

**15.** $200 \text{ m} \times \dfrac{1.09 \text{ yd}}{1 \text{ m}} \approx 218 \text{ yd}$

**17.** $65 \text{ in.} \times \dfrac{2.54 \text{ cm}}{1 \text{ in.}} \approx 165.1 \text{ cm}$

**19.** $280 \text{ gal} \times \dfrac{3.79 \text{ L}}{1 \text{ gal}} \approx 1061.2 \text{ L}$

**21.** $23 \text{ qt} \times \dfrac{0.946 \text{ L}}{1 \text{ qt}} \approx 21.76 \text{ L}$

**23.** $19 \text{ L} \times \dfrac{0.264 \text{ gal}}{1 \text{ L}} \approx 5.02 \text{ gal}$

**25.** $4.5 \text{ L} \times \dfrac{1.06 \text{ qt}}{1 \text{ L}} \approx 4.77 \text{ qt}$

**27.** $130 \text{ lb} \times \dfrac{0.454 \text{ kg}}{1 \text{ lb}} \approx 59.02 \text{ kg}$

**29.** $126 \text{ g} \times \dfrac{0.0353 \text{ oz}}{1 \text{ g}} \approx 4.45 \text{ oz}$

**31.** $1260 \text{ cm} \times \dfrac{0.394 \text{ in.}}{1 \text{ cm}} \times \dfrac{1 \text{ ft}}{12 \text{ in.}} \approx 41.37 \text{ ft}$

**33.** $\dfrac{800 \text{ mi}}{1 \text{ hr}} \times \dfrac{1 \text{ hr}}{3600 \text{ sec}} \times \dfrac{5280 \text{ ft}}{1 \text{ mi}} \approx 1173 \text{ ft/sec}$

**35.** $\dfrac{400 \text{ ft}}{1 \text{ sec}} \times \dfrac{3600 \text{ sec}}{1 \text{ hr}} \times \dfrac{1 \text{ mi}}{5280 \text{ ft}} \approx 273 \text{ mi/hr}$

**37.**     $13 \text{ mm} = 1.3 \text{ cm}$

$1.3 \text{ cm} \times \dfrac{0.394 \text{ in.}}{1 \text{ cm}} \approx 0.51 \text{ in.}$

**39.** $F = 1.8 \times C + 32$
$= 1.8 \times 85 + 32$
$= 185$
$185°\text{F}$

**41.** $F = 1.8 \times C + 32$
$= 1.8 \times 12 + 32$
$= 53.6$
$53.6°\text{F}$

**43.** $C = \dfrac{5 \times F - 160}{9}$
$= \dfrac{5 \times 140 - 160}{9}$
$= \dfrac{540}{9}$
$= 60$
$60°\text{C}$

**45.** $C = \dfrac{5 \times F - 160}{9}$
$= \dfrac{5 \times 40 - 160}{9}$
$= \dfrac{40}{9}$
$\approx 4.44$
$4.44°\text{C}$

**47.** $67 \text{ mi} \times \dfrac{1.61 \text{ km}}{1 \text{ mi}} \approx 107.87 \text{ km}$

$\text{Total} = 36 + 107.87$
$= 143.87 \text{ km}$

**49.** $15 \text{ gal} \times \dfrac{3.79 \text{ L}}{1 \text{ gal}} \approx 56.85 \text{ L}$

$\text{Difference} = 56.85 - 38$
$= 18.85 \text{ L}$

**51.** $635 \text{ kg} \times \dfrac{2.2 \text{ lb}}{1 \text{ kg}} \approx 1397 \text{ lb}$

**53.** $F = 1.8 \times C + 32$
4 A.M.: $F = 1.8 \times 19 + 32 = 66.2$
7 A.M.: $F = 1.8 \times 45 + 32 = 113$

It is $66.2°\text{F}$ at 4 A.M.
The temperature may reach $113°\text{F}$
after 7 A.M.

**55.** $96{,}550 \text{ km} \times \dfrac{0.62 \text{ mi}}{1 \text{ km}} \approx 59{,}861 \text{ mi}$

**57.** $28 \times 2.54 \times 2.54 \approx 180.6448 \text{ sq cm}$

**59.**   American: $8 \times 4 \times 28 = \$896$
Germany: $8 \text{ yd} \times \dfrac{0.914 \text{ m}}{1 \text{ yd}} \approx 7.312 \text{ m}$

$4 \text{ yd} \times \dfrac{0.914 \text{ m}}{1 \text{ yd}} \approx 3.656 \text{ m}$

$7.312 \times 3.656 \times 30 \approx \$802$
German carpet is cheaper by
$896 - 802 = \$94.$

## Cumulative Review Problems

**61.** $2^3 \times 6 - 4 + 3 = 8 \times 6 - 4 + 3$
$\qquad = 48 - 4 + 3$
$\qquad = 44 + 3$
$\qquad = 47$

**63.** $\dfrac{1}{2} \cdot \dfrac{3}{4} - \dfrac{1}{5}\left(\dfrac{1}{2}\right)^2 = \dfrac{1}{2} \cdot \dfrac{3}{4} - \dfrac{1}{5}\left(\dfrac{1}{4}\right)$

$\qquad = \dfrac{3}{8} - \dfrac{1}{20}$

$\qquad = \dfrac{3}{8} \times \dfrac{5}{5} - \dfrac{1}{20} \times \dfrac{2}{2}$

$\qquad = \dfrac{15}{40} - \dfrac{2}{40}$

$\qquad = \dfrac{13}{40}$

## Putting Your Skills To Work

**1.** $32.5 \text{ knots} \times \dfrac{38 \text{ miles}}{33 \text{ knots}}$
$\qquad \approx 37.4 \text{ miles per hour}$

## 6.5   Exercises

**1.** $14\dfrac{1}{3} + 8\dfrac{2}{3} + 13 = 36 \text{ in.}$

$36 \text{ in.} \times \dfrac{1 \text{ ft}}{12 \text{ in.}} = 3 \text{ ft}$

**3.**   $86 \text{ yd} + 77 \text{ yd} = 163 \text{ yd}$

$522 \text{ ft} \times \dfrac{1 \text{ yd}}{3 \text{ ft}} = 174 \text{ yd}$

$174 \text{ yd} - 163 \text{ yd} = 11 \text{ yd}$

11 yards left over

**5.** Perimeter $= 2 \times 87 + 2 \times 152$
$\qquad = 174 \times 304$
$\qquad = 478 \text{ cm}$
$\qquad = 4.78 \text{ m}$

$\text{Cost} = 4.78 \text{ m} \times \dfrac{\$7.00}{1 \text{ m}}$

$\qquad = \$33.46$

**7.** $12.4 \text{ m} = 1240 \text{ cm}$

$\dfrac{1240}{4} = 310$

$310 \text{ cm/piece}$

**9.**   $400 \times \dfrac{1}{10} = 40 \text{ m}$

$40 \text{ m} \times \dfrac{3.28 \text{ ft}}{1 \text{ m}} \approx 131.2 \text{ ft}$

$131.2 \text{ ft}$

**11.**   $340 \text{ mg} = 0.34 \text{ g}$
$0.34 \times 7 = 2.38 \text{ g}$

**13.**   $F = 1.8 \times 27 + 32$
$\qquad = 80.6°\text{F}$

$85 - 80.6 = 4.4$

The discrepancy is $4.4°\text{F}$. The sign in the store is $4.4°\text{F}$ less than it should be.

**15.**
$$F = 1.8 \times 180 + 32$$
$$= 356°F$$
$$356 - 350 = 6$$
The difference is 6°F. The temperature reading of 180°C is hotter.

**17. a.**
$$\frac{520 \text{ mi}}{8 \text{ hr}} = 65 \text{ mi/hr}$$
$$\frac{65 \text{ mi}}{1 \text{ hr}} \times \frac{1.61 \text{ km}}{1 \text{ mi}} \approx 105 \text{ km/hr}$$

**b.** Probably not, but we cannot be sure. They may have gone faster or slower than their average speed.

**19.** $2 \times \dfrac{1 \text{ pt}}{1 \text{ min}} \times \dfrac{1 \text{ qt}}{2 \text{ pt}} \times \dfrac{1 \text{ gal}}{4 \text{ qt}} \times \dfrac{60 \text{ min}}{1 \text{ hr}}$
$$= 15 \text{ gal/hr}$$

**21.** $3.4 \text{ tons} \times \dfrac{2000 \text{ lb}}{1 \text{ ton}} \times \dfrac{\$0.015}{1 \text{ lb}} = \$102$

**23.** $16 \text{ oz} - 5 \text{ oz} = 11 \text{ oz of raisins}$
$$11 \text{ oz} \times \frac{28.35 \text{ g}}{1 \text{ oz}} \approx 311.85 \text{ grams of raisins}$$

**25. a.** $11 \text{ L} \times \dfrac{1.06 \text{ qt}}{1 \text{ L}} \approx 11.66 \text{ qt}$
$$16 - 11.66 = 4.34 \text{ qt}$$
She bought 4.34 qt extra.

**b.** $11.66 \times 2.89 \approx \$33.70$

**27. a.** $392 \text{ km} \times \dfrac{1 \text{ L}}{6 \text{ km}} \times \dfrac{\$0.78}{1 \text{ L}} \approx \$5.46$

**b.** $56 \text{ km/L} = \dfrac{56 \text{ km}}{1 \text{ L}} \times \dfrac{0.62 \text{ m}}{1 \text{ km}} \times \dfrac{1 \text{ L}}{0.264 \text{ gal}}$
$$\approx \frac{132 \text{ mi}}{1 \text{ gal}}$$
132 mi/gal

**29.** $\dfrac{240,000 \text{ gal}}{1 \text{ hr}} \times \dfrac{8 \text{ pt}}{1 \text{ gal}} \times \dfrac{1 \text{ hr}}{3600 \text{ sec}}$
$$= 533\frac{1}{3} \text{ pt/sec}$$
Yes

## Cumulative Review Problems

**31.** $\dfrac{n}{16} = \dfrac{2}{50}$
$$50n = 16 \times 2$$
$$\frac{50n}{50} = \frac{32}{50}$$
$$n = 0.64$$

**33.** $6 \text{ in.} \times \dfrac{7.75 \text{ mi}}{3 \text{ in.}} = 15.5 \text{ mi}$

## Chapter 6 Review Problems

**1.** $33 \text{ ft} \times \dfrac{1 \text{ yd}}{3 \text{ ft}} = 11 \text{ yd}$

**2.** $27 \text{ ft} \times \dfrac{1 \text{ yd}}{3 \text{ ft}} = 9 \text{ yd}$

**3.** $5 \text{ mi} \times \dfrac{1760 \text{ yd}}{1 \text{ mi}} = 8000 \text{ yd}$

**4.** $6 \text{ mi} \times \dfrac{1760 \text{ yd}}{1 \text{ mi}} = 10,560 \text{ yd}$

**5.** $90 \text{ in.} \times \dfrac{1 \text{ ft}}{12 \text{ in.}} = 7.5 \text{ ft}$

**6.** $78 \text{ in.} \times \dfrac{1 \text{ ft}}{12 \text{ in.}} = 6.5 \text{ ft}$

**7.** $15,840 \text{ ft} \times \dfrac{1 \text{ mi}}{5280 \text{ ft}} = 3 \text{ mi}$

**8.** $10,560 \text{ ft} \times \dfrac{1 \text{ mi}}{5280 \text{ ft}} = 2 \text{ mi}$

**9.** $7 \text{ tons} \times \dfrac{2000 \text{ lb}}{1 \text{ ton}} = 14,000 \text{ lb}$

**10.** $4 \text{ tons} \times \dfrac{2000 \text{ lb}}{1 \text{ ton}} = 8000 \text{ lb}$

**11.** $92 \text{ oz} \times \dfrac{1 \text{ lb}}{16 \text{ oz}} = 5.75 \text{ lb}$

**12.** $100 \text{ oz} \times \dfrac{1 \text{ lb}}{16 \text{ oz}} = 6.25 \text{ lb}$

**13.** $15 \text{ gal} \times \dfrac{4 \text{ qt}}{1 \text{ gal}} = 60 \text{ qt}$

**14.** $21 \text{ gal} \times \dfrac{4 \text{ qt}}{1 \text{ gal}} = 84 \text{ qt}$

**15.** $31 \text{ pt} \times \dfrac{1 \text{ qt}}{2 \text{ pt}} = 15.5 \text{ qt}$

**16.** $27 \text{ pt} \times \dfrac{1 \text{ qt}}{2 \text{ pt}} = 13.5 \text{ qt}$

**17.** $56 \text{ cm} = 560 \text{ mm}$

**18.** $29 \text{ cm} = 290 \text{ mm}$

**19.** $1763 \text{ mm} = 176.3 \text{ mm}$

**20.** $2598 \text{ mm} = 259.8 \text{ cm}$

**21.** $9.2 \text{ m} = 920 \text{ cm}$

**22.** $7.4 \text{ m} = 740 \text{ cm}$

**23.** $9 \text{ km} = 9000 \text{ m}$

**24.** $8 \text{ km} = 8000 \text{ m}$

**25.** $6.2 \text{ m} + 121 \text{ cm} + 0.52 \text{ m}$
$= 6.2 \text{ m} + 1.21 \text{ m} + 0.52 \text{ m}$
$= 7.93 \text{ m}$

**26.** $9.8 \text{ m} + 673 \text{ cm} + 0.48 \text{ m}$
$= 9.8 \text{ m} + 6.73 \text{ m} + 0.48 \text{ m}$
$= 17.01 \text{ m}$

**27.** $0.024 \text{ km} + 1.8 \text{ m} + 983 \text{ cm}$
$= 24 \text{ m} + 1.8 \text{ m} + 9.83 \text{ m}$
$= 35.63 \text{ m}$

**28.** $0.078 \text{ km} + 5.5 \text{ m} + 609 \text{ cm}$
$= 78 \text{ m} + 5.5 \text{ m} + 6.09 \text{ m}$
$= 89.59 \text{ m}$

**29.** $17 \text{ kL} = 17,000 \text{ L}$

**30.** $23 \text{ kL} = 23.000 \text{ L}$

**31.** $196 \text{ kg} = 196,000 \text{ g}$

**32.** $721 \text{ kg} = 721,000 \text{ g}$

**33.** $778 \text{ mg} = 0.778 \text{ g}$

**34.** $459 \text{ mg} = 0.459 \text{ g}$

**35.** $76 \text{ kg} = 76,000 \text{ g}$

**36.** $41 \text{ kg} = 41,000 \text{ g}$

**37.** $765 \text{ cm}^3 = 765 \text{ mL}$

**38.** $423 \text{ cm}^3 = 423 \text{ mL}$

**39.** 2.43 L = 2430 mL = 2430 cm³

**40.** 1.93 L = 1930 mL = 1930 cm³

**41.** $42 \text{ kg} \times \dfrac{2.2 \text{ lb}}{1 \text{ kg}} \approx 92.4 \text{ lb}$

**42.** $9 \text{ ft} \times \dfrac{0.305 \text{ m}}{1 \text{ ft}} \approx 2.75 \text{ m}$

**43.** $13 \text{ oz} \times \dfrac{28.35 \text{ g}}{1 \text{ oz}} \approx 368.55 \text{ g}$

**44.** $1.3 \text{ ft} \times \dfrac{12 \text{ in.}}{1 \text{ ft}} \times \dfrac{2.54 \text{ cm}}{1 \text{ in.}} \approx 39.62 \text{ cm}$

**45.** $14 \text{ cm} \times \dfrac{0.394 \text{ in.}}{1 \text{ cm}} \approx 5.52 \text{ in.}$

**46.** $18 \text{ cm} \times \dfrac{0.394 \text{ in.}}{1 \text{ cm}} \approx 7.09 \text{ in.}$

**47.** $20 \text{ lb} \times \dfrac{0.454 \text{ kg}}{1 \text{ lb}} \approx 9.08 \text{ kg}$

**48.** $30 \text{ lb} \times \dfrac{0.454 \text{ kg}}{1 \text{ lb}} \approx 13.62 \text{ kg}$

**49.** $12 \text{ yd} \times \dfrac{0.914 \text{ m}}{1 \text{ yd}} \approx 10.97 \text{ m}$

**50.** $14 \text{ yd} \times \dfrac{0.914 \text{ m}}{1 \text{ yd}} \approx 12.80 \text{ m}$

**51.** $\dfrac{80 \text{ km}}{1 \text{ hr}} \times \dfrac{0.62 \text{ mi}}{1 \text{ km}} \approx 49.6 \text{ mi/hr}$

**52.** $\dfrac{70 \text{ km}}{1 \text{ hr}} \times \dfrac{0.62 \text{ mi}}{1 \text{ km}} \approx 43.4 \text{ mi/hr}$

**53.** $F = 1.8 \times C + 32$
$\phantom{F} = 1.8 \times 15 + 32$
$\phantom{F} = 59$
$59°F$

**54.** $F = 1.8 \times C + 32$
$\phantom{F} = 1.8 \times 25 + 32$
$\phantom{F} = 77$
$77°F$

**55.** $C = \dfrac{5 \times F - 160}{9}$
$\phantom{C} = \dfrac{5 \times 221 - 160}{9}$
$\phantom{C} = 105$
$105°C$

**56.** $C = \dfrac{5 \times F - 160}{9}$
$\phantom{C} = \dfrac{5 \times 185 - 160}{9}$
$\phantom{C} = 85$
$85°C$

**57.** $C = \dfrac{5 \times F - 160}{9}$
$\phantom{C} = \dfrac{5 \times 32 - 160}{9}$
$\phantom{C} = 0$
$0°C$

**58.** 1.76 cm + 4.32 mm + 0.93 cm
$\phantom{58.} = 1.76 \text{ cm} + 0.432 \text{ cm} + 0.93 \text{ cm}$
$\phantom{58.} = 3.12 \text{ cm}$

**59. a.** $7\dfrac{2}{3} \text{ ft} + 4\dfrac{1}{3} \text{ ft} + 5 \text{ ft} = 16\dfrac{3}{3} \text{ feet}$
$\phantom{59. a. 7\frac{2}{3} ft} = 17 \text{ feet}$

**b.** $17 \text{ feet} \times \dfrac{12 \text{ in.}}{1 \text{ foot}} = 204 \text{ inches}$

**60. a.** $16 \text{ m} + 84 \text{ m} + 16 \text{ m} + 84 \text{ m} = 200 \text{ m}$

**b.** $200 \text{ m} = 0.2 \text{ km}$

**61.** $450 \text{ g} \times \dfrac{0.0353 \text{ oz}}{1 \text{ g}} \approx 15.89 \text{ oz}$

$\dfrac{\$0.14}{1 \text{ oz}} \times 15.89 \text{ oz} \approx \$2.22$

**62.** $12.6 \text{ m} \times \dfrac{3.28 \text{ ft}}{1 \text{ m}} \approx 41.328 \text{ ft}$

Yes

$43 - 41.328 = 1.672 \text{ ft extra}$

**63.** $\qquad 90 \text{ km/hr} \times 3 = 270 \text{ km}$

$270 \text{ km} \times \dfrac{0.62 \text{ mi}}{1 \text{ km}} \approx 167.4 \text{ mi}$

Difference $= 200 - 167.4 = 32.6 \text{ mi}$

**64.** $F = 1.8 \times C + 32$

$\quad = 1.8 \times 185 + 32$

$\quad = 365°$

$390° - 365° = 25°F$

**65.** $\qquad 19 \text{ m} = 1900 \text{ cm}$

Bottom part $= 1900 \times \dfrac{1}{5}$

$\qquad\qquad\quad = 380 \text{ cm}$

**66.** $\dfrac{100 \text{ m}}{13.0 \text{ sec}} \times \dfrac{3.28 \text{ ft}}{1 \text{ m}} \approx 25.2 \text{ ft/sec}$

**67.** $\dfrac{80 \text{ km}}{1 \text{ hr}} \times \dfrac{0.62 \text{ mi}}{1 \text{ km}} \approx 49.6 \text{ mi/hr}$

**68.** $2.2 + 1.4 + 3.8 = 7.4 \text{ kg}$

$7.4 \text{ kg} \times \dfrac{2.2 \text{ lb}}{1 \text{ kg}} \approx 16.28 \text{ lb}$

They are slightly over the weight limit of 16 lb

**69.** $\dfrac{\$0.87}{1 \text{ L}} \times \dfrac{1 \text{ L}}{0.264 \text{ gal}} \approx \$3.30/\text{gal}$

**70.** $1.88 \text{ m} \times \dfrac{3.28 \text{ ft}}{1 \text{ m}} \approx 6.1664 \text{ ft}$

Now, 6 ft 2 in. $= 6 \text{ ft } \dfrac{2}{12} \text{ ft}$

$\qquad\qquad\qquad\quad = 6\dfrac{1}{6} \text{ ft}$

$\qquad\qquad\qquad\quad \approx 6.1667 \text{ ft}$

The person would be comfortable in the car.

**71.** $4 \text{ m} \times \dfrac{3.28 \text{ ft}}{1 \text{ m}} \approx 13.12 \text{ ft}$

$12 \text{ m} \times \dfrac{3.28 \text{ ft}}{1 \text{ m}} \approx 39.36 \text{ ft}$

Area $= 13.12 \times 39.36 \approx 516.4 \text{ sq ft}$

**72.** $4 \times \dfrac{\$1.23}{1 \text{ kg}} \times \dfrac{1 \text{ kg}}{2.2 \text{ lb}} \approx \$2.24$

The cost is about \$2.24.

## Chapter 6 Test

**1.** $1.6 \text{ tons} \times \dfrac{2000 \text{ lb}}{1 \text{ ton}} = 3200 \text{ lb}$

**2.** $19 \text{ ft} \times \dfrac{12 \text{ in.}}{1 \text{ ft}} = 228 \text{ in.}$

**3.** $21 \text{ gal} \times \dfrac{4 \text{ qt}}{1 \text{ gal}} = 84 \text{ qt}$

**4.** $26{,}960 \text{ ft} \times \dfrac{1 \text{ mi}}{5280 \text{ ft}} = 7 \text{ mi}$

**5.** $1800 \text{ sec} \times \dfrac{1 \text{ min}}{60 \text{ sec}} = 30 \text{ min}$

**6.** $3 \text{ cups} \times \dfrac{1 \text{ qt}}{4 \text{ cups}} = 0.75 \text{ qt}$

**7.** $9.2 \text{ km} = 9200 \text{ m}$

**8.** $27.3 \text{ cm} = 0.273 \text{ m}$

**9.** $9.88 \text{ cm} = 0.0988 \text{ m}$

**10.** $46 \text{ mm} = 4.6 \text{ cm}$

**11.** $12.7 \text{ m} = 1270 \text{ cm}$

**12.** $0.936 \text{ cm} = 9.36 \text{ mm}$

**13.** $46 \text{ L} = 0.046 \text{ kL}$

**14.** $127 \text{ L} = 127{,}000 \text{ mL}$

**15.** $28.9 \text{ mg} = 0.0289 \text{ g}$

**16.** $983 \text{ g} = 0.983 \text{ kg}$

**17.** $0.92 \text{ L} = 920 \text{ mL}$

**18.** $9.42 \text{ g} = 9420 \text{ mg}$

**19.** $42 \text{ mi} \times \dfrac{1.61 \text{ km}}{1 \text{ mi}} \approx 67.62 \text{ km}$

**20.** $1.78 \text{ yd} \times \dfrac{0.914 \text{ m}}{1 \text{ yd}} \approx 1.63 \text{ m}$

**21.** $9 \text{ cm} \times \dfrac{0.394 \text{ in.}}{1 \text{ cm}} \approx 3.55 \text{ in.}$

**22.** $38 \text{ L} \times \dfrac{0.264 \text{ gal}}{1 \text{ L}} \approx 10.03 \text{ gal}$

**23.** $7.3 \text{ kg} \times \dfrac{2.2 \text{ lb}}{1 \text{ kg}} \approx 16.06 \text{ lb}$

**24.** $3 \text{ oz} \times \dfrac{28.35 \text{ g}}{1 \text{ oz}} \approx 85.05 \text{ g}$

**25. a.** $3 \text{ m} + 7 \text{ m} + 3 \text{ m} + 7 \text{ m} = 20 \text{ m}$

**b.** $20 \text{ m} \times \dfrac{1.09 \text{ yd}}{1 \text{ m}} \approx 21.8 \text{ yd}$

**26. a.**
$$F = 1.8 \times C + 32$$
$$= 1.8 \times 35 + 32$$
$$= 95$$
$$95°F - 80°F = 15°F$$

**b.** Yes; $80°F < 95°F$

**27.** $\dfrac{5.5 \text{ qt}}{1 \text{ min}} \times \dfrac{1 \text{ gal}}{4 \text{ qt}} \times \dfrac{60 \text{ min}}{1 \text{ hr}} = 82.5 \text{ gal/hr}$

**28. a.** $100 \times 3 = 300 \text{ km}$

**b.** $300 \text{ km} \times \dfrac{0.62 \text{ mi}}{1 \text{ km}} \approx 186 \text{ mi}$

$200 - 186 = 14 \text{ mi farther}$

**29.**
$$C = \dfrac{5 \times F - 160}{9}$$
$$= \dfrac{5(-16) - 160}{9}$$
$$= \dfrac{-240}{9}$$
$$\approx -27$$
-27°C

**30.**
$$F = 1.8 \times C + 32$$
$$= 1.8 \times 40 + 32$$
$$= 104$$
104°F

## Chapters 1–6 Cumulative Test

**1.**
$$\begin{array}{r} 9824 \\ -\ 3796 \\ \hline 6028 \end{array}$$

**2.**
$$\begin{array}{r} 608 \\ \times\ 305 \\ \hline 3\ 040 \\ 182\ 40\ \ \\ \hline 185{,}440 \end{array}$$

**3.**
$$\begin{array}{r} 69 \\ 28\overline{)1932} \\ \underline{168}\phantom{00} \\ 252 \\ \underline{252} \\ 0 \end{array}$$

**4.** $\dfrac{1}{7} + \dfrac{3}{14} + \dfrac{2}{21} = \dfrac{6}{42} + \dfrac{9}{42} + \dfrac{4}{42}$

$$= \frac{6 + 9 + 4}{42}$$

$$= \frac{19}{42}$$

**5.**
$$\begin{array}{r} 3\frac{1}{8} \\ -\ 1\frac{3}{4} \end{array} \qquad \begin{array}{r} 2\frac{9}{8} \\ -\ 1\frac{6}{8} \\ \hline 1\frac{3}{8} \end{array}$$

**6.** $\dfrac{21}{35} \overset{?}{=} \dfrac{12}{20}$

$21 \times 20 \overset{?}{=} 35 \times 12$

$420 = 420 \quad$ True

**7.** $\dfrac{0.4}{n} = \dfrac{2}{30}$

$0.4 \times 30 = n \times 2$

$\dfrac{12}{2} = \dfrac{n \times 2}{2}$

$n = 6$

**8.** $\dfrac{6.5\text{ cm}}{68\text{ g}} = \dfrac{20\text{ cm}}{n\text{ g}}$

$6.5 \times n = 68 \times 20$

$6.5 \times n = 1360$

$\dfrac{6.5 \times n}{6.5} = \dfrac{1360}{6.5}$

$n \approx 209.23\text{ g}$

**9.** What percent of 66 is 165?

$n \times 66 = 165$

$\dfrac{n \times 66}{66} = \dfrac{165}{66}$

$n = 2.5$

$= 250\%$

**10.** $n = 26\% \times 7500$

$n = 0.26 \times 7500$

$n = 1950$

**11.** 0.5% of what number is 100?

$0.5\% \times n = 100$

$\dfrac{0.005 \times n}{0.005} = \dfrac{100}{0.005}$

$n = 20{,}000$

**12.** $38\text{ qt} \times \dfrac{1\text{ gal}}{4\text{ qt}} = 9.5\text{ gal}$

**13.** $2.5\text{ tons} \times \dfrac{2000\text{ lb}}{1\text{ ton}} = 5000\text{ lb}$

**14.** $7\text{ pt} \times \dfrac{1\text{ qt}}{2\text{ pt}} = 3.5\text{ qt}$

**15.** $25\text{ feet} \times \dfrac{12\text{ in.}}{1\text{ ft}} = 300\text{ in.}$

**16.** $3.7\text{ km} = 3700\text{ m}$

**17.** $62.8\text{ g} = 0.0628\text{ kg}$

**18.** 0.79 L = 790 mL

**19.** 5 cm = 0.05 m

**20.** $42 \text{ lb} \times \dfrac{16 \text{ oz}}{1 \text{ lb}} = 672 \text{ oz}$

**21.** $28 \text{ gal} \times \dfrac{3.79 \text{ L}}{1 \text{ gal}} \approx 106.12 \text{ L}$

**22.** $96 \text{ lb} \times \dfrac{0.454 \text{ kg}}{1 \text{ lb}} \approx 43.58 \text{ kg}$

**23.** $7.87 \text{ m} \times \dfrac{3.28 \text{ ft}}{1 \text{ m}} \approx 25.81 \text{ feet}$

**24.** $9 \text{ mi} \times \dfrac{1.61 \text{ km}}{1 \text{ mi}} \approx 14.49 \text{ km}$

**25.** $6 \text{ yd} + 4 \text{ yd} + 3 \text{ yd} = 13 \text{ yd}$

$13 \text{ yd} \times \dfrac{0.914 \text{ m}}{1 \text{ yd}} \approx 11.88 \text{ m}$

**26. a.** $F = 1.8 \times C + 32$

$= 11.8 \times 15 + 32$

$= 59$

59°F

59°F − 15°F = 44°F

**27.** $\dfrac{100 \text{ km}}{1 \text{ hr}} \times 1\frac{1}{2} \text{ hr} = 150 \text{ km}$

$150 \text{ km} \times \dfrac{0.62 \text{ mi}}{1 \text{ km}} \approx 93 \text{ mi}$

$100 - 93 = 7$

He needs to travel 7 miles farther.

**28.** 0.72 cm + 0.98 cm + 0.38 mm

0.72 cm + 0.98 cm + 0.038 cm

1.738 cm

# GEOMETRY

## Pretest Chapter 7

**1.** $90° - 72° = 18°$

**2.** $180° - 117° = 63°$

**3.** $\angle b = 136°$
$\angle a = \angle c$
$\quad = 180° - 136°$
$\quad = 44°$

**4.** $P = 2\ell + 2w$
$\quad = 2\,(6.5 \text{ m}) + 2\,(2.5 \text{ m})$
$\quad = 13 \text{ m} + 5 \text{ m}$
$\quad = 18 \text{ m}$

**5.** $P = 4s$
$\quad = 4\,(3.5 \text{ m})$
$\quad = 14 \text{ m}$

**6.** $A = s^2$
$\quad = (4.8 \text{ cm})^2$
$\quad = 23.04 \text{ cm}^2$
$\quad \approx 23.0 \text{ cm}^2$

**7.** $A = \ell w$
$\quad = (2.7 \text{ cm})\,(0.9 \text{ cm})$
$\quad = 2.43 \text{ cm}^2$
$\quad \approx 2.4 \text{ cm}^2$

**8.** $P = 2\ell + 2w$
$\quad = 2\,(9.2 \text{ yd}) + 2\,(3.6 \text{ yd})$
$\quad = 18.4 \text{ yd} + 7.2 \text{ yd}$
$\quad = 25.6 \text{ yd}$

**9.** $P = 17 \text{ ft} + 15 \text{ ft} + 25 \text{ ft} + 21 \text{ ft}$
$\quad = 78 \text{ ft}$

**10.** $A = bh$
$\quad = (27 \text{ in.})\,(13 \text{ in.})$
$\quad = 351 \text{ in.}^2$

**11.** $A = \dfrac{h\,(b + B)}{2}$
$\quad = \dfrac{9 \text{ in.}\,(16 \text{ in.} + 22 \text{ in.})}{2}$
$\quad = \dfrac{9 \text{ in.}\,(38 \text{ in.})}{2}$
$\quad = 171 \text{ in.}^2$

**12.** $A = \ell w + \dfrac{h\,(b + B)}{2}$
$\quad = (7 \text{ m})\,(9 \text{ m}) + \dfrac{4 \text{ m}\,(7 \text{ m} + 10 \text{ m})}{2}$
$\quad = 63 \text{ m}^2 + \dfrac{4 \text{ m}\,(17 \text{ m})}{2}$
$\quad = 63 \text{ m}^2 + 34 \text{ m}^2$
$\quad = 97 \text{ m}^2$

**13.** $180° - (39° + 118°)$
$\quad = 180° - 157°$
$\quad = 23°$

**14.** $P = 7.2 \text{ m} + 4.3 \text{ m} + 3.8 \text{ m}$
$= 15.3 \text{ m}$

**15.** $A = \dfrac{bh}{2}$

$= \dfrac{(16 \text{ m})(9 \text{ m})}{2}$

$= 72 \text{ m}^2$

**16.** $\sqrt{64} = 8$

**17.** $\sqrt{4} + \sqrt{100} = 2 + 10$
$= 12$

**18.** $\sqrt{46} \approx 6.782$

**19.** $\text{leg} = \sqrt{10^2 - 6^2}$
$= \sqrt{100 - 36}$
$= \sqrt{64}$
$= 8 \text{ ft}$

**20.** $\text{leg} = \sqrt{13^2 - 5^2}$
$= \sqrt{169 - 25}$
$= \sqrt{144}$
$= 12 \text{ ft}$

**21.** $d = 2r$
$= 2(14 \text{ in.})$
$= 28 \text{ in.}$

**22.** $c = \pi d$
$= 3.14(30 \text{ cm})$
$= 94.2 \text{ cm}$

**23.** $A = \pi r^2$
$= 3.14(9 \text{ m})^2$
$= 3.14(81 \text{ m}^2)$
$= 254.34 \text{ m}^2$
$\approx 254.3 \text{ m}^2$

**25.** $V = \ell wh$
$= (6 \text{ yd})(5 \text{ yd})(8 \text{ yd})$
$= (30 \text{ yd}^2)(8 \text{ yd})$
$= 240 \text{ yd}^3$

**26.** $V = \dfrac{4\pi r^3}{3}$

$= \dfrac{4(3.14)(3 \text{ ft})^3}{3}$

$= \dfrac{4(3.14)(27 \text{ ft}^3)}{3}$

$= 113.04 \text{ ft}^3$

$\approx 113.0 \text{ ft}^3$

**27.** $V = \pi r^2 h$
$= 3.14(7 \text{ in.})^2(12 \text{ in.})$
$= 3.14(49 \text{ in.}^2)(12 \text{ in.})$
$= 1846.32 \text{ in.}^3$
$\approx 1846.3 \text{ in.}^3$

**28.** $B = (25 \text{ m})(25 \text{ m}) = 625 \text{ m}^2$

$V = \dfrac{hB}{3}$

$= \dfrac{(21 \text{ m})(625 \text{ m}^2)}{3}$

$= 4375 \text{ m}^3$

**29.** $V = \dfrac{\pi r^2 h}{3}$

$\qquad = \dfrac{3.14\,(6\text{ m})^2\,(30\text{ m})}{3}$

$\qquad = \dfrac{3.14\,(36\text{ m}^2)\,(30\text{ m})}{3}$

$\qquad = 1130.4\text{ m}^3$

**30.** $\dfrac{n}{16} = \dfrac{30}{4}$

$\qquad 4n = (16)\,(30)$

$\qquad 4n = 480$

$\qquad \dfrac{4n}{4} = \dfrac{480}{4}$

$\qquad n = 120$

$120\text{ cm}$

**31.** $\dfrac{n}{8} = \dfrac{52.5}{15}$

$\qquad 15n = (8)\,(52.5)$

$\qquad 15n = 420$

$\qquad \dfrac{15n}{15} = \dfrac{420}{15}$

$\qquad n = 28$

$28\text{ m}$

**32. a.** $A = \ell w + \pi r^2$

$\qquad = (100\text{ yd})\,(30\text{ yd}) + 3.14\,(15\text{ yd})^2$

$\qquad = 3000\text{ yd}^2 + 706.5\text{ yd}^2$

$\qquad = 3706.5\text{ yd}^2$

**b.** $\text{Cost} = 3706.5\text{ yd}^2 \times \dfrac{\$0.22}{1\text{ yd}^2}$

$\qquad = \$815.43$

## 7.1    Exercises

**1.** An acute angle is an angle whose measure is between 0° and 90°.

**3.** Complementary angles are two angles that have a sum of 90°.

**5.** When two lines intersect, the two angles that are opposite each other are called vertical angles.

**7.** A transversal is a line that intersects two or more other lines at different points.

**9.** $\angle ABD$, $\angle CBE$

**11.** $\angle ABD$ and $\angle CBE$
$\angle DBC$ and $\angle ABE$

**13.** None exist

**15.** $\angle PQR = 50°$

**17.** $\angle RQT = 35° + 25°$
$\qquad = 60°$

**19.** $\angle TQR = 25° + 35°$
$\qquad = 60°$

**21.** $\angle RQS = 35°$

**23.** $\angle TQV = 180° - 50° - 35° - 25°$
$\qquad = 70°$

**25.** $90° - 31° = 59°$

**27.** $180° - 127° = 53°$

**29.** $\angle a = 90° - 74°$
$= 16°$

**31.** $\angle a = 87° - 43°$
$= 44°$

**33.** $\angle a = 115° - 44° - 26°$
$= 45°$

**35.** $\angle b = 102°$
$\angle a = \angle c$
$= 180° - 102°$
$= 78°$

**37.** $\angle b = 38°$
$\angle a = \angle c$
$= 180° - 38°$
$= 142°$

**39.** $\angle a = \angle c$
$= 48°$
$\angle b = 180° - 48°$
$= 132°$

**41.** $\angle c = \angle d$
$= \angle a$
$= 123°$
$\angle b = \angle c$
$= \angle f$
$= \angle g$
$= 180° - 123°$
$= 57°$

**43.** $\angle x = 180° - 121°$
$= 59°$

**45.** New angle is
$56° - 7° = 49°$ north of east.

## Cumulative Review Problems

**47.** $3.2 + 4.3 + 5.8 = 13.3$ miles
Then, $23 - 13.3 = 9.7$ miles remain.

**49.** Value
$= 4(1000) + 3(500) + 1(300) - 5(400)$
$= 4000 + 1500 + 300 - 2000$
$= 5500 + 300 - 2000$
$= 5800 - 2000$
$= 3800$
$3800 increase

## 7.2   Exercises

**1. a.** perpendicular
   **b.** equal

**3.** multiply

**5.** $P = 2\ell + 2w$
$= 2(5.5 \text{ mi}) + 2(2 \text{ mi})$
$= 11 \text{ mi} + 4 \text{ mi}$
$= 15 \text{ mi}$

**7.** $P = 2\ell + 2w$
$= 2(2.5 \text{ ft}) + 2(9.3 \text{ ft})$
$= 5.0 \text{ ft} + 18.6 \text{ ft}$
$= 23.6 \text{ ft}$

**9.** $P = 2\ell + 2w$
$= 2(4.2 \text{ ft}) + 2(12.8 \text{ ft})$
$= 8.4 \text{ ft} + 25.6 \text{ ft}$
$= 34 \text{ ft}$

**11.** $P = 2\ell + 2w$

$\quad = 2(0.84 \text{ mm}) + 2(0.12 \text{ mm})$

$\quad = 1.68 \text{ mm} + 0.24 \text{ mm}$

$\quad = 1.92 \text{ mm}$

**13.** $P = 2\ell + 2w$

$\quad = 2(4.28 \text{ km}) + 2(4.28 \text{ mm})$

$\quad = 8.56 \text{ km} + 8.56 \text{ km}$

$\quad = 17.12 \text{ km}$

**15.** $3.2 \text{ ft} \times \dfrac{12 \text{ in.}}{1 \text{ ft}} = 38.4 \text{ in.}$

$\quad P = 2\ell + 2w$

$\quad = 2(38.4 \text{ in.}) + 2(48 \text{ in.})$

$\quad = 76.8 \text{ in.} + 96 \text{ in.}$

$\quad = 172.8 \text{ in.}$

**17.** $P = 4s$

$\quad = 4(0.068 \text{ mm})$

$\quad = 0.272 \text{ mm}$

**19.** $P = 4s$

$\quad = 4(7.96 \text{ cm})$

$\quad = 31.84 \text{ cm}$

**21.**

$$
\begin{array}{r}
16 \text{ m} \\
7 \text{ m} \\
20 \text{ m} \\
4 \text{ m} \\
36 \text{ m} \\
+ 11 \text{ m} \\
\hline
94 \text{ m}
\end{array}
$$

$P = 94 \text{ m}$

**23.**

$$
\begin{array}{r}
9 \text{ cm} \\
13 \text{ cm} \\
11 \text{ cm} \\
13 \text{ cm} \\
16 \text{ cm} \\
41 \text{ cm} \\
36 \text{ cm} \\
+ 41 \text{ cm} \\
\hline
180 \text{ cm}
\end{array}
$$

$P = 180 \text{ cm}$

**25.** $A = \ell w$

$\quad = (0.96 \text{ m})(0.3 \text{ m})$

$\quad = 0.288 \text{ m}^2$

**27.** $39 \text{ yd} \times \dfrac{3 \text{ ft}}{1 \text{ yd}} = 117 \text{ ft}$

$\quad A = \ell w$

$\quad = (117 \text{ ft})(9 \text{ ft})$

$\quad = 1053 \text{ ft}^2$

**29.** $A = \ell w + \ell w$

$\quad = (12 \text{ m})(21 \text{ m}) + (6 \text{ m})(7 \text{ m})$

$\quad = 252 \text{ m}^2 + 42 \text{ m}^2$

$\quad = 294 \text{ m}^2$

**31.** $A = 220 \text{ ft} \times 50 \text{ ft}$

$\quad = 11{,}000 \text{ ft}^2$

$\quad \text{Cost} = 11{,}000 \text{ ft}^2 \times \dfrac{\$12.00}{\text{ft}^2}$

$\quad = \$132{,}000$

**33.** $P = 2\ell + 2w$

$\quad = 2(12.5 \text{ ft}) + 2(9.5 \text{ ft})$

$\quad = 25 \text{ ft} + 19 \text{ ft}$

$\quad = 44 \text{ ft}$

$\quad \text{Cost} = 44 \text{ ft} \times \dfrac{\$1.35}{\text{ft}}$

$\quad = \$59.40$

**35.**  $A = \ell w + \ell w$

$= (24 \text{ ft})(12 \text{ ft})(7 \text{ ft})(8 \text{ ft})$

$= 288 \text{ ft}^2 + 56 \text{ ft}^2$

$= 344 \text{ ft}^2$

Cost of carpet

$= 344 \text{ ft}^2 \times \dfrac{\$14.50}{\text{yd}^2} \times \dfrac{1 \text{ yd}^2}{9 \text{ ft}^2}$

$\approx \$554.22$

$P = 12 \text{ ft} + 17 \text{ ft} + 8 \text{ ft} + 7 \text{ ft} + 20 \text{ ft} + 24 \text{ ft}$

$= 88 \text{ ft}$

Cost of binding

$= 88 \text{ ft} \times \dfrac{\$1.50}{\text{yd}} \times \dfrac{1 \text{ yd}}{3 \text{ ft}}$

$= \$44$

Total cost

$= \$544.22 + \$44$

$= \$598.22$

**37.**  $11 \text{ ft} \times \dfrac{12 \text{ in.}}{1 \text{ ft}} = 132 \text{ in.}$

$11 \text{ ft} - 3 \text{ ft} = 8 \text{ ft}$

$8 \text{ ft} \times \dfrac{12 \text{ in.}}{1 \text{ ft}} = 96 \text{ in.}$

Large area $= (132 \text{ in.})(96 \text{ in.})$

$= 12{,}672 \text{ in.}^2$

Smaller area $= (129 \text{ in.})(94 \text{ in.})$

$= 12{,}126 \text{ in.}^2$

Difference $= 12{,}672 \text{ in.}^2 - 12{,}126 \text{ in.}^2$

$= 546 \text{ in.}^2$

---

## Cumulative Review Problems

**39.**
$$
\begin{array}{r}
156.8 \\
27.2 \\
+\ 39.3 \\
\hline
223.3
\end{array}
$$

**41.**
$$
\begin{array}{r}
1076 \\
\times\ 20.3 \\
\hline
322\ 8 \\
21\ 520 \\
\hline
21{,}842.8
\end{array}
$$

**43. a.**  Value $= 11 \times 72\dfrac{3}{4}$

$= 11 \times \dfrac{291}{4}$

$= \dfrac{3201}{4}$

$= 800\dfrac{1}{4}$

$\$800\dfrac{1}{4}$

**b.**
$$
\begin{array}{r}
72\frac{3}{4} \\
-\ 2\frac{3}{16} \\
\hline
\end{array}
\qquad
\begin{array}{r}
72\frac{12}{16} \\
-\ 2\frac{3}{16} \\
\hline
70\frac{9}{16}
\end{array}
$$

Value $= 11 \times 70\dfrac{9}{16}$

$= 11 \times \dfrac{1129}{16}$

$= \dfrac{12{,}419}{16}$

$= 776\dfrac{3}{16}$

$\$776\dfrac{3}{16}$

---

## 7.3  Exercises

**1.**  adding

**3.**  perpendicular

**5.**  $P = 2(2.8 \text{ m}) + 2(17.3 \text{ m})$

$= 5.6 \text{ m} + 34.6 \text{ m}$

$= 40.2 \text{ m}$

**7.** $P = 2\,(12.3 \text{ in.}) + 2\,(2.6 \text{ in.})$

$\qquad = 24.6 \text{ in.} + 5.2 \text{ in.}$

$\qquad = 29.8 \text{ in.}$

**9.** $A = bh$

$\qquad = (14.2 \text{ m})\,(21.25 \text{ m})$

$\qquad = 301.75 \text{ m}^2$

**11.** $A = bh$

$\qquad = (126 \text{ yd})\,(28 \text{ yd})$

$\qquad = 3528 \text{ yd}^2$

**13.** $P = 4\,(12 \text{ m})$

$\qquad = 48 \text{ m}$

$\quad A = bh$

$\qquad = (12 \text{ m})\,(6 \text{ m})$

$\qquad = 72 \text{ m}^2$

**15.** $P = 4\,(2.4 \text{ ft})$

$\qquad = 9.6 \text{ ft}$

$\quad A = bh$

$\qquad = (2.4 \text{ ft})\,(1.5 \text{ ft})$

$\qquad = 3.6 \text{ ft}^2$

**17.** $P = 13 \text{ m} + 20 \text{ m} + 15 \text{ m} + 34 \text{ m}$

$\qquad = 82 \text{ m}$

**19.** $P = 130 \text{ cm} + 70 \text{ cm} + 40 \text{ cm} + 260 \text{ cm}$

$\qquad = 500 \text{ cm}$

**21.** $A = \dfrac{h\,(b + B)}{2}$

$\qquad = \dfrac{(12 \text{ yd})\,(9.6 \text{ yd} + 10.2 \text{ yd})}{2}$

$\qquad = \dfrac{(12 \text{ yd})\,(19.8 \text{ yd})}{2}$

$\qquad = 118.8 \text{ yd}^2$

**23.** $A = \dfrac{h\,(b + B)}{2}$

$\qquad = \dfrac{(20 \text{ km})\,(24 \text{ km} + 31 \text{ km})}{2}$

$\qquad = \dfrac{(20 \text{ km})\,(55 \text{ km})}{2}$

$\qquad = 550 \text{ km}^2$

**25.** $\quad \text{Top area} = \ell w$

$\qquad\qquad\quad = (28 \text{ m})\,(16 \text{ m})$

$\qquad\qquad\quad = 448 \text{ m}^2$

$\quad \text{Bottom area} = \dfrac{h\,(b + B)}{2}$

$\qquad\qquad\quad = \dfrac{(9 \text{ m})\,(28 \text{ m} + 32 \text{ m})}{2}$

$\qquad\qquad\quad = \dfrac{(9 \text{ m})\,(60 \text{ m})}{2}$

$\qquad\qquad\quad = 270 \text{ m}^2$

$\quad \text{Total area} = 448 \text{ m}^2 + 270 \text{ m}^2$

$\qquad\qquad\quad = 718 \text{ m}^2$

**27.** $\quad \text{Top area} = bh$

$\qquad\qquad\quad = (12 \text{ ft})\,(5 \text{ ft})$

$\qquad\qquad\quad = 60 \text{ ft}^2$

$\quad \text{Bottom area} = \dfrac{h\,(b + B)}{2}$

$\qquad\qquad\quad = \dfrac{(18 \text{ ft})\,(12 \text{ ft} + 21 \text{ ft})}{2}$

$\qquad\qquad\quad = \dfrac{(18 \text{ ft})\,(33 \text{ ft})}{2}$

$\qquad\qquad\quad = 297 \text{ ft}^2$

$\quad \text{Total area} = 60 \text{ ft}^2 + 297 \text{ ft}^2$

$\qquad\qquad\quad = 357 \text{ ft}^2$

**29.**   Top area $= bh$
$$= (46 \text{ yd})\,(49 \text{ yd})$$
$$= 2254 \text{ yd}^2$$

Bottom area $= bh$
$$= (46 \text{ yd})\,(31 \text{ yd})$$
$$= 1426 \text{ yd}^2$$

Total area $= 2254 \text{ yd}^2 + 1426 \text{ yd}^2$
$$= 3680 \text{ yd}^2$$

$$\text{Cost} = 3680 \text{ yd}^2 \times \frac{\$22}{\text{yd}^2}$$
$$= \$80,900$$

**31.**   Top area $= \dfrac{h\,(b + B)}{2}$
$$= \frac{(0.71 \times b)\,(b + 2.65 \times b)}{2}$$
$$= \frac{(0.71b)\,(3.65b)}{2}$$
$$= 1.29575b^2$$

Bottom area $= 1.29575b^2$

Middle area $= \ell w$
$$= (2.65 \times b)\,(b)$$
$$= 2.65b^2$$

Total area $= 2\,(1.29575b^2) + 2.65b^2$
$$= 2.5915b^2 + 2.65b^2$$
$$= 5.2415b^2$$

## Cumulative Review Problems

**33.**  $10 \text{ yd} \times \dfrac{3 \text{ ft}}{1 \text{ yd}} = 30 \text{ ft}$

**35.**  $18 \text{ m} = 1800 \text{ cm}$

## 7.4   Exercises

**1.** right

**3.** Add the measures of the two known angles and subtract that value from $180°$.

**5.** You could conclude that the lengths of all three sides of the triangle are equal.

**7.** True

**9.** True

**11.** False

**13.** True

**15.** $180° - (36° + 74°) = 180° - 110°$
$$= 70°$$

**17.** $180° - (44° + 8°) = 180° - 52°$
$$= 128°$$

**19.** $P = 36 \text{ m} + 27 \text{ m} + 41 \text{ m}$
$$= 104 \text{ m}$$

**21.** $P = 45.25 \text{ in.} + 35.75 \text{ in.} + 35.75 \text{ in.}$
$$= 116.75 \text{ in.}$$

**23.** $P = 3\,(3.5 \text{ mi})$

$= 10.5 \text{ mi}$

**25.** $A = \dfrac{bh}{2}$

$= \dfrac{(4.5 \text{ in.})\,(7 \text{ in.})}{2}$

$= 15.75 \text{ in}^2$

**27.** $A = \dfrac{bh}{2}$

$= \dfrac{(17.5 \text{ cm})\,(9.5 \text{ cm})}{2}$

$= 83.125 \text{ cm}^2$

**29.** $30 \text{ ft} \times \dfrac{1 \text{ yd}}{3 \text{ ft}} = 10 \text{ yd}$

$A = \dfrac{bh}{2}$

$= \dfrac{(3.5 \text{ yd})\,(10 \text{ yd})}{2}$

$= 17.5 \text{ yd}^2$

**31.** $A = \dfrac{bh}{2}$

$= \dfrac{(7 \text{ ft})\,(8.5 \text{ ft})}{2}$

$= 29.75 \text{ ft}^2$

**33.**    Top area $= \dfrac{bh}{2}$

$= \dfrac{(16 \text{ yd})\,(4.5 \text{ yd})}{2}$

$= 36 \text{ yd}^2$

Bottom area $= \ell w$

$= (16 \text{ yd})\,(9.5 \text{ yd})$

$= 152 \text{ yd}^2$

Total area $= 36 \text{ yd}^2 + 152 \text{ yd}^2$

$= 188 \text{ yd}^2$

**35.**  Area of side

$= \ell w$

$= (45 \text{ ft})\,(20 \text{ ft})$

$= 900 \text{ ft}^2$

Area of front (or back)

$= \ell w + \dfrac{bh}{2}$

$= (35 \text{ ft})\,(20 \text{ ft}) + \dfrac{(35 \text{ ft})\,(5 \text{ ft})}{2}$

$= 700 \text{ ft}^2 + 87.5 \text{ ft}^2$

$= 787.5 \text{ ft}^2$

Area of 4 sides (total area)

$= 2\,(900 \text{ ft}^2) + 2\,(787.5 \text{ ft}^2)$

$= 1800 \text{ ft}^2 + 1575 \text{ ft}^2$

$= 3375 \text{ ft}^2$

**37.** Draw a vertical line through the middle of the wing to produce two identical triangles. For each triangle,

$b = 22 \text{ yd and } h = \dfrac{29 \text{ yd}}{2} = 14.5 \text{ yd.}$

$A = \dfrac{bh}{2}$

$= \dfrac{(22 \text{ yd})\,(14.5 \text{ yd})}{2}$

$= 159.5 \text{ yd}^2$

Total area $= 2\,(159.5 \text{ yd}^2)$

$= 319 \text{ yd}^2$

Cost $= 319 \text{ yd}^2 \times \dfrac{\$90}{\text{yd}^2}$

$= \$28{,}710$

**39.** Area of smallest triangle

$$= \frac{bh}{2}$$

$$= 0.625h \text{ meters}$$

Area of largest triangle

$$= \frac{bh}{2}$$

$$= \frac{(20 \text{ m})(h \text{ m})}{2}$$

$$= 10h \text{ meters}$$

Percent

$$= \frac{0.625h}{10h}$$

$$= 0.0625$$

$$= 6.25\%$$

## Cumulative Review Problems

**41.**
$$\frac{5}{n} = \frac{7.5}{18}$$

$$5 \times 18 = 7.5 \times n$$

$$\frac{90}{7.5} = \frac{7.5 \times n}{7.5}$$

$$12 = n$$

**43.**
$$\frac{4}{15} = \frac{n}{2685}$$

$$4 \times 2685 = 15 \times n$$

$$\frac{10{,}746}{15} = \frac{15 \times n}{15}$$

$$716 = n$$

716 waitstaff

**45.**
$$\begin{array}{cc} 6\frac{1}{2} & 6\frac{3}{6} \\ + 14\frac{2}{3} & + 14\frac{4}{6} \\ \hline & 20\frac{7}{6} = 21\frac{1}{6} \end{array}$$

$$52 - 21\frac{1}{6} = 51\frac{6}{6} - 21\frac{1}{6}$$

$$= 30\frac{5}{6} \text{ ft}$$

## 7.5   Exercises

**1.** $\sqrt{25} = 5$ because $(5)(5) = 25$

**3.** 32 is not a perfect square because no whole number multiplied by itself equals 32.

**5.** $\sqrt{25} = 5$

**7.** $\sqrt{49} = 7$

**9.** $\sqrt{121} = 11$

**11.** $\sqrt{0} = 0$

**13.** $\sqrt{49} + \sqrt{9} = 7 + 3$
$$= 10$$

**15.** $\sqrt{100} + \sqrt{1} = 10 + 1$
$$= 11$$

**17.** $\sqrt{225} - \sqrt{144} = 15 - 12$
$$= 3$$

**19.** $\sqrt{169} - \sqrt{121} + \sqrt{36} = 13 - 11 + 6$
$$= 2 + 6$$
$$= 8$$

**21.** $\sqrt{4} \times \sqrt{121} = 2 \times 11$
$$= 22$$

**23. a.** Yes because $16 \times 16 = 256$.

  **b.** $\sqrt{256} = 16$

**25.** $\sqrt{18} \approx 4.243$

**27.** $\sqrt{76} \approx 8.718$

**29.** $\sqrt{194} \approx 13.928$

**31.** $\sqrt{136 \text{ m}^2} \approx 11.662$ m

**33.** $\sqrt{10,964}$ ft $\approx 104.7$ ft

**35. a.** $\sqrt{4} = 2$

  **b.** $\sqrt{0.04} = 0.2$

  **c.** Each answer is obtained from the previous answer by divding by 10.

  **d.** No because 0.004 is not a perfect square

**37.** $\sqrt{456} + \sqrt{322} \approx 21.3542 + 17.9443$
$$= 39.2985$$
which rounds to 32.299.

## Cumulative Review Problems

**39.** $A = \ell w$
$$= (60 \text{ in.})\,(80 \text{ in.})$$
$$= 4800 \text{ sq in.}$$

**41.** 92 cm $= 0.92$ m

## 7.6   Exercises

**1.** Square the length of each leg and add those two results.  Then take the square root of the remaining number.

**3.** hypotenuse $= \sqrt{8^2 + 11^2}$
$$= \sqrt{64 + 121}$$
$$= \sqrt{185}$$
$$\approx 13.601 \text{ yd}$$

**5.** leg $= \sqrt{16^2 - 5^2}$
$$= \sqrt{256 - 25}$$
$$= \sqrt{231}$$
$$\approx 15.199 \text{ ft}$$

**7.** hypotenuse $= \sqrt{5^2 + 4^2}$
$$= \sqrt{25 + 16}$$
$$= \sqrt{41}$$
$$\approx 6.403 \text{ m}$$

**9.** hypotenuse $= \sqrt{7^2 + \sqrt{7}^2}$
$$= \sqrt{49 + 49}$$
$$= \sqrt{98}$$
$$\approx 9.899 \text{ m}$$

**11.** $\text{leg} = \sqrt{13^2 - 11^2}$

$= \sqrt{169 - 121}$

$= \sqrt{48}$

$\approx 6.928 \text{ yd}$

**13.** $\text{hypotenuse} = \sqrt{12^2 + 5^2}$

$= \sqrt{144 + 25}$

$= \sqrt{169}$

$= 13 \text{ ft}$

**15.** $\text{hypotenuse} = \sqrt{9^2 + 4^2}$

$= \sqrt{81 + 16}$

$= \sqrt{97}$

$\approx 9.8 \text{ cm}$

**17.** $\text{leg} = \sqrt{32^2 - 30^2}$

$= \sqrt{1024 - 900}$

$= \sqrt{124}$

$\approx 11.1 \text{ yd}$

**19.** Side opposite $30°$ angle $= \dfrac{1}{2}(8) = 4 \text{ in.}$

$\text{leg} = \sqrt{8^2 - 4^2}$

$= \sqrt{64 - 16}$

$= \sqrt{48}$

$\approx 6.9 \text{ in.}$

**21.** $\text{hypotenuse} = \sqrt{2} \times \text{leg}$

$= \sqrt{2} \times 6$

$\approx 1.414 \times 6$

$\approx 8.5 \text{ m}$

**23.** The other leg $= 25 \text{ cm}$

$\text{hypotenuse} = \sqrt{2} \times \text{leg}$

$= \sqrt{2} \times 25$

$\approx 1.414 \times 25$

$\approx 35.4 \text{ cm}$

**25.** $\text{leg} = \sqrt{10^2 - 7^2}$

$= \sqrt{100 - 49}$

$= \sqrt{51}$

$\approx 7.1 \text{ in.}$

**27.** $\text{hypotenuse} = \sqrt{2^2 + 5^2}$

$= \sqrt{4 + 25}$

$= \sqrt{29}$

$\approx 5.39 \text{ ft}$

**29.** $\text{hypotenuse} = \sqrt{7^2 + 11^2}$

$= \sqrt{49 + 121}$

$= \sqrt{170}$

$\approx 13.038 \text{ yd}$

## Cumulative Review Problems

**31.** $A = \dfrac{bh}{2}$

$= \dfrac{(31 \text{ m})(22 \text{ m})}{2}$

$= 341 \text{ m}^2$

**33.** $A = s^2$

$= (21 \text{ in.})^2$

$= 441 \text{ in.}^2$

## 7.7   Exercises

**1.** circumference

**3.** radius

**5.** You need to multiply the radius by 2 and then use the formula $C = \pi d$

**7.** $d = 2r$
$= 2\,(29\text{ in.})$
$= 58\text{ in.}$

**9.** $d = 2r$
$= 2\,(8.5\text{ mm})$
$= 17\text{ mm}$

**11.** $r = \dfrac{d}{2}$
$= \dfrac{45\text{ yd}}{2}$
$= 22.5\text{ yd}$

**13.** $r = \dfrac{d}{2}$
$= \dfrac{19.84\text{ cm}}{2}$
$= 9.92\text{ cm}$

**15.** $C = \pi d$
$= 3.14\,(32\text{ cm})$
$= 100.48\text{ cm}$

**17.** $C = \pi r$
$= 2\,(3.14)\,(18.5\text{ in.})$
$= 116.18\text{ in.}$

**19.** $C = \pi d$
$= 3.14\,(32\text{ in.})$
$= 100.48\text{ in}$
Distance
$= (100.48\text{ in})\,(5\text{ rev}) \times \dfrac{1\text{ ft}}{12\text{ in.}}$
$\approx 41.9\text{ ft}$

**21.** $A = \pi r^2$
$= 3.14\,(5\text{ yd})^2$
$= 3.14\,(25\text{ yd}^2)$
$= 78.5\text{ yd}^2$

**23.** $r = \dfrac{d}{2} = \dfrac{32\text{ cm}}{2} = 16\text{ cm}$
$A = \pi r^2$
$= 3.14\,(16\text{ cm})^2$
$= 3.14\,(256\text{ cm}^2)$
$= 803.84\text{ cm}^2$

**25.** $A = \pi r^2$
$= 3.14\,(8\text{ ft})^2$
$= 3.14\,(64\text{ ft}^2)$
$= 200.96\text{ ft}^2$

**27.** $r = \dfrac{120\text{ mi}}{2} = 60\text{ mi}$
$A = \pi r^2$
$= 3.14\,(60\text{ mi})^2$
$= 3.14\,(3600\text{ mi}^2)$
$\approx 11{,}304\text{ mi}^2$

**29.** $A = \pi r^2 - \pi r^2$
$= 3.14\,(13\text{ m})^2 - 3.14\,(9\text{ m})^2$
$= 3.14\,(169\text{ m}^2) - 3.14\,(81\text{ m}^2)$
$= 520.66\text{ m}^2 - 254.34\text{ m}^2$
$= 276.32\text{ m}^2$

**31.**  $r = \dfrac{d}{2} = \dfrac{10 \text{ m}}{2} = 5 \text{ m}$

$A = \dfrac{1}{2}\pi r^2 + \ell w$

$= \dfrac{1}{2}(3.14)(5 \text{ m})^2 + (15 \text{ m})(10 \text{ m})$

$= \dfrac{1}{2}(3.14)(25 \text{ m}^2) + 150 \text{ m}^2$

$= 39.25 \text{ m}^2 + 150 \text{ m}^2$

$= 189.25 \text{ m}^2$

**33.**  $A = \ell w = \pi r^2$

$= (12 \text{ m})(12 \text{ m}) - (3.14)(6 \text{ m})^2$

$= 144 \text{ m}^2 - 3.14 (36 \text{ m}^2)$

$= 144 \text{ m}^2 - 113.04 \text{ m}^2$

$= 30.96 \text{ m}^2$

**35.**  $r = \dfrac{40 \text{ yd}}{2} = 20 \text{ yd}$

$A = \pi r^2 + \ell w$

$= 3.14 (20 \text{ yd})^2 + (120 \text{ yd})(40 \text{ yd})$

$= 3.14 (400 \text{ yd}^2) + 4800 \text{ yd}^2$

$= 1256 \text{ yd}^2 + 4800 \text{ yd}^2$

$= 6056 \text{ yd}^2$

$\text{Cost} = 6056 \text{ yd}^2 \times \dfrac{\$0.20}{1 \text{ yd}^2} = \$1211.20$

**37.**  $C = \pi d$

$= 3.14 (2 \text{ ft})$

$= 6.28 \text{ ft}$

**39.**  $C = 2\pi r$

$= 2 (3.14)(14 \text{ in.})$

$= 87.92 \text{ in.}$

Distance

$= (87.92 \text{ in.})(35 \text{ rev}) \times \dfrac{1 \text{ ft}}{12 \text{ in.}}$

$\approx 256.43 \text{ ft}$

**41.**  $1 \text{ mi} = 5280 \text{ ft} \times \dfrac{12 \text{ in.}}{1 \text{ ft}}$

$= 63{,}360 \text{ in.}$

$C = 2\pi r$

$= 2 (3.14)(16 \text{ in.})$

$= 100.48 \text{ in.}$

$\text{rev} = \dfrac{63{,}360 \text{ in.}}{100.48 \text{ in.}}$

$\approx 630.57$

**43.**  $r = \dfrac{d}{2} = \dfrac{6 \text{ ft}}{2} = 3 \text{ ft}$

$A = \pi r^2$

$= 3.14 (3 \text{ ft})^2$

$= 3.14 (9 \text{ ft}^2)$

$= 28.26 \text{ ft}^2$

$\text{Cost} = 28.26 \text{ ft}^2 \times \dfrac{\$72}{\text{yd}^2} \times \dfrac{1 \text{ yd}^2}{9 \text{ ft}^2}$

$= \$226.08$

**45. a.**  $\text{Cost} = \dfrac{6.00}{8} = 0.75 = \$0.75$

$r = \dfrac{d}{2} = \dfrac{15 \text{ in.}}{2} = 7.5 \text{ in}$

$A = \pi r^2$

$= 3.14 (7.5 \text{ in.})^2$

$= 3.14 (56.25 \text{ in.}^2)$

$= 176.625 \text{ in}^2$

One slice

$= \dfrac{175.625 \text{ in.}^2}{8} \approx 22.1 \text{ in.}^2$

**b.** $\text{Cost} = \dfrac{4.00}{6} \approx 0.67 = \$0.67$

$$r = \dfrac{d}{2} = \dfrac{12 \text{ in.}}{2} = 6 \text{ in}$$

$$A = \pi r^2$$
$$= 3.14 \, (6 \text{ in.})^2$$
$$= 3.14 \, (36 \text{ in.}^2)$$
$$= 113.04 \text{ in}^2$$

One slice

$$= \dfrac{113.04 \text{ in.}^2}{6} \approx 18.84 \text{ in.}^2$$

**c.** Cost per square inch for 15 in. pizza is

$$\dfrac{\$0.75}{22.1} \approx \$0.034.$$

Cost per square inch for 12 in. pizza is

$$\dfrac{\$0.67}{18.84} \approx \$0.035.$$

The 15 in. pizza is a better value.

## Cumulative Review Problems

**47.** $n = 16\% \text{ of } 87$
$n = 16\% \times 87$
$n = 0.16 \times 87$
$n = 13.92$

**49.** $\dfrac{12}{100} = \dfrac{720}{b}$

$12 \times b = 100 \times 720$

$\dfrac{12 \times b}{12} = \dfrac{72{,}000}{12}$

$b = 6000$

## 7.8   Exercises

**1.** box

**3.** sphere

**5.** pyramid

**7.** $V = \ell w h$
$= (20 \text{ mm})\,(14 \text{ mm})\,(2.5 \text{ mm})$
$= 700 \text{ mm}^3$

**9.** $V = \pi r^2 h$
$= 3.14 \, (2 \text{ m})^2 \, (7 \text{ m})$
$= 3.14 \, (4 \text{ m}^2) \, (7 \text{ m})$
$= 87.89 \text{ m}^3$
$\approx 87.9 \text{ m}^3$

**11.** $r = \dfrac{d}{2} = \dfrac{30 \text{ m}}{2} = 15 \text{ m}$

$V = \pi r^2 h$
$= 3.14 \, (15 \text{ m})^2 \, (9 \text{ m})$
$= 3.14 \, (225 \text{ m}^2) \, (9 \text{ m})$
$= 6358.5 \text{ m}^3$

**13.** $V = \dfrac{4 \pi r^3}{3}$

$$= \dfrac{4 \, (3.14) \, (9 \text{ yd})^3}{3}$$

$$= \dfrac{4 \, (3.14) \, (729 \text{ yd}^3)}{3}$$

$$\approx 3052.1 \text{ yd}^3$$

**15.** $V = \dfrac{1}{2} \times \dfrac{4\pi r^3}{3}$

$\quad = \dfrac{1}{2} \times \dfrac{4\,(3.14)\,(7\text{ m})^3}{3}$

$\quad = \dfrac{1}{2} \times \dfrac{4\,(3.14)\,(343\text{ m}^3)}{3}$

$\quad \approx \dfrac{1}{2} \times 1436.027\text{ m}^3$

$\quad \approx 718.0\text{ m}^3$

**17.** $V = \dfrac{\pi r^2 h}{3}$

$\quad = \dfrac{3.14\,(9\text{ cm})^2\,(12\text{ cm})}{3}$

$\quad = \dfrac{3.14\,(81\text{ cm}^2)\,(12\text{ cm})}{3}$

$\quad \approx 1017.4\text{ cm}^3$

**19.** $V = \dfrac{\pi r^2 h}{3}$

$\quad = \dfrac{3.14\,(6\text{ ft})^2\,(12\text{ ft})}{3}$

$\quad = \dfrac{3.14\,(36\text{ ft}^2)\,(12\text{ ft})}{3}$

$\quad \approx 452.2\text{ ft}^3$

**21.** $B = (7\text{ m})\,(7\text{ m}) = 49\text{ m}^2$

$\quad V = \dfrac{Bh}{3}$

$\quad = \dfrac{(49\text{ m}^2)\,(10\text{ m})}{3}$

$\quad \approx 163.3\text{ m}^3$

**23.** $B = (8\text{ m})\,(14\text{ m}) = 112\text{ m}^2$

$\quad V = \dfrac{Bh}{3}$

$\quad = \dfrac{(112\text{ m}^2)\,(10\text{ m})}{3}$

$\quad \approx 373.3\text{ m}^3$

**25.** $4\text{ in.} \times \dfrac{1\text{ ft}}{12\text{ in.}} \times \dfrac{1\text{ yd}}{3\text{ ft}} = \dfrac{1}{9}\text{ yd}$

$\quad V = \ell w h$

$\quad = (120\text{ yd})\,(7\text{ yd})\left(\dfrac{1}{9}\text{ yd}\right)$

$\quad = (840\text{ yd}^2)\left(\dfrac{1}{9}\text{ yd}\right)$

$\quad \approx 93.3\text{ yd}^3$

**27.** $\text{Outer} = \pi r^2 h$

$\quad = 3.14\,(5\text{ in.})^2\,(20\text{ in.})$

$\quad = 3.14\,(25\text{ in.}^2)\,(20\text{ in.})$

$\quad = 1570\text{ in.}^3$

$\quad \text{Inner} = \pi r^2 h$

$\quad = 3.14\,(3\text{ in.})^2\,(20\text{ in.})$

$\quad = 3.14\,(9\text{ in.}^2)\,(20\text{ in.})$

$\quad = 565.2\text{ in.}^3$

$\text{Difference} = 1570\text{ in.}^3 - 565.2\text{ in.}^3$

$\quad = 1004.8\text{ in.}^3$

**29.** Jupiter:

$\quad V = \dfrac{4\pi r^3}{3}$

$\quad = \dfrac{4\,(3.14)\,(45{,}000\text{ mi})^3}{3}$

$\quad = 381{,}510{,}000{,}000{,}000\text{ mi}^3$

Earth:

$\quad V = \dfrac{4\pi r^3}{3}$

$\quad = \dfrac{4\,(3.14)\,(3950\text{ mi})^3}{3}$

$\quad \approx 258{,}023{,}743{,}333\text{ mi}^3$

Difference

$\quad = \begin{array}{r} 381{,}510{,}000{,}000{,}000\text{ mi}^3 \\ -\ 258{,}023{,}743{,}333\text{ mi}^3 \\ \hline 381{,}251{,}976{,}256{,}667\text{ mi}^3 \end{array}$

**31.** Smaller:

$$V = \ell wh$$
$$= (18 \text{ in.}) (6 \text{ in.}) (12 \text{ in.})$$
$$= (108 \text{ in.}^2) (12 \text{ in.})$$
$$= 1296 \text{ in.}^3$$

Larger:

$$V = \ell wh$$
$$= (12 \text{ in.}) (22 \text{ in.}) (16 \text{ in.})$$
$$= (264 \text{ in.}^2) (16 \text{ in.})$$
$$= 4224 \text{ in.}^3$$

Difference

$$= 4224 \text{ in.}^3 - 1296 \text{ in.}^3$$
$$= 2928 \text{ in.}^3$$

**33.**   $V = \dfrac{\pi r^2 h}{3}$

$$= \dfrac{3.14 \,(5 \text{ cm})^2 (9 \text{ cm})}{3}$$

$$= \dfrac{3.14 \,(25 \text{ cm}^2) (9 \text{ cm})}{3}$$

$$= 235.5 \text{ cm}^3$$

$$\text{Cost} = 235.5 \text{ cm}^3 \times \dfrac{\$4.00}{1 \text{ cm}^3}$$

$$= \$942$$

**35.**   $B = (87 \text{ yd}) (130 \text{ yd})$

$$= 11{,}310 \text{ yd}^3$$

$$V = \dfrac{Bh}{3}$$

$$= \dfrac{(11{,}310 \text{ yd}^2) (70 \text{ yd})}{3}$$

$$= 263{,}900 \text{ yd}^3$$

**37.** Many possibilities are possible.

## Cumulative Review Problems

**39.**

$$\begin{array}{r} 7\frac{1}{3} \\ + \ 2\frac{1}{4} \\ \hline \end{array} \qquad \begin{array}{r} 7\frac{4}{12} \\ + \ 2\frac{3}{12} \\ \hline 9\frac{7}{12} \end{array}$$

**41.** $2\dfrac{1}{4} \times 3\dfrac{3}{4} = \dfrac{9}{4} \times \dfrac{15}{4}$

$$= \dfrac{135}{16}$$

$$= 8\dfrac{7}{16}$$

## Putting Your Skills To Work

**1. a.**   $P = 2\ell + 2w$

$$= 2\,(27 \text{ in.}) + 2\,(22.5 \text{ in.})$$
$$= 54 \text{ in.} + 45 \text{ in.}$$
$$= 99 \text{ in.}$$

$$A = \ell w$$
$$= (27 \text{ in.}) (22.5 \text{ in.})$$
$$= 607.5 \text{ in.}^2$$

**b.**   $P = 2\ell + 2w$

$$= 2\,(25 \text{ in.}) + 2\,(22.5 \text{ in.})$$
$$= 50 \text{ in.} + 45 \text{ in.}$$
$$= 95 \text{ in.}$$

$$A = \ell w$$
$$= (25 \text{ in.}) (22.5 \text{ in.})$$
$$= 562.5 \text{ in.}^2$$

**c.**   $\text{Difference} = 607.5 \text{ in.}^2 - 562.5 \text{ in.}^2$

$$= 45 \text{ in.}^2$$

**3. a.** $10{,}000 \text{ tons} \times \dfrac{2000 \text{ lb}}{1 \text{ ton}}$

$$= 20{,}000{,}000 \text{ pounds}$$

**b.** $\dfrac{20{,}000{,}000}{365} \approx 54{,}795 \text{ pounds}$

Recall that 1 year = 365 days.

## 7.9   Exercises

**1.** size; shape

**3.** sides

**5.**  $\dfrac{n}{2} = \dfrac{12}{3}$

$3n = (2)(12)$

$3n = 24$

$\dfrac{3n}{3} = \dfrac{24}{3}$

$n = 8$

8 m

**7.**  $\dfrac{n}{6} = \dfrac{9}{21}$

$21n = (6)(9)$

$21n = 54$

$\dfrac{21n}{21} = \dfrac{54}{21}$

$n \approx 2.6$

2.6 ft

**9.**  $\dfrac{n}{7} = \dfrac{5}{18}$

$18n = (7)(5)$

$18n = 35$

$\dfrac{18n}{18} = \dfrac{35}{18}$

$n \approx 1.9$

1.9 yd

**11.** a corresponds to f

b corresponds to e

c corresponds to d

**13.**  $\dfrac{n}{8} = \dfrac{10.5}{25}$

$25n = (8)(10.5)$

$25n = 84$

$\dfrac{25n}{25} = \dfrac{84}{25}$

$n \approx 3.4$

3.4 m

**15.** $3 \text{ in.} \times \dfrac{1 \text{ ft}}{12 \text{ in.}} = \dfrac{1}{4} \text{ ft}$

$5 \text{ in.} \times \dfrac{1 \text{ ft}}{12 \text{ in.}} = \dfrac{5}{12} \text{ ft}$

$\dfrac{n}{\frac{1}{4}} = \dfrac{3.5}{\frac{5}{12}}$

$\dfrac{5}{12}n = \left(\dfrac{1}{4}\right)(3.5)$

$\dfrac{5}{12}n = \dfrac{7}{8}$

$\dfrac{5}{12} \cdot \dfrac{12}{5}n = \dfrac{7}{8} \cdot \dfrac{12}{5}$

$n \approx 2.1$

2.1 ft

**17.** $1\dfrac{1}{2} \text{ ft} = \dfrac{3}{2} \text{ ft} \times \dfrac{12 \text{ in.}}{1 \text{ ft}} = 18 \text{ in.}$

$\dfrac{n}{25} = \dfrac{18}{6}$

$6n = (25)(18)$

$6n = 450$

$\dfrac{6n}{6} = \dfrac{450}{6}$

$n = 75$

75 in.

**19.**   $\dfrac{n}{6} = \dfrac{24}{4}$

$4n = (6)(24)$

$4n = 144$

$\dfrac{6n}{6} = \dfrac{144}{4}$

$n = 36$

36 ft

**21.**   $\dfrac{n}{96} = \dfrac{5.5}{6.5}$

$6.5n = (96)(5.5)$

$6.5n = 528$

$\dfrac{6.5n}{6.5} = \dfrac{528}{6.5}$

$n \approx 81$

81 ft

**23.**   $\dfrac{n}{15} = \dfrac{5}{9}$

$9n = (15)(5)$

$9n = 75$

$\dfrac{9n}{9} = \dfrac{75}{9}$

$n \approx 8.3$

8.3 ft

**25.**   $\dfrac{n}{7} = \dfrac{28}{12}$

$12n = (7)(28)$

$12n = 196$

$\dfrac{12n}{12} = \dfrac{196}{12}$

$n \approx 16.3$

16.3 cm

**27.**   $\left(\dfrac{2}{3}\right)^2 = \dfrac{4}{9}$

Proportion:

$\dfrac{A}{26} = \dfrac{4}{9}$

$9A = (26)(4)$

$9A = 104$

$\dfrac{9A}{9} = \dfrac{104}{9}$

$A \approx 11.6$

11.6 yd$^2$

## Cumulative Review Problems

**29.**  $2 \times 3^2 + 4 - 2 \times 5$

$\quad = 2 \times 9 + 4 - 2 \times 5$

$\quad = 18 + 4 - 10$

$\quad = 22 - 10$

$\quad = 12$

**31.**  $(5)(9) - (21 + 3) \div 8$

$\quad = (5)(9) - (24) \div 8$

$\quad = 45 - 3$

$\quad = 42$

## 7.10   Exercises

**1. a.**   Trip $= 13$ km $+ 17$ km

$\qquad = 30$ km

Speed $= \dfrac{30 \text{ km}}{0.4 \text{ hr}}$

$\qquad\quad = 75$ km/hr

**b.**   Trip $= 12$ km $+ 15$ km $+ 11$ km

$\qquad = 38$ km

Speed $= \dfrac{38 \text{ km}}{0.5 \text{ hr}}$

$\qquad\quad = 76$ km/hr

**c.** Trip through Woodville and Palermo has higher speed.

**3.**  $A = (7 \text{ ft}) (10 \text{ ft}) = 70 \text{ ft}^2$

$A = (7 \text{ ft}) (14 \text{ ft}) = 98 \text{ ft}^2$

$A = (6 \text{ ft}) (10 \text{ ft}) = 60 \text{ ft}^2$

$A = (6 \text{ ft}) (8 \text{ ft}) = 48 \text{ ft}^2$

Total $A$

$= 70 \text{ ft}^2 + 98 \text{ ft}^2 + 60 \text{ ft}^2 + 48 \text{ ft}^2$

$= 276 \text{ ft}^2$

Time

$= 276 \text{ ft}^2 \times \dfrac{15 \text{ min}}{120 \text{ ft}^2} = 34.5 \text{ min}$

34.5 minutes

**5.**  $A = \ell w - \dfrac{bh}{2}$

$= (21 \text{ ft}) (15 \text{ ft}) - \dfrac{(3 \text{ ft}) (6 \text{ ft})}{2}$

$= 315 \text{ ft}^2 - 9 \text{ ft}^2$

$= 306 \text{ ft}^2$

Cost $= 306 \text{ ft}^2 \times \dfrac{1 \text{ yd}^2}{9 \text{ ft}^2} \times \dfrac{\$15}{\text{yd}^2}$

$= \$510$

$\$510$

**7.**  $r = \dfrac{d}{2} = \dfrac{2 \text{ m}}{2} = 1 \text{ m}$

$V = \ell w h - \pi r^2 h$

$= (7 \text{ m}) (4 \text{ m}) (3 \text{ m})$

$\quad - (3.14) (1 \text{ m})^2 (7 \text{ m})$

$= 84 \text{ m}^3 - 21.98 \text{ m}^3$

$= 62.02 \text{ m}^3$

Cost $= 62.02 \text{ m}^3 \times \dfrac{\$1.20}{\text{m}^3}$

$\approx \$74.42$

$\$74.42$

**9. a.**  $C = 2\pi r$

$= 2 (3.14) (6500 \text{ km})$

$= 40{,}820 \text{ km}$

**b.**  $S = \dfrac{40{,}820 \text{ km}}{2 \text{ hr}}$

$= 20{,}410 \text{ km/hr}$

**11.**  $V = 400 \times \pi r^2 h$

$= 400 \times (3.14) (2 \text{ in.})^2 (10 \text{ in.})$

$= 400 \times (3.14) (4 \text{ in.}^2) (10 \text{ in.})$

$= 400 \times 125.6 \text{ in.}^3$

$= 50{,}240 \text{ in.}^3$

**13.**  $P = 2\ell + 2w$

$= 2(456 \text{ ft}) + 2 (625 \text{ ft})$

$= 912 \text{ ft} + 1250 \text{ ft}$

$= 2162 \text{ ft}$

Cost $= 2162 \text{ ft} \times \dfrac{1 \text{ yd}}{3 \text{ ft}} \times \dfrac{\$4.57}{\text{yd}}$

$\approx \$3293.45$

$\$3293.45$

**15.**  $r = \dfrac{d}{2} = \dfrac{28 \text{ in.}}{2} = 14 \text{ in.}$

Convert to feet:

$14 \text{ in.} \times \dfrac{1 \text{ ft}}{12 \text{ in.}} = \dfrac{14}{12} \text{ ft} = \dfrac{7}{6} \text{ ft}$

$C = 2\pi r = 2 (3.14) \left( \dfrac{7}{6} \text{ ft} \right)$

$= \dfrac{21.98}{3} \text{ ft}$

Speed

$= (\text{rev}) (C)$

$= (200 \text{ rev/min}) \left( \dfrac{21.98}{3} \text{ ft} \right)$

$\approx 1465.33 \text{ ft/min}$

$1465.33 \text{ ft/min}$

## Cumulative Review Problems

**17.**
$$
\begin{array}{r}
128 \\
16\overline{)2048} \\
\underline{16} \\
44 \\
\underline{32} \\
128 \\
\underline{128} \\
0
\end{array}
$$

**19.**
$$
\begin{array}{r}
0.25 \\
1.3\overline{)0.325} \\
\underline{26} \\
65 \\
\underline{65} \\
0
\end{array}
$$

## Chapter 7 Review Problems

**1.** $90° - 76° = 14°$

**2.** $180° - 12° = 168°$

**3.** $\angle b = 146°$
   $\angle a = \angle c$
   $= 180° - 146°$
   $= 44°$

**4.** $\angle t = \angle x = \angle y = 65°$
   $\angle s = \angle u = \angle w = \angle z$
   $= 180° - 65°$
   $= 115$

**5.** $P = 2(8.3 \text{ m}) + 2(1.6 \text{ m})$
   $= 16.6 \text{ m} + 3.2 \text{ m}$
   $= 19.8 \text{ m}$

**6.** $P = 4s$
   $= 4(2.4 \text{ yd})$
   $= 9.6 \text{ yd}$

**7.** $A = (5.9 \text{ cm})(2.8 \text{ cm})$
   $= 16.52 \text{ cm}^2$
   $\approx 16.5 \text{ cm}^2$

**8.** $A = s^2$
   $= (7.2 \text{ in.})^2$
   $= 51.84 \text{ in.}^2$
   $\approx 51.8 \text{ in.}^2$

**9.** $P = 3(8 \text{ ft}) + 2(2 \text{ ft}) + 4 \text{ ft} + 2(3 \text{ ft})$
   $= 24 \text{ ft} + 4 \text{ ft} + 4 \text{ ft} + 6 \text{ ft}$
   $= 38 \text{ ft}$

**10.** $P = 3(11 \text{ ft}) + 2(7 \text{ ft}) + 2(3.5 \text{ ft}) + 4 \text{ ft}$
   $= 33 \text{ ft} + 14 \text{ ft} + 7 \text{ ft} + 4 \text{ ft}$
   $= 58 \text{ ft}$

**11.** $A = (14 \text{ m})(5 \text{ m}) - 2(1 \text{ m})^2$
   $= 70 \text{ m}^2 - 2 \text{ m}^2$
   $= 68 \text{ m}^2$

**12.** $A = (9 \text{ m})^2 - (2.7 \text{ m})(6.5 \text{ m})$
   $= 81 \text{ m}^2 - 17.55 \text{ m}^2$
   $= 63.45 \text{ m}^2$
   $\approx 63.5 \text{ m}^2$

**13.** $P = 2(52 \text{ m}) + 2(20.6 \text{ m})$
   $= 104 \text{ m} + 41.2 \text{ m}$
   $= 145.2 \text{ m}$

**14.** $P = 5 \text{ mi} + 22 \text{ mi} + 5 \text{ mi} + 30 \text{ mi}$
   $= 62 \text{ mi}$

**15.** $A = (90 \text{ m})(30 \text{ m})$
   $= 2700 \text{ m}^2$

**16.** $A = \dfrac{36 \text{ yd} (17 \text{ yd} + 23 \text{ yd})}{2}$

$= \dfrac{36 \text{ yd} (40 \text{ yd})}{2}$

$= 720 \text{ yd}^2$

**17.** $A = \dfrac{(8 \text{ cm}) (33 \text{ cm} + 20 \text{ cm})}{2}$

$+ \dfrac{(20 \text{ cm}) (9 \text{ cm} + 20 \text{ cm})}{2}$

$= \dfrac{(8 \text{ cm}) (33 \text{ cm})}{2} + \dfrac{(20 \text{ cm}) (29 \text{ cm})}{2}$

$= 132 \text{ cm}^2 + 290 \text{ cm}^2$

$= 422 \text{ cm}^2$

**18.** $A = (15 \text{ m}) (17 \text{ m}) + (17 \text{ m}) (6 \text{ m})$

$= 255 \text{ m}^2 + 102 \text{ m}^2$

$= 357 \text{ m}^2$

**19.** $P = 10 \text{ ft} + 5 \text{ ft} + 7 \text{ ft}$

$= 22 \text{ ft}$

**20.** $P = 5.5 \text{ ft} + 3 \text{ ft} + 5.5 \text{ ft}$

$= 14 \text{ ft}$

**21.** Angle is

$180° - (15° + 12°) = 180° - 27°$

$= 153°$

**22.** Angle is

$180° - (37° + 96°) = 180° - 133°$

$= 47°$

**23.** $A = \dfrac{(8.5 \text{ m}) (12.3 \text{ m})}{2}$

$= \dfrac{104.55 \text{ m}^2}{2}$

$= 52.275 \text{ m}^2$

$\approx 52.3 \text{ m}^2$

**24.** $A = \dfrac{(12.5 \text{ m}) (9.5 \text{ m})}{2}$

$= \dfrac{118.75 \text{ m}^2}{2}$

$= 59.375 \text{ m}^2$

$\approx 59.4 \text{ m}^2$

**25.** $A = (18 \text{ m}) (22 \text{ m}) + \dfrac{(18 \text{ m}) (6 \text{ m})}{2}$

$= 396 \text{ m}^2 + 54 \text{ m}^2$

$= 450 \text{ m}^2$

**26.** $A = (12 \text{ m}) (6 \text{ m}) + \dfrac{(6 \text{ m}) (3 \text{ m})}{2}$

$+ \dfrac{(6 \text{ m}) (2 \text{ m})}{2}$

$= 72 \text{ m}^2 + 9 \text{ m}^2 + 6 \text{ m}^2$

$= 87 \text{ m}^2$

**27.** $\sqrt{81} = 9$

**28.** $\sqrt{64} = 8$

**29.** $\sqrt{121} = 11$

**30.** $\sqrt{36} + \sqrt{0} = 6 + 0$

$= 6$

**31.** $\sqrt{9} + \sqrt{4} = 3 + 2$
$\qquad\qquad = 5$

**32.** $\sqrt{35} \approx 5.916$

**33.** $\sqrt{45} \approx 6.708$

**34.** $\sqrt{165} \approx 12.845$

**35.** $\sqrt{180} \approx 13.416$

**36.** hypotenuse $= \sqrt{3^2 + 4^2}$
$\qquad\qquad = \sqrt{9 + 16}$
$\qquad\qquad = \sqrt{25}$
$\qquad\qquad = 5$ m

**37.** leg $= \sqrt{13^2 - 12^2}$
$\qquad = \sqrt{169 - 144}$
$\qquad = \sqrt{25}$
$\qquad = 5$ yd

**38.** leg $= \sqrt{18^2 - 16^2}$
$\qquad = \sqrt{324 - 256}$
$\qquad = \sqrt{68}$
$\qquad \approx 8.25$ cm

**39.** hypotenuse $= \sqrt{4^2 + 7^2}$
$\qquad\qquad = \sqrt{16 + 49}$
$\qquad\qquad = \sqrt{65}$
$\qquad\qquad \approx 8.06$ m

**40.** hypotenuse $= \sqrt{5^2 + 4^2}$
$\qquad\qquad = \sqrt{25 + 16}$
$\qquad\qquad = \sqrt{41}$
$\qquad\qquad \approx 6.4$ cm

**41.** hypotenuse $= \sqrt{9^2 + 2^2}$
$\qquad\qquad = \sqrt{81 + 4}$
$\qquad\qquad = \sqrt{85}$
$\qquad\qquad \approx 9.2$ ft

**42.** leg $= \sqrt{11^2 - 9^2}$
$\qquad = \sqrt{121 - 81}$
$\qquad = \sqrt{40}$
$\qquad \approx 6.3$ ft

**43.** leg $= \sqrt{7^2 - 6^2}$
$\qquad = \sqrt{49 - 36}$
$\qquad = \sqrt{13}$
$\qquad \approx 3.6$ ft

**44.** $d = 2r$
$\qquad = 2\,(53 \text{ cm})$
$\qquad = 106$ cm

**45.** $r = \dfrac{d}{2}$
$\qquad = \dfrac{126 \text{ cm}}{2}$
$\qquad = 63$ cm

**46.** $C = \pi d$
$\qquad = 3.14\,(14 \text{ in.})$
$\qquad = 43.96$ in.
$\qquad \approx 44.0$ in.

**47.** $C = 2\pi r$
$\qquad = 2\,(3.14)\,(9 \text{ in.})$
$\qquad = 56.52$ in.
$\qquad \approx 56.5$ in.

**48.**  $A = \pi r^2$

$\quad = 3.14 \, (9 \text{ m})^2$

$\quad = 3.14 \, (81 \text{ m}^2)$

$\quad = 254.34 \text{ m}^2$

$\quad \approx 254.3 \text{ m}^2$

**49.**  $r = \dfrac{d}{2} = \dfrac{16 \text{ ft}}{2} = 8 \text{ ft}$

$A = \pi r^2$

$\quad = 3.14 \, (8 \text{ ft})^2$

$\quad = 3.14 \, (64 \text{ ft}^2)$

$\quad = 200.96 \text{ ft}^2$

$\quad \approx 201.0 \text{ ft}^2$

**50.**  $A = \pi r^2 - \pi r^2$

$\quad = 3.14 \, (11 \text{ in.})^2 - 3.14 \, (7 \text{ in.})^2$

$\quad = 3.14 \, (121 \text{ in.}^2) - 3.14 \, (49 \text{ in}^2)$

$\quad = 379.94 \text{ in.}^2 - 153.86 \text{ in}^2$

$\quad = 226.08 \text{ in}^2$

$\quad \approx 226.1 \text{ in.}^2$

**51.**  $A = \pi r^2 - \pi r^2$

$\quad = 3.14 \, (10 \text{ m})^2 - 3.14 \, (6 \text{ m})^2$

$\quad = 3.14 \, (100 \text{ m}^2) - 3.14 \, (36 \text{ m}^2)$

$\quad = 314 \text{ m}^2 - 113.04 \text{ m}^2$

$\quad = 200.96 \text{ m}^2$

$\quad \approx 201.0 \text{ m}^2$

**52.**  $A = \ell w + \pi r^2$

$\quad = (24 \text{ ft}) \, (10 \text{ ft}) + 3.14 \, (5 \text{ ft})^2$

$\quad = 240 \text{ ft}^2 + 78.5 \text{ ft}^2$

$\quad = 318.5 \text{ ft}^2$

**53.**  $A = \ell w - \pi r^2$

$\quad = (20 \text{ m}) \, (14 \text{ m}) - 3.14 \, (7 \text{ m})^2$

$\quad = 280 \text{ m}^2 - 153.86 \text{ m}^2$

$\quad = 126.14 \text{ m}^2$

$\quad \approx 126.1 \text{ m}^2$

**54.**  $A = bh - \pi r^2$

$\quad = (12 \text{ ft}) \, (10 \text{ ft}) - 3.14 \, (2 \text{ ft})^2$

$\quad = 120 \text{ ft}^2 - 12.56 \text{ ft}^2$

$\quad = 107.44 \text{ ft}^2$

$\quad \approx 107.4 \text{ ft}^2$

**55.**  $A = \dfrac{h \, (b + B)}{2} + \dfrac{1}{2} \times \pi r^2$

$\quad = \dfrac{5 \text{ m} \, (8 \text{ m} + 14 \text{ m})}{2}$

$\quad \quad + \dfrac{1}{2} \times (3.14) \, (4 \text{ m})^2$

$\quad = \dfrac{5 \text{ m} \, (22 \text{ m})}{2} + \dfrac{1}{2} \times (3.14) \, (16 \text{ m}^2)$

$\quad = 55 \text{ m} + 25.12 \text{ m}^2$

$\quad = 80.12 \text{ m}^2$

$\quad \approx 80.1 \text{ m}^2$

**56.**  $V = \ell w h$

$\quad = (3 \text{ ft}) \, (6 \text{ ft}) \, (2.5 \text{ ft})$

$\quad = (18 \text{ ft}^2) \, (2.5 \text{ ft})$

$\quad = 45 \text{ ft}^3$

**57.**  $V = \dfrac{4 \pi r^3}{3}$

$\quad = \dfrac{4 \, (3.14) \, (1.2 \text{ ft})^3}{3}$

$\quad = \dfrac{4 \, (3.14) \, (1.728 \text{ ft}^3)}{3}$

$\quad = \dfrac{21.70368 \text{ ft}^3}{3}$

$\quad = 7.23456 \text{ ft}^3$

$\quad \approx 7.2 \text{ ft}^3$

**58.**  $V = \pi r^2 h$

$\quad = 3.14 \, (7 \text{ m})^2 \, (2 \text{ m})$

$\quad = 3.14 \, (49 \text{ m}^2) \, (2 \text{ m})$

$\quad = 307.72 \text{ m}^3$

$\quad \approx 307.7 \text{ m}^3$

**59.** $B = (16 \text{ m}) (18 \text{ m}) = 288 \text{ m}^2$

$$V = \frac{hB}{3}$$

$$= \frac{(18 \text{ m}) (288 \text{ m}^2)}{3}$$

$$= \frac{5184 \text{ m}^3}{3}$$

$$= 1728 \text{ m}^3$$

**60.** $B = (7 \text{ m}) (7 \text{ m}) = 49 \text{ m}^2$

$$V = \frac{hB}{3}$$

$$= \frac{(15 \text{ m}) (49 \text{ m}^2)}{3}$$

$$= \frac{735 \text{ m}^3}{3}$$

$$= 245 \text{ m}^3$$

**61.** $V = \frac{\pi r^2 h}{3}$

$$= \frac{3.14 (20 \text{ ft})^2 (9 \text{ ft})}{3}$$

$$= \frac{3.14 (400 \text{ ft}^2) (9 \text{ ft})}{3}$$

$$= \frac{11{,}304 \text{ ft}^3}{3}$$

$$= 3768 \text{ ft}^3$$

**62.** $V = \frac{\pi r^2 h}{3}$

$$= \frac{3.14 (17 \text{ yd})^2 (30 \text{ yd})}{3}$$

$$= \frac{3.14 (289 \text{ yd}^2) (30 \text{ yd})}{3}$$

$$= \frac{27{,}203.8 \text{ yd}^3}{3}$$

$$= 9074.6 \text{ yd}^3$$

**63.** $\dfrac{n}{2} = \dfrac{45}{3}$

$$3n = (2) (45)$$

$$3n = 90$$

$$\frac{3n}{3} = \frac{90}{3}$$

$$n = 30$$

30 m

**64.** $\dfrac{n}{20} = \dfrac{6}{36}$

$$36n = (20) (6)$$

$$36n = 120$$

$$\frac{36n}{36} = \frac{120}{36}$$

$$n \approx 3.3$$

3.3 m

**65.** Small figure:

$$P = 18 \text{ cm} + 26 \text{ cm} + 2 (5 \text{ cm})$$

$$= 54 \text{ cm}$$

$$\frac{n}{54} = \frac{108}{18}$$

$$18n = (54) (108)$$

$$18n = 5832$$

$$\frac{18n}{18} = \frac{5832}{18}$$

$$n = 324$$

324 cm

**66.** Small figure:

$$P = 20 \text{ ft} + 13 \text{ ft} + 25 \text{ ft} + 12 \text{ ft}$$

$$= 70 \text{ ft}$$

$$\frac{n}{70} = \frac{32.5}{13}$$

$$13n = (70)(32.5)$$

$$13n = 2275$$

$$\frac{13n}{13} = \frac{2275}{13}$$

$$n = 175$$

175 ft

**67.** $3\frac{1}{2} \times 3\frac{1}{2} = \frac{7}{2} \times \frac{7}{2} = \frac{49}{4}$

$$\frac{n}{\frac{49}{4}} = \frac{12}{1}$$

$$n = \left(\frac{49}{4}\right)(12)$$

$$= 147$$

147 yd²

**68.** $A = \ell w - \ell w$

$$= (14 \text{ yd})(8 \text{ yd}) - (4 \text{ yd})(5 \text{ yd})$$

$$= 112 \text{ yd}^2 - 20 \text{ yd}^2$$

$$= 92 \text{ yd}^2$$

$$\text{Cost} = 92 \text{ yd}^2 \times \frac{\$8}{\text{yd}^2}$$

$$= \$736$$

**69.** $V = \dfrac{\pi r^2 h}{3}$

$$= \frac{3.14 \, (9 \text{ in.})^2 (24 \text{ in.})}{3}$$

$$= \frac{3.14 \, (81 \text{ in.}^2)(24 \text{ in.})}{3}$$

$$\approx 2034.7 \text{ in.}^3$$

$$W = 2034.7 \text{ in.}^3 \times \frac{16 \text{ g}}{1 \text{ in}^3}$$

$$= 32{,}555.2 \text{ g}$$

**70. a.** $V = \pi r^2 h + \dfrac{1}{2} \times \dfrac{4\pi r^3}{3}$

$$= 3.14 \, (9 \text{ ft})^2 (80 \text{ ft})$$

$$+ \frac{1}{2} \times \frac{4 \, (3.14)(9 \text{ ft})^3}{3}$$

$$= 3.14 \, (81 \text{ ft}^2)(80 \text{ ft})$$

$$+ \frac{1}{2} \times \frac{4 \, (3.14)(729 \text{ ft}^3)}{3}$$

$$= 20{,}347.2 \text{ ft}^3 + \frac{1}{2} \times 3052.08 \text{ ft}^3$$

$$= 20{,}347.2 \text{ ft}^3 + 1526.04 \text{ ft}^3$$

$$= 21{,}873.24 \text{ ft}^3$$

$$\approx 21{,}873.2 \text{ ft}^3$$

**b.** $B = 21{,}873.2 \text{ ft}^3 \times \dfrac{0.8 \text{ bushel}}{1 \text{ ft}^3}$

$$\approx 17{,}498.6 \text{ bushels}$$

**71. a.** $\text{Trip} = 32 \text{ km} + 18 \text{ km}$

$$= 50 \text{ km}$$

$$\text{Speed} = \frac{50 \text{ km}}{0.5 \text{ hr}}$$

$$= 100 \text{ km/hr}$$

**b.**   Trip = 26 km + 14 km + 16 km

   = 56 km

   Speed = $\dfrac{56 \text{ km}}{0.8 \text{ km}}$

   = 70 km/hr

**c.** through Ipswich

**72.** 2.757 billion × 1.244 ft$^3$

   = 3.429708 billion ft$^3$

   = 3,429,708,000 ft$^3$

**73.**   $h = \dfrac{V}{\ell w}$

   $= \dfrac{3,429,108,000 \text{ ft}^3}{(10,000 \text{ ft})(20,000 \text{ ft})}$

   $= \dfrac{3,429,108,000 \text{ ft}^3}{200,000,000 \text{ ft}^2}$

   $\approx 17.1$ ft

**74.**   $P = 2(21 \text{ ft}) + 2(30 \text{ ft})$

   = 42 ft + 60 ft

   = 102 ft

   Cost $= 102 \text{ ft} \times \dfrac{\$7}{1 \text{ ft}}$

   = \$714

**75.** Posts $= 102 \text{ ft} \times \dfrac{1 \text{ post}}{3 \text{ ft}}$

   = 34 posts

**76.**   $A = \ell w - \ell w$

   $= (8.5 \text{ in.})(11 \text{ in.}) - (7 \text{ in.})(9 \text{ in.})$

   = 93.5 in.$^2$ − 63 in.$^2$

   = 30.5 in.$^2$

**77.** $A = \ell w + \pi r^2, \quad r = \dfrac{d}{2} = \dfrac{16 \text{ yd}}{2} = 8$ yd

   $= (16 \text{ yd})(20 \text{ yd}) + 3.14(8 \text{ yd})^2$

   = 320 yd$^2$ + 200.96 yd$^2$

   = 520.96 yd$^2$

   $\approx 521.0$ yd$^2$

**78.**   $A \approx 521.0$ yd$^2$

   Cost $= 521 \text{ yd}^2 \times \dfrac{\$50}{1 \text{ yd}^2}$

   = \$26,050

**79.**   leg $= \sqrt{33^2 - 30^2}$

   $= \sqrt{1089 - 900}$

   $= \sqrt{189}$

   $\approx 13.7$ ft

**80.**   $r = \dfrac{d}{2} = \dfrac{90 \text{ m}}{2} = 45$ m

   $V = \dfrac{4\pi r^3}{3}$

   $= \dfrac{4(3.14)(45 \text{ m})^3}{2}$

   $= \dfrac{4(3.14)(91,125 \text{ m}^3)}{3}$

   = 381,510 m$^3$

**81.** 18 in. $\times \dfrac{1 \text{ ft}}{12 \text{ in.}} = \dfrac{3}{2}$ ft

   $r = \dfrac{d}{2} = \dfrac{\frac{3}{2}}{2} = \dfrac{3}{4}$ ft = 0.75 ft

   $V = \pi r^2 h$

   $= 3.14(0.75 \text{ ft})^2(5 \text{ ft})$

   $= 3.14(0.5625 \text{ ft}^2)(5 \text{ ft})$

   = 8.83125 ft$^3$

   $\approx 8.8$ ft$^3$

**82.**     $V \approx 8.8 \text{ ft}^3$

gallons $= 8.8 \text{ ft}^3 \times \dfrac{7.5 \text{ gal}}{1 \text{ ft}^3}$

$= 66 \text{ gal}$

**83.**  $A = \dfrac{(35 \text{ ft}) (45 \text{ ft} + 50 \text{ ft})}{2}$

$= \dfrac{(35 \text{ ft}) (95 \text{ ft})}{2}$

$= \dfrac{3325 \text{ ft}^2}{2}$

$= 1662.5 \text{ ft}^2$

**84.**   $A = 1662.5 \text{ ft}^2$

Cost $= 1662.5 \text{ ft}^2 \times \dfrac{\$0.50}{1 \text{ ft}^3} \times 3 \text{ times/yr}$

$= \$2493.75$

## Chapter 7 Test

**1.**  $P = 2 (9 \text{ yd}) + 2 (11 \text{ yd})$
$= 18 \text{ yd} + 22 \text{ yd}$
$= 40 \text{ yd}$

**2.**  $P = 4 (6.3 \text{ ft})$
$= 25.2 \text{ ft}$

**3.**  $P = 2 (6.5 \text{ m}) + 2 (3.5 \text{ m})$
$= 13 \text{ m} + 7 \text{ m}$
$= 20 \text{ m}$

**4.**  $P = 2 (13 \text{ m}) + 22 \text{ m} + 32 \text{ m}$
$= 26 \text{ m} + 22 \text{ m} + 32 \text{ m}$
$= 80 \text{ m}$

**5.**  $P = 58.6 \text{ m} + 32.9 \text{ m} + 45.5 \text{ m}$
$= 137 \text{ m}$

**6.**  $A = (10 \text{ yd}) (18 \text{ yd})$
$= 180 \text{ yd}^2$

**7.**  $A = (10.2 \text{ m})^2$
$= 104.04 \text{ m}^2$
$\approx 104.0 \text{ m}^2$

**8.**  $A = (13 \text{ m}) (6 \text{ m})$
$= 78 \text{ m}^2$

**9.**  $A = \dfrac{(9 \text{ m}) (7 \text{ m} + 25 \text{ m})}{2}$

$= \dfrac{(9 \text{ m}) (32 \text{ m})}{2}$

$= \dfrac{288 \text{ m}^2}{2}$

$= 144 \text{ m}^2$

**10.**  $A = \dfrac{(4 \text{ cm}) (6 \text{ cm})}{2}$

$= \dfrac{24 \text{ cm}^2}{2}$

$= 12 \text{ cm}^2$

**11.**  $A = \dfrac{(15 \text{ m}) (7 \text{ m})}{2}$

$= \dfrac{105 \text{ m}^2}{2}$

$= 52.5 \text{ m}^2$

**12.**  $\sqrt{81} = 9$

**13.**  $\sqrt{121} = 11$

**14.**  $90° - 63° = 27°$

**15.**  $180° - 107° = 73°$

**16.** $\sqrt{54} \approx 7.348$

**17.** $\sqrt{135} \approx 11.619$

**18.** $\sqrt{187} \approx 13.675$

**19.** hypotenuse $= \sqrt{7^2 + 6^2}$
$\qquad = \sqrt{49 + 36}$
$\qquad = \sqrt{85}$
$\qquad \approx 9.220$

**20.** leg $= \sqrt{26^2 - 24^2}$
$\qquad = \sqrt{676 - 576}$
$\qquad = \sqrt{100}$
$\qquad = 10$

**21.** hypotenuse $= \sqrt{5^2 + 3^2}$
$\qquad = \sqrt{25 + 9}$
$\qquad = \sqrt{34}$
$\qquad \approx 5.83$ cm

**22.** hypotenuse $= \sqrt{15^2 - 12^2}$
$\qquad = \sqrt{225 - 144}$
$\qquad = \sqrt{81}$
$\qquad = 9$ ft

**23.** $C = 2\pi r$
$\qquad \approx 2\,(3.14)\,(6 \text{ in.})$
$\qquad = 37.68$ in.
$\qquad \approx 37.7$ in.

**24.** $r = \dfrac{d}{2} = \dfrac{18 \text{ ft}}{2} = 9$ ft

$A = \pi r^2$
$\quad = 3.14\,(9 \text{ ft})^2$
$\quad = 3.14\,(81 \text{ ft}^2)$
$\quad = 254.34 \text{ ft}^2$
$\quad \approx 254.3 \text{ ft}^2$

**25.** $A = bh - \pi r^2$
$\quad = (15 \text{ in.})\,(8 \text{ in.}) - (3.14)\,(2 \text{ in.})^2$
$\quad = 120 \text{ in.}^2 - 12.56 \text{ in.}^2$
$\quad = 107.44 \text{ in.}^2$
$\quad \approx 107.4 \text{ in.}^2$

**26.** $A = \dfrac{h\,(b + B)}{2} + \dfrac{1}{2} \times \pi r^2$

$\quad = \dfrac{(7 \text{ in.})\,(10 \text{ in.} + 20 \text{ in.})}{2}$
$\qquad + \dfrac{1}{2} \times (3.14)\,(5 \text{ in.})^2$

$\quad = \dfrac{(7 \text{ in.})\,(30 \text{ in.})}{2}$
$\qquad + \dfrac{1}{2} \times (3.14)\,(25 \text{ in.}^2)$

$\quad = 105 \text{ in.}^2 + 39.25 \text{ in.}^2$
$\quad = 144.25 \text{ in.}^2$
$\quad \approx 144.3 \text{ in.}^2$

**27.** $V = \ell wh$
$\quad = (7 \text{ m})\,(12 \text{ m})\,(10 \text{ m})$
$\quad = 840 \text{ m}^3$

**28.**  $V = \dfrac{\pi r^2 h}{3}$

$= \dfrac{3.14\,(8\text{ m})^2\,(12\text{ m})}{3}$

$= \dfrac{3.14\,(64\text{ m}^2)\,(12\text{ m})}{3}$

$= 803.84\text{ m}^3$

$\approx 803.8\text{ m}^3$

**29.**  $V = \dfrac{4\pi r^3}{3}$

$= \dfrac{4\,(3.14)\,(3\text{ m})^3}{3}$

$= \dfrac{4\,(3.14)\,(27\text{ m})^3}{3}$

$= 113.04\text{ m}^3$

$\approx 113.0\text{ m}^3$

**30.**  $V = \pi r^2 h$

$= 3.14\,(9\text{ ft})^2\,(2\text{ ft})$

$= 3.14\,(81\text{ ft}^2)\,(2\text{ ft})$

$= 508.68\text{ ft}^3$

$\approx 508.7\text{ ft}^3$

**31.**  $B = (4\text{ m})\,(3\text{ m}) = 12\text{ m}^2$

$V = \dfrac{hB}{3}$

$= \dfrac{(14\text{ m})\,(12\text{ m}^2)}{3}$

$= 56\text{ m}^3$

**32.**  $\dfrac{n}{18} = \dfrac{12}{5}$

$5n = (18)\,(12)$

$5n = 216$

$\dfrac{3n}{5} = \dfrac{216}{5}$

$n = 43.2$

43.2 m

**33.**  $\dfrac{n}{7} = \dfrac{60}{9}$

$9n = (7)\,(60)$

$9n = 420$

$\dfrac{9n}{9} = \dfrac{420}{9}$

$n = 46.\overline{6}$

$n \approx 46.7$

46.7 ft

**34.**  $r = \dfrac{d}{2} = 20\text{ yd}$

$A = \ell w + \pi r^2$

$= (130\text{ yd})\,(40\text{ yd}) + 3.14\,(20\text{ yd})^2$

$= 5200\text{ yd}^2 + 1256\text{ yd}^2$

$= 6456\text{ yd}^2$

**35.**  $A = 6456\text{ yd}^2$

$\text{Cost} = 6456\text{ yd}^2 \times \dfrac{\$0.40}{1\text{ yd}^2}$

$= \$2582.40$

## Chapters 1–7 Cumulative Test

**1.**  126,350
278,120
+ 531,290
935,760

**2.**
$$\begin{array}{r} 163 \\ \times\ 205 \\ \hline 815 \\ 32\ 60 \\ \hline 33{,}415 \end{array}$$

**3.**
$$\frac{17}{18} \qquad \frac{85}{90}$$
$$-\frac{11}{30} \qquad -\frac{33}{90}$$
$$\qquad\qquad \frac{52}{90} = \frac{26}{45}$$

**4.** $\dfrac{3}{7} \div 2\dfrac{1}{4} = \dfrac{3}{7} \div \dfrac{9}{4}$
$$= \frac{3}{7} \times \frac{4}{9}$$
$$= \frac{4}{21}$$

**5.** $56.1279 \approx 56.13$

**6.**
$$\begin{array}{r} 9.034 \\ \times\ 0.8 \\ \hline 7.2272 \end{array}$$

**7.**
$$0.021\overline{)1.743}$$
with quotient $83$, $1\ 68$, $63$, $63$, $0$

**8.**
$$\frac{3}{n} = \frac{2}{18}$$
$$3 \times 18 = n \times 2$$
$$54 = n \times 2$$
$$\frac{54}{2} = \frac{n \times 2}{2}$$
$$27 = n$$

**9.**
$$\frac{7}{100} = \frac{56}{n}$$
$$7 \times n = 100 \times 56$$
$$\frac{7 \times n}{7} = \frac{5600}{7}$$
$$n = 800 \text{ students}$$

**10.** $\dfrac{18}{24} = 0.75 = 75\%$

**11.** $0.8\%$ of what number is 16?
$$0.8\% \times n = 16$$
$$\frac{0.008 \times n}{0.008} = \frac{16}{0.008}$$
$$n = 2000$$

**12.**
$$n = 18.5\% \times 220$$
$$n = 0.185 \times 220$$
$$n = 40.7$$

**13.** $586 \text{ cm} \times \dfrac{1 \text{ m}}{100 \text{ cm}} = 5.86 \text{ cm}$

**14.** $42 \text{ yd} \times \dfrac{36 \text{ in.}}{1 \text{ yd}} = 1512 \text{ in.}$

**15.** $88 \text{ km} \times \dfrac{0.62 \text{ mi}}{1 \text{ km}} = 54.56 \text{ mi}$

**16.** $P = 2(17 \text{ m}) + 2(8 \text{ m})$
$$= 34 \text{ m} + 16 \text{ m}$$
$$= 50 \text{ m}$$

**17.** $P = 86 \text{ cm} + 13 \text{ cm} + 96 \text{ cm} + 13 \text{ cm}$
$$= 208 \text{ cm}$$

18. $C = \pi d$

$= 3.14\,(18\ \text{yd})$

$= 56.52\ \text{yd}$

$\approx 56.5\ \text{yd}$

19. $A = \dfrac{bh}{2}$

$= \dfrac{(1.2\ \text{cm})\,(2.4\ \text{cm})}{2}$

$= \dfrac{2.88\ \text{cm}^2}{2}$

$= 1.44\ \text{cm}^2$

$\approx 1.4\ \text{cm}^2$

20. $A = \dfrac{h\,(b + B)}{2}$

$= \dfrac{(18\ \text{m})\,(26\ \text{m} + 34\ \text{m})}{2}$

$= \dfrac{(18\ \text{m})\,(60\ \text{m})}{2}$

$= 540\ \text{m}^2$

21. $A = \ell w + bh$

$= (12\ \text{m})\,(12\ \text{m}) + (12\ \text{m})\,(4\ \text{m})$

$= 144\ \text{m}^2 + 48\ \text{m}^2$

$= 192\ \text{m}^2$

22. $A = \ell w - \ell w$

$= (35\ \text{yd})\,(20\ \text{yd}) - (6\ \text{yd})\,(6\ \text{yd})$

$= 700\ \text{yd}^2 - 36\ \text{yd}^2$

$= 664\ \text{yd}^2$

23. $A = \pi r^2$

$= 3.14\,(4\ \text{m})^2$

$= 3.14\,(16\ \text{m}^2)$

$= 50.24\ \text{m}^2$

$\approx 50.2\ \text{m}^2$

24. $V = \pi r^2 h$

$= 3.14\,(8\ \text{m})^2\,(12\ \text{m})$

$= 3.14\,(64\ \text{m}^2)\,(12\ \text{m})$

$= 2411.52\ \text{m}^3$

$\approx 2411.5\ \text{m}^3$

25. $V = \dfrac{4\pi r^3}{3}$

$= \dfrac{4\,(3.14)\,(9\ \text{cm})^3}{3}$

$= \dfrac{4\,(3.14)\,(729\ \text{cm}^3)}{3}$

$= 3052.08\ \text{cm}^3$

$\approx 3052.1\ \text{cm}^3$

26. $B = (14\ \text{cm})\,(21\ \text{cm})$

$= 294\ \text{cm}^2$

$V = \dfrac{hB}{3}$

$= \dfrac{(32\ \text{cm})\,(294\ \text{cm}^2)}{3}$

$= 3136\ \text{cm}^2$

27. $V = \dfrac{\pi r^2 h}{3}$

$= \dfrac{3.14\,(12\ \text{m})^2\,(18\ \text{m})}{3}$

$= \dfrac{3.14\,(144\ \text{m}^2)\,(18\ \text{m})}{3}$

$= 2712.96\ \text{m}^3$

$\approx 2713.0\ \text{m}^3$

**28.**   $\dfrac{n}{9} = \dfrac{26}{7}$

$7n = (9)(26)$

$7n = 234$

$\dfrac{7n}{7} = \dfrac{234}{7}$

$n \approx 33.4$

33.4 m

**29.**   $\dfrac{n}{11} = \dfrac{1.5}{4}$

$4n = (11)(1.5)$

$4n = 16.5$

$\dfrac{4n}{4} = \dfrac{16.5}{4}$

$n = 4.125$

$n \approx 4.1$

4.1 ft

**30. a.**  $A = (14 \text{ yd})(6 \text{ yd}) + (5 \text{ yd})(5 \text{ yd})$

$\qquad + \dfrac{(5 \text{ yd})(6 \text{ yd})}{2}$

$= 84 \text{ yd}^2 + 25 \text{ yd}^2 + 15 \text{ yd}^2$

$= 124 \text{ yd}^2$

**b.** $\text{Cost} = 124 \text{ yd}^2 \times \dfrac{\$8}{1 \text{ yd}^2}$

$= \$992$

**31.** $\sqrt{36} + \sqrt{25} = 6 + 5$

$= 11$

**32.** $\sqrt{57} \approx 7.550$

**33.**  hypotenuse $= \sqrt{10^2 + 3^2}$

$= \sqrt{100 + 9}$

$= \sqrt{109}$

$\approx 10.440$ in.

**34.**  leg $= \sqrt{7^2 - 5^2}$

$= \sqrt{49 - 25}$

$= \sqrt{24}$

$\approx 4.899$ m

**35.**  hypotenuse $= \sqrt{12^2 + 7^2}$

$= \sqrt{144 + 49}$

$= \sqrt{193}$

$\approx 13.9$ mi

**36.**   $20 \text{ ft} \times \dfrac{12 \text{ in.}}{1 \text{ ft}} = 240 \text{ in.}$

Number needed $= 240 \div 7\dfrac{1}{2}$

$= 240 \div \dfrac{15}{2}$

$= 240 \times \dfrac{2}{15}$

$= 32$ paintbrushes

# STATISTICS

## Pretest Chapter 8

**1.** Under age 18

**2.** $33\% + 10\% = 43\%$

**3.** What percent of the students are 25 or older?

Age 25–27 + Over age 27
$$= 10\% + 7\%$$
$$= 17\%$$

**4.** $33\% \times 5000 = 0.33 \times 5000$
$$= 1650 \text{ students}$$

**5.** How many students are over age 27?

$7\% \times 5000 = 0.07 \times 5000$
$$= 350 \text{ students}$$

**6.** 300 housing starts

**7.** 450 housing starts

**8.** During the 2nd quarter of 2000

**9.** During the 3rd quarter of 2001

**10.** $550 - 250 = 300$

**11.** $450 - 400 = 50$

**12.** August and December

**13.** December

**14.** November

**15.** 20,000 sets

**16.** 30,000 sets

**17.** 55,000 cars

**18.** 60,000 cars

**19.** $20,000 + 5000 = 25,000$ cars

**20.** $5000 + 15,000 = 20,000$ cars

**21.** $\dfrac{38 + 42 + 44 + 27 + 32 + 27}{6}$
$$= \dfrac{210}{6}$$
$$= 35 \text{ pages}$$

**22.** $\dfrac{38 + 32}{2} = \dfrac{70}{2} = 35$ pages

**23.** 27 pages for mode

---

## 8.1 Exercises

**1.** rent

**3.** $200 for utilities

**5.** $650 + $150 = $800

**7.** $\dfrac{\$650}{\$200} = \dfrac{650 \div 50}{200 \div 50} = \dfrac{13}{4}$

**9.** $\dfrac{\$1000}{\$2700} = \dfrac{1000 \div 100}{2700 \div 100} = \dfrac{10}{27}$

**11.** Hit batters is the category with the least number of pitches.

**13.** 294 pitches were balls.

**15.** 85 hits + 294 balls = 379 pitches

**17.** $\dfrac{144}{650} = \dfrac{144 \div 2}{650 \div 2} = \dfrac{72}{325}$

**19.** $\dfrac{294}{144} = \dfrac{294 \div 6}{144 \div 6} = \dfrac{49}{24}$

**21.** $11\% + 50\% = 61\%$

**23.** $100\% - (25\% + 50\%) = 100\% - 75\%$
$$= 25\%$$

**25.** Number $= 0.50 \times 198{,}000{,}000$
$$= 99{,}000{,}000 \text{ people}$$

**27.** $0.36 \times 6{,}000{,}000{,}000$
$$= 2{,}160{,}000{,}000 \text{ Christians}$$

**29.** $20\% + 19\% = 39\%$

**31.** $100\% - 20\% = 80\%$

**33.** $\text{percent} = \dfrac{69{,}000{,}000}{6{,}000{,}000{,}000}$

$$= 0.0115$$

$$\approx 1.2\%$$

## Cumulative Review Problems

**35.** $A = \dfrac{bh}{2}$

$$= \dfrac{(6 \text{ in.})\,(14 \text{ in.})}{2}$$

$$= \dfrac{84 \text{ in.}^2}{2}$$

$$= 42 \text{ in.}^2$$

**37.** $A = 2lw + 2lw$

$$= 2\,(7 \text{ yd})\,(12 \text{ yd}) + 2\,(7 \text{ yd})\,(20 \text{ yd})$$

$$= 2\,(84 \text{ yd}^2) + 2\,(140 \text{ yd}^2)$$

$$= 168 \text{ yd}^2 + 280 \text{ yd}^2$$

$$= 448 \text{ yd}^2$$

$$448 \text{ yd}^2 \times \dfrac{1 \text{ gal}}{28 \text{ yd}^2} = 16 \text{ gal}$$

16 gallons of paint

## 8.2  Exercises

**1.** 20 million or 20,000,000 people

**3.** 14 million or 14,000,000 people

**5.** 1960–1970 with 1 million or 1,000,000 people

**7.** 22 quadrillion Btu

**9.** 1970

**11.** $18 - 14 = 4$ quadrillion Btu

**13.** $22 - 16 = 6$ quadrillion Btu

**15.** from 1975 to 1980 and from 1985 to 1990 with an increase of 3 quadrillion Btu

**17.** $24 - 18 = 6$ quadrillion Btu
For 2020: $24 + 6 = 30$ quadrillion Btu

**19.** $3.5 million or $3,500,000

**21.** 1999

**23.** $4.5 - $3.5 = $1 million or $1,000,000

**25.** 2.5 inches of rainfall

**27.** October, November, and December

**29.** $4.0 - 2.5 = 1.5$ inches

**31.**

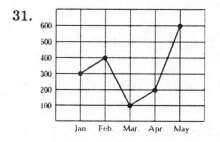

Jan.   Feb.   Mar.   Apr.   May

## Cumulative Review Problems

**33.** $7 \times 6 + 3 - 5 \times 2 = 42 + 3 - 10$
$$= 45 - 10$$
$$= 35$$

**35.** $\dfrac{1}{5} + \left( \dfrac{1}{5} - \dfrac{1}{6} \right) \times \dfrac{2}{3} = \dfrac{1}{5} + \left( \dfrac{6}{30} - \dfrac{5}{30} \right) \times \dfrac{2}{3}$

$$= \dfrac{1}{5} + \dfrac{1}{30} \times \dfrac{2}{3}$$

$$= \dfrac{1}{5} + \dfrac{1}{45}$$

$$= \dfrac{9}{45} + \dfrac{1}{45}$$

$$= \dfrac{10}{45}$$

$$= \dfrac{2}{9}$$

## Putting Your Skills To Work

**1.** $3{,}021{,}000 - 1{,}771{,}000 = 1{,}250{,}000$

**3.** 2030 to 2040

## 8.3   Exercises

**1.** 10 cars achieve between 28 and 30.9 mi per gallon

**3.** 35 cars achieve between 25 and 27.9 mi per gallon

**5.**     35 cars between 25 and 27.9
   + 10 cars between 28 and 30.9
    45 cars more than 24.9

**7.** 50 cars between 19 and 21.9
60 cars between 22 and 24.9
35 cars between 25 and 27.9
145 cars achieve between 19 and 27.9 mpg

**9.** 8,000 books

**11.** books costing $5.00 - $7.99

**13.**     3,000
            8,000
         + 17,000
         ─────────
           28,000  books

**15.**    17,000
           10,000
           12,000
         + 13,000
         ─────────
           52,000  books

**17.**    13,000
         + 7,000
         ─────────
           20,000  books

$$\text{percent} = \frac{20,000}{70,000}$$

$$\approx 0.286$$

$$= 28.6\%$$

**19.** Tally: |||
    Frequency: 3

**21.** Tally: ||||||
    Frequency: 6

**23.** Tally: |||
    Frequency: 3

**25.** Tally: ||
    Frequency: 2

**27.**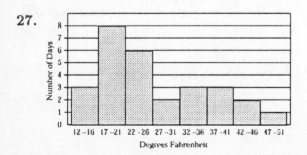

**29.** $3 + 2 + 1 = 6$ days

**31.** Tally: |||
    Frequency: 3

**33.** Tally: ||
    Frequency: 2

**35.** Tally: |
    Frequency: 1

**37.** Tally: ||
    Frequency: 2

**39.** $6 + 3 + 3 = 12$ prescriptions

───────────────────────────────

## Cumulative Review Problems

**41.**     $$\frac{126}{n} = \frac{36}{17}$$

$$125 \times 17 = n \times 36$$

$$2142 = n \times 36$$

$$\frac{2142}{36} = \frac{n \times 36}{36}$$

$$59.5 = n$$

**43.**     $$\frac{3}{5} = \frac{n}{20}$$

$$3 \times 20 = 5 \times n$$

$$60 = 5 \times n$$

$$\frac{60}{5} = \frac{5 \times n}{5}$$

$$n = 12 \text{ pounds}$$

## 8.4    Exercises

1. The median of a set of numbers when they are arranged in order from smallest to largest is that value that has the same number of values above it as below it. The mean of a set of values is the sum of the values divided by the number of values. The mean is most likely to be not typical of the values you would expect if there are many extremely low values or many extremely high values. The median is more likely to be typical of the value you would expect.

3. Mean $= \dfrac{30 + 29 + 28 + 35 + 34 + 37 + 31}{7}$

   $= \dfrac{224}{7}$

   $= 32$

5. Mean $= \dfrac{6 + 2 + 3 + 3.5 + 2.5 + 1 + 0}{7}$

   $= \dfrac{18}{7}$

   $\approx 2.6$ hours

7. Mean

   $= \dfrac{\begin{array}{c}67{,}000 + 86{,}000 + 107{,}000 \\ + 134{,}000 + 152{,}000\end{array}}{5}$

   $= \dfrac{546{,}000}{5}$

   $= 109{,}200$

9. Batting Average

   $= \dfrac{\#\ \text{Hits}}{\#\ \text{At Bats}}$

   $= \dfrac{0 + 2 + 3 + 2 + 2}{5 + 4 + 6 + 5 + 4}$

   $= \dfrac{9}{24}$

   $= 0.375$

11. Avg miles/gallon

    $= \dfrac{\text{Miles driven}}{\text{Gallons used}}$

    $= \dfrac{276 + 350 + 391 + 336}{12 + 14 + 17 + 14}$

    $= \dfrac{1353}{57}$

    $\approx 23.7$ miles per gallon

13. 865, 968, 999, 1023, 1052, 1152

    Median $= \dfrac{999 + 1023}{2} = 1011$

15. 0.34, 0.52, 0.58, 0.69, 0.71

    Median $= 0.58$

17. $11,600, $15,700, $17,000, $23,500, $26,700, $31,500

    Median $= \dfrac{\$17{,}000 + \$23{,}500}{2} = \$20{,}250$

19. 19, 20, 21, 24, 28, 30, 32, 33, 35, 44

    Median $= \dfrac{28 + 30}{2} = 29$ years old

21. $97, $109, $185, $207, $218, $330, $420

    Median $= \$207$

**23.**   1.8, 1.9, 2.0, 2.0, 2.4, 3.1, 3.1, 3.7

$$\text{Median} = \frac{2.0 + 2.4}{2} = 2.2$$

**25.**
$$
\begin{array}{r}
\$\ 30{,}000 \\
74{,}500 \\
47{,}890 \\
89{,}000 \\
57{,}645 \\
78{,}090 \\
110{,}370 \\
+\quad 65{,}800 \\
\hline
\$553{,}295
\end{array}
$$

$$\text{Mean} = \frac{\$553{,}295}{8} \approx \$69{,}161.88$$

**27.**   1987, 2576, 3700, 4700, 5000, 7200
8764, 9365

$$\text{Median} = \frac{4700 + 5000}{2} = 4850$$

**29.**  $\text{Mean} = \dfrac{18 + 27 + 101 + 93 + 111}{5}$

$$= \frac{350}{5}$$

$$= 70$$

18, 27, 93, 101, 111

$$\text{Median} = 93$$

**31. a.** Mean

$$= \frac{\begin{array}{c}1500 + 1700 + 1650 + 1300 + 1440 \\ +1580 + 1820 + 1380 + 2900 + 6300\end{array}}{10}$$

$$= \frac{21{,}570}{10}$$

$$= \$2157$$

**b.** 1300, 1380, 1440, 1500, 1580, 1650,
1700, 1820, 2900, 6300

$$\text{Median} = \frac{1580 + 1650}{2} = \$1615$$

**c.** The median since the mean is affected
by the score of \$6300.

**33.** The mode is 60 since this occurs twice.

**35.** The modes are 121 and 150 which both
occur twice.

**37.** The mode is \$249 which occurs twice.

**39.**  $\text{Mean} = \dfrac{\begin{array}{c}869 + 992 + 482 + 791 \\ +\ 399 + 855 + 869\end{array}}{7}$

$$= \frac{5257}{7}$$

$$= 751$$

399, 482, 791, 855, 869, 869, 992

$$\text{Median} = 855$$

$$\text{Mode} = 869 \text{ since it occurs twice.}$$

**41.**  $\text{GPA} = \dfrac{4\,(3) + 3\,(4) + 2\,(3) + 3\,(3)}{3 + 4 + 3 + 3}$

$$= \frac{12 + 12 + 6 + 9}{3 + 4 + 3 + 3}$$

$$= \frac{39}{13}$$

$$= 3.0$$

## Cumulative Review Problems

**43.**  $A = \dfrac{bh}{2}$

$$= \frac{(7 \text{ in.})\,(5.5 \text{ in.})}{2}$$

$$= \frac{38.5 \text{ in.}^2}{2}$$

$$= 19.25 \text{ in.}^2$$

$$\approx 19.3 \text{ in.}^2$$

**45.**
$$A = bh$$
$$= (5 \text{ ft}) (4 \text{ ft})$$
$$= 20 \text{ ft}^2$$

$$\text{Cost} = 20 \text{ ft}^2 \times \frac{\$16.50}{1 \text{ ft}^2}$$
$$= \$330$$

## Chapter 8 Review Problems

**1.** 13 computers

**2.** 32 computers

**3.** $43 + 25 = 68$ computers

**4.** $21 + 6 = 27$ computers

**5.** $\dfrac{13}{21}$

**6.** $\dfrac{43}{32}$

**7.** $\dfrac{25}{140} \approx 0.179 = 17.9\%$

**8.** $\dfrac{32}{140} \approx 0.229 = 22.9\%$

**9.** $100\% - 36\% = 64\%$

**10.** $100\% - 16\% = 84\%$

**11.** 6% in the black color

**12.** blue at 16%

**13.** purple at 9%

**14.** $18\% + 15\% = 33\%$
$$33\% \times 500 = 0.33 \times 500$$
$$= 165 \text{ students}$$

**15.** $100\% - 6\% = 94\%$
$$94\% \times 500 = 0.94 \times 500$$
$$= 470 \text{ students}$$

**16.** $16\% + 9\% + 6\% = 31\%$
$$31\% \times 500 = 0.31 \times 500$$
$$= 155 \text{ students}$$

**17.** 36

**18.** 26

**19.** 10–13 years

**20.** 6–9 years

**21.** $18 - 6 = 12$ glasses

**22.** $30 - 8 = 22$ glasses

**23.** $\dfrac{10}{2} = \dfrac{5}{1}$

**24.** $\dfrac{36}{18} = \dfrac{2}{1}$

**25.** $31,000

**26.** $58,000

**27.** between 1985 and 1990 with a difference of
$$\$44,000 - \$33,000 = \$11,000$$

**28.** between 1980 and 1985 with a difference of
$$\$24{,}000 - \$16{,}000 = \$8000$$

**29.** $\$33{,}000 - \$24{,}000 = \$9000$

**30.** $\$52{,}000 - \$37{,}000 = \$15{,}000$

**31.** 2000 with a difference of
$$\$58{,}000 - \$41{,}000 = \$17{,}000$$

**32.** 1980 with a difference of
$$\$23{,}000 - \$16{,}000 = \$7000$$

**33.** Difference $= \$58{,}000 - \$23{,}000$
$$= \$35{,}000$$
$$\text{Average} = \frac{\$35{,}000}{4}$$
$$= \$8750$$

**34.** Difference $= \$41{,}000 - \$16{,}000$
$$= \$25{,}000$$
$$\text{Average} = \frac{\$25{,}000}{4}$$
$$= \$6250$$

**35.** Difference $= \$41{,}000 - \$37{,}000$
$$= \$4000$$
$$\text{Year } 2005 = \$41{,}000 + \$4000$$
$$= \$45{,}000$$

**36.** Difference $= \$58{,}000 - \$44{,}000$
$$= \$14{,}000$$
$$\text{Year } 2005 = \$58{,}000 + \$14{,}000$$
$$= \$72{,}000$$

**37.** 400 students

**38.** 500 students

**39.** 650 students

**40.** 450 students

**41.** $300 - 200 = 100$ students

**42.** $500 - 450 = 50$ students

**43.** 1999–2000

**44.** 2000–2001

**45.** 45,000 cones

**46.** 30,000 cones

**47.** $20{,}000 - 10{,}000 = 10{,}000$ cones

**48.** $60{,}000 - 30{,}000 = 30{,}000$ cones

**49.** $55{,}000 - 30{,}000 = 25{,}000$ cones

**50.** $60{,}000 - 30{,}000 = 30{,}000$ cones

**51.** The sharp drop in the number of ice cream cones purchased from July 2000 to August 2000 is probably directly related to the weather. Since August was cold and rainy, significantly fewer people wanted ice cream during August.

**52.** The sharp increase in the number of ice cream cones purchased from June 2001 to July 2001 is probably directly related to the weather. Since June was cold and rainy, significantly more people wanted ice cream during July.

**53.** 14,000,000

**54.** 18,000,000

**55.** Between 1996 and 1998

**56.** Between 1990 and 1992

**57.** $18,000,000 - 17,000,000 = 1,000,000$

**58.** $33,000,000 - 21,000,000 = 12,000,000$

**59.** 1992

**60.** 2000

**61.** $\dfrac{33,000,000 - 14,000,000}{5}$

$= \dfrac{19,000,000}{5}$

$= 3,800,000$

**62.** $\dfrac{21,000,000 - 10,000,000}{5}$

$= \dfrac{11,000,000}{5}$

$= 2,200,000$

**63.** $21,000,000 - 18,000,000 = 3,000,000$

Year 2004:

$21,000,000 + 3,000,000 = 24,000,000$

**64.** $33,000,000 - 17,000,000 = 16,000,000$

Year 2006:

$33,000,000 + 16,000,000 = 49,000,000$

**65.** 50 bridges

**66.** 25 bridges

**67.** 20 and 39 years old

**68.** $20 + 5 = 25$ bridges

**69.** $80 + 70 = 150$ bridges

**70.** $25 - 20 = 5$ bridges

**71.** Tally: ||||||||||
Frequency: 10

**72.** Tally: ||||||||
Frequency: 8

**73.** Tally: |||
Frequency: 3

**74.** Tally: |||||
Frequency: 5

**75.** Tally: ||
Frequency: 2

**76.**

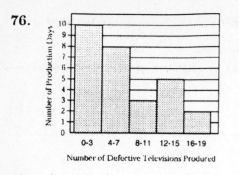

**77.** $10 + 8 = 18$ times

**78.** Mean

$$= \frac{86 + 83 + 88 + 95 + 97 + 100 + 81}{7}$$

$$= \frac{630}{7}$$

$$= 90$$

$90°$

**79.** Mean $= \dfrac{\begin{array}{c}\$87 + \$105 + \$89 + \$120 \\ + \$139 + \$160 + \$98\end{array}}{7}$

$$= \frac{\$798}{7}$$

$$= \$114$$

**80.** Mean

$$= \frac{1327 + 1567 + 1429 + 1307 + 1481}{5}$$

$$= \frac{7105}{5}$$

$$= 1421 \text{ cars}$$

**81.** Mean $= \dfrac{\begin{array}{c} 882 + 913 + 1017 + 1592 \\ + 1778 + 1936 \end{array}}{7}$

$$= \frac{8118}{6}$$

$$= 1353 \text{ employees}$$

**82.** $21,690, $28,500, $29,300, $35,000, $37,000, $38,600, $43,600, $45,300

$$\text{Median} = \frac{\$35,000 + \$37,000}{2}$$

$$= \$36,000$$

**83.** $98,000, $120,000, $126,000, $135,000, $139,000, $144,000, $150,000, $154,000, $156,000, $170,000

$$\text{Median} = \frac{\$139,000 + \$144,000}{2}$$

$$= \$141,500$$

**84.** 0, 1, 4, 5, 9, 18, 19, 19, 20, 21, 22, 22, 27, 36, 38, 43

$$\text{Median} = \frac{19 + 20}{2} = 19.5 \text{ cups}$$

$$\text{Mode} = 19 \text{ cups and } 22 \text{ cups}$$

**85.** 3, 9, 13, 14, 15, 15, 16, 18, 19, 21, 24, 25, 26, 28, 31, 36

$$\text{Median} = \frac{18 + 19}{2}$$

$$= 18.5 \text{ deliveries}$$

$$\text{Mode} = 15 \text{ deliveries}$$

**86.** The median, because of one low score, 31.

**87.** The median is a better measure of the usual sales because of the one very high data item, 39.

## Chapter 8 Test

**1.** 37%

**2.** 21%

**3.** $6\% + 2\% + 4\% = 12\%$

**4.** $30\% \times 300{,}000$

$\qquad = 0.30 \times 300{,}000$

$\qquad = 90{,}000$ automobiles

**5.** $\qquad 21\% + 6\% = 27\%$

$27\% \times 300{,}000 = 0.27 \times 300{,}000$

$81{,}000$ automobiles

**6.** 350 cars

**7.** 500 cars

**8.** 3rd quarter 2001

**9.** 1st quarter

**10.** $400 - 350 = 50$ cars

**11.** $450 - 300 = 150$ cars

**12.** 20 years

**13.** 26 years

**14.** $48 - 36 = 12$ years

**15.** age 35

**16.** age 65

**17.** 60,000 televisions

**18.** 25,000 televisions

**19.** $15{,}000 + 5000 = 20{,}000$ televisions

**20.** $45{,}000 + 15{,}000 = 60{,}000$ televisions

**21.** Mean $= \dfrac{\begin{array}{c}19 + 16 + 15 + 12 + 18 + 17 \\ + 14 + 10 + 16 + 20\end{array}}{10}$

$\qquad\qquad = \dfrac{157}{10}$

$\qquad\qquad = 15.7$

**22.** 10, 12, 14, 15, 16, 16, 17, 18, 19, 20

$\qquad$ Median $= 16$

**23.** 16

# Chapters 1–8 Cumulative Test

**1.** $\quad\begin{array}{r}1\,376 \\ 2\,804 \\ 9\,003 \\ +\ 7\,642 \\ \hline 20{,}825\end{array}$

**2.** $\quad\begin{array}{r}2008 \\ \times\ \ 37 \\ \hline 14\,056 \\ 60\,24\phantom{0} \\ \hline 74{,}296\end{array}$

**3.** $\begin{array}{r}7\frac{1}{5} \\ -\ 3\frac{3}{8} \\ \hline \end{array}$ $\qquad \begin{array}{r}6\frac{48}{40} \\ -\ 3\frac{15}{40} \\ \hline 3\frac{33}{40}\end{array}$

**4.** $10\dfrac{4}{5} \div 3\dfrac{1}{2} = \dfrac{54}{5} \div \dfrac{7}{2}$

$\qquad\qquad = \dfrac{54}{5} \times \dfrac{2}{7}$

$\qquad\qquad = \dfrac{108}{35}$

$\qquad\qquad = 3\dfrac{3}{35}$

**5.** 1796.4289 rounds to 1796.43

**6.**    200.58
       − 127.93
       ─────────
         72.65

**7.**
$$
\begin{array}{r}
72.23 \\
0.72\overline{)52.0056} \\
\underline{50\ 4} \\
1\ 60 \\
\underline{1\ 44} \\
165 \\
\underline{144} \\
216 \\
\underline{216} \\
0
\end{array}
$$

**8.**    $\dfrac{7}{n} = \dfrac{35}{3}$

   $7 \times 3 = n \times 35$

   $\dfrac{21}{35} = \dfrac{n \times 35}{35}$

   $0.6 = n$

**9.**   $\dfrac{n}{26,390} = \dfrac{3}{2030}$

   $2030n = 3 \times 26,390$

   $\dfrac{2030n}{2030} = \dfrac{79,170}{2030}$

   $n = 39$ defects

**10.**   $n = 1.3\% \times 25$

   $n = 0.013 \times 25$

   $n = 0.325$

**11.**   72% of what number is 252?

   $72\% \times n = 252$

   $\dfrac{0.72 \times n}{0.72} = \dfrac{252}{0.72}$

   $n = 350$

**12.** 198 cm = 1.98 m (move 2 places left)

**13.** 18 yd $\times \dfrac{3 \text{ ft}}{1 \text{ yd}} = 54$ ft

**14.**   $A = \pi r^2$

   $= 3.14 \,(3 \text{ in.})^2$

   $= 3.14 \,(9 \text{ in.}^2)$

   $\approx 28.3 \text{ in.}^2$

**15.** $P = 4s = 4\,(17 \text{ in.}) = 68$ in.

**16.** $18\% + 16\% = 34\%$

**17.** Number of Freshmen

   $= 32\% \times 12,000$

   $= 0.32 \times 12,000$

   $= 3840$ students

**18.** 3 million dollars

**19.**

   $7,000,000 profit in 2nd quarter 2001
   − $6,000,000 profit in 2nd quarter 2000
   ────────────────────────────────────
   $1,000,000 more profit in 2nd quarter 2001

**20.** 16 in.

**21.** In 1960 and 1970 the annual rainfall in
   Weston was greater than Dixville.

**22.** 8 students

**23.**    4 students ages 17–19
        7 students ages 20–22
      + 5 students ages 23–25
      ───────────────────────────────
       16 students less than 26 years of age

**24.** $\text{Mean} = \dfrac{\begin{array}{c}\$5.00 + \$4.50 + \$3.95 + \$4.90 \\ + \$12.15 + \$4.50 + \$6.00\end{array}}{2}$

$= \dfrac{\$48.00}{8}$

$= \$6.00$

**25.** $3.95, $4.50, $4.50, $4.90, $5.00, $6.00, $7.00, $12.15

$\text{Median} = \dfrac{\$4.90 + \$5.00}{2}$

$= \$4.95$

**26.** $\text{Mode} = \$4.50$

# SIGNED NUMBERS

## Pretest Chapter 9

**1.** $-7 + (-12) = -19$

**2.** $-23 + 19 = -4$

**3.** $7.6 + (-3.1) = 4.5$

**4.** $8 + (-5) + 6 + (-9)$
$= 3 + 6 + (-9)$
$= 9 + (-9)$
$= 0$

**5.** $\dfrac{5}{12} + \left(-\dfrac{3}{4}\right) = \dfrac{5}{12} + \left(-\dfrac{9}{12}\right)$
$= -\dfrac{4}{12}$
$= -\dfrac{1}{3}$

**6.** $-\dfrac{5}{6} + \left(-\dfrac{1}{3}\right) = -\dfrac{5}{6} + \left(-\dfrac{2}{6}\right)$
$= -\dfrac{7}{6}$
$= -1\dfrac{1}{6}$

**7.** $-2.8 + (-4.2) = -7$

**8.** $-3.7 + 5.4 = 1.7$

**9.** $13 - 21 = 13 + (-21)$
$= -8$

**10.** $-26 - 15 = -26 + (-15)$
$= -41$

**11.** $\dfrac{5}{17} - \left(-\dfrac{9}{17}\right) = \dfrac{5}{17} + \dfrac{9}{17}$
$= \dfrac{14}{17}$

**12.** $-19 - (-7) = -19 + 7$
$= -12$

**13.** $-4.9 - (-6.3) = -4.9 + 6.3$
$= 1.4$

**14.** $2.8 - 5.6 = 2.8 + (-5.6)$
$= -2.8$

**15.** $21 - (-21) = 21 + 21$
$= 42$

**16.** $\dfrac{2}{3} - \left(-\dfrac{3}{5}\right) = \dfrac{2}{3} + \dfrac{3}{5}$
$= \dfrac{10}{15} + \dfrac{9}{15}$
$= \dfrac{19}{15}$
$= 1\dfrac{4}{15}$

**17.** $(-3)(-8) = 24$

**18.** $-48 \div (-12) = 4$

**19.** $-72 \div 9 = -8$

**20.** $(5)(-4)(2)(-1)\left(-\dfrac{1}{4}\right)$

$$= -20(2)(-1)\left(-\dfrac{1}{4}\right)$$

$$= -40(-1)\left(-\dfrac{1}{4}\right)$$

$$= 40\left(-\dfrac{1}{4}\right)$$

$$= -10$$

**21.** $\dfrac{72}{-3} = -24$

**22.** $\dfrac{-\frac{3}{4}}{-\frac{4}{5}} = -\dfrac{3}{4} \div \left(-\dfrac{4}{5}\right)$

$$= -\dfrac{3}{4} \times \left(-\dfrac{5}{4}\right)$$

$$= \dfrac{15}{16}$$

**23.** $(-8)(-2)(-4) = (16)(-4)$

$$= -64$$

**24.** $120 \div (-12) = -10$

**25.** $24 \div (-4) + 28 \div (-7)$

$$= -6 + (-4)$$

$$= -10$$

**26.** $18 \div 3(5) + (-5) \div (-5)$

$$= 6(5) + 1$$

$$= 30 + 1$$

$$= 31$$

**27.** $7 + (-9) + 2(-5)$

$$= 7 + (-9) + (-10)$$

$$= -2 + (-10)$$

$$= -12$$

**28.** $8(-6) \div (-10) = -48 \div (-10)$

$$= 4.8$$

**29.** $5 - (-6) + 18 \div (-3)$

$$= 5 + 6 + (-6)$$

$$= 11 + (-6)$$

$$= 5$$

**30.** $8(-3) + 4(-2) - (-6)$

$$= -27 + (-8) + 6$$

$$= -35 + 6$$

$$= -29$$

**31.** $\dfrac{12 + 8 - 4}{(-4)(3)(4)} = \dfrac{12 + 8 + (-4)}{-48}$

$$= \dfrac{20 + (-4)}{-48}$$

$$= \dfrac{16}{-48}$$

$$= -\dfrac{1}{3}$$

**32.** $\dfrac{56 \div (-7) - 1}{3(-9) + 6(-3)} = \dfrac{-8 - 1}{-27 - 18}$

$$= \dfrac{-9}{-45}$$

$$= \dfrac{1}{5}$$

**33.** $80,000 = 8 \times 10^4$

**34.** $0.0005 = 5 \times 10^{-4}$

**35.** $128,000,000 = 1.28 \times 10^8$

**36.** $6.7 \times 10^{-3} = 0.0067$

**37.** $1.32 \times 10^6 = 1,320,000$

**38.** $1.5678 \times 10^{-8} = 0.000000015678$

## 9.1    Exercises

27. $1.5 + (-2.2) = -0.7$

1. Find the absolute value of each number. Then add those two absolute values. Use the common sign in the answer.

29. $-14.6 + (-5.7) = -20.3$

3. $-6 + (-11) = -17$

31. $13 + (-9) = 4$

5. $-4.9 + (-2.1) = -7$

33. $-\dfrac{1}{3} + \left(-\dfrac{2}{7}\right) = -\dfrac{7}{21} + \left(-\dfrac{6}{21}\right)$

7. $8.9 + 7.6 = 16.5$

$\qquad\qquad = -\dfrac{13}{21}$

9. $\dfrac{1}{5} + \dfrac{2}{7} = \dfrac{7}{35} + \dfrac{10}{35} = \dfrac{17}{35}$

35. $-7.3 + 13.8 = 6.5$

11. $-\dfrac{1}{12} + \left(-\dfrac{5}{6}\right) = -\dfrac{1}{12} + \left(-\dfrac{10}{12}\right)$

37. $-5 + \left(-\dfrac{1}{2}\right) = -\dfrac{10}{2} + \left(-\dfrac{1}{2}\right)$

$\qquad\qquad = -\dfrac{11}{12}$

$\qquad\qquad = -\dfrac{11}{2}$

13. $14 + (-5) = 9$

$\qquad\qquad = -5\dfrac{1}{2}$

15. $-17 + 12 = -5$

17. $-36 + 58 = 22$

39. $6 + (-14) + 4 = -8 + 4$

$\qquad\qquad\qquad = -4$

19. $-9.3 + 6.5 = -2.8$

41. $11 + (-9) + (-10) + 8$

21. $\dfrac{1}{12} + \left(-\dfrac{3}{4}\right) = \dfrac{1}{12} + \left(-\dfrac{9}{12}\right)$

$\qquad = 2 + (-10) + 8$

$\qquad = -8 + 8$

$\qquad = 0$

$\qquad\qquad = -\dfrac{8}{12}$

$\qquad\qquad = -\dfrac{2}{3}$

43. $-7 + 6 + (-2) + 5 + (-3) + (-5)$

$\qquad = -1 + (-2) + 5 + (-3) + (-5)$

$\qquad = -3 + 5 + (-3) + (-5)$

23. $\dfrac{7}{9} + \left(-\dfrac{2}{9}\right) = \dfrac{5}{9}$

$\qquad = 2 + (-3) + (-5)$

$\qquad = -1 + (-5)$

25. $-18 + (-4) = -22$

$\qquad = -6$

**45.** $-\dfrac{1}{5} + \left(-\dfrac{2}{3}\right) + \dfrac{4}{25}$

$= -\dfrac{15}{75} + \left(-\dfrac{50}{75}\right) + \dfrac{12}{75}$

$= -\dfrac{65}{75} + \dfrac{12}{75}$

$= -\dfrac{53}{75}$

**47.** $-\$43{,}000 + (-\$51{,}000) = -\$94{,}000$

**49.** $\$28{,}000 + (-\$19{,}000) = \$9000$

**51.** $-\$35{,}000 + \$17{,}000 + (-\$20{,}000)$

$= -\$18{,}000 + (-\$20{,}000)$

$= -\$38{,}000$

**53.** $-1° + (-17°) = -18°\text{F}$

**55.** $-5° + 4° = -1°\text{F}$

**57.** $15° + (-6°) + 2° + 1° + (-3°)$

$= 9° + 2° + 1° + (-3°)$

$= 11° + 1° + (-3°)$

$= 12° + (-3°)$

$= 9$

**59.** $-8 + 13 + (-6)$

$= 5 + (-6)$

$= -1$ yard or a loss of 1 yard.

**61.** $-\$28 + \$30 + (-\$15)$

$= \$2 + (-\$15)$

$= -\$13$

**63.** $\$89.50 + (-\$50.00) + (-\$2.50)$

$= \$39.50 + (-\$2.50)$

$= \$37.00$

## Cumulative Review Problems

**65.** $v = \dfrac{4\pi r^3}{3}$

$= \dfrac{4\,(3.14)\,(16 \text{ in.})^3}{3}$

$\approx 904.3 \text{ cu ft}$

## 9.2   Exercises

**1.** $3 - 9 = 3 + (-9)$

$= -6$

**3.** $-14 - 3 = -14 + (-3)$

$= -17$

**5.** $-12 - (-10) = -12 + 10$

$= -2$

**7.** $3 - (-21) = 3 + 21$

$= 24$

**9.** $12 - 30 = 12 + (-30)$

$= -18$

**11.** $-12 - (-15) = -12 + 15$

$= 3$

**13.** $150 - 210 = 150 + (-210)$

$= -60$

**15.** $300 - (-256) = 300 + 256$

$= 556$

**17.** $-2.5 - 4.2 = -2.5 + (-4.2)$

$= -6.7$

**19.** $4.2 - 10.7 = 4.2 + (-10.7)$
$= -6.5$

**21.** $-10.9 - (-2.3) = -10.9 + 2.3$
$= -8.6$

**23.** $20.23 - (-12.71) = 20.33 + 12.71$
$= 32.94$

**25.** $\dfrac{1}{4} - \left(-\dfrac{3}{4}\right) = \dfrac{1}{4} + \dfrac{3}{4}$
$= \dfrac{4}{4}$
$= 1$

**27.** $-\dfrac{5}{6} - \dfrac{1}{3} = -\dfrac{5}{6} + \left(-\dfrac{1}{3}\right)$
$= -\dfrac{5}{6} + \left(-\dfrac{2}{6}\right)$
$= -\dfrac{7}{6}$
$= -1\dfrac{1}{6}$

**29.** $-\dfrac{5}{12} - \left(-\dfrac{1}{4}\right) = -\dfrac{5}{12} + \dfrac{1}{4}$
$= -\dfrac{5}{12} + \dfrac{3}{12}$
$= -\dfrac{2}{12}$
$= -\dfrac{1}{6}$

**31.** $\dfrac{5}{9} - \dfrac{2}{7} = \dfrac{5}{9} + \left(-\dfrac{2}{7}\right)$
$= \dfrac{35}{63} + \left(-\dfrac{18}{63}\right)$
$= \dfrac{17}{63}$

**33.** $2 - (-8) + 5 = 2 + 8 + 5$
$= 10 + 5$
$= 15$

**35.** $-5 - 6 - (-11)$
$= -5 + (-6) + 11$
$= -11 + 11$
$= 0$

**37.** $7 - (-2) - (-18)$
$= 7 + 2 + 8$
$= 9 + 8$
$= 17$

**39.** $-16 - (-6) - 12$
$= -16 + 6 + (-12)$
$= -10 + (-12)$
$= -22$

**41.** $9 - 3 - 2 - 6$
$= 9 + (-3) + (-2) + (-6)$
$= 6 + (-2) + (-6)$
$= 4 + (-6)$
$= -2$

**43.** $-16 + 9 - (-2) - 8$
$= -16 + 9 + 2 + (-8)$
$= -7 + 2 + (-8)$
$= -5 + (-8)$
$= -13$

**45.** $5277 - (-844) = 5277 + 844$
$= 6121 \text{ feet}$

**47.** $23° - (-19°) = 23° + 19°$
$= 42°\text{F}$

**49.** $-29° + 16° = -13°\text{F}$

**51.** $18,750 + (-\$32,800)$

$\qquad = \$18,750 + (-\$32,800)$

$\qquad = -\$14,100$

**53.** $-\$6300 + \$43,500 = \$37,200$

**55.** $\$15\frac{1}{2} - \$1\frac{1}{2} + \$2\frac{3}{4} - \$3\frac{1}{4}$

$\qquad = \$15\frac{1}{2} + \left(-\$1\frac{1}{2}\right) + \$2\frac{3}{4} + \left(-\$3\frac{1}{4}\right)$

$\qquad = \$14 + \$2\frac{3}{4} + \left(-\$3\frac{1}{4}\right)$

$\qquad = \$16\frac{3}{4} + \left(-\$3\frac{1}{4}\right)$

$\qquad = \$13\frac{2}{4}$

$\qquad = \$13\frac{1}{2}$

**57.** The bank finds that a customer has $50 in a checking account. However, the bank must remove an erroneous debit of $80 from the account. When the bank makes the correction, what will the new balance be?

---

## Cumulative Review Problems

**59.** $20 \times 2 \div 10 + 4 - 3$

$\qquad = 40 \div 10 + 4 - 3$

$\qquad = 4 + 4 - 3$

$\qquad = 8 - 3$

$\qquad = 8 + (-3)$

$\qquad = 5$

**61.** $r = 6 \div 2 = 3$

$\qquad A = \pi r^2$

$\qquad = 3.14 \, (3 \text{ in.})^2$

$\qquad = 28.26 \text{ square inches}$

---

## 9.3   Exercises

**1.** To multiply two numbers with the same sign, multiply the absolute values. The sign of the result is positive.

**3.** $(12)(3) = 36$

**5.** $(-20)(-3) = -60$

**7.** $(-20)(8) = -160$

**9.** $(2.5)(-0.6) = -1.5$

**11.** $(-12.5)(-2.25) = 28.125$

**13.** $\left(-\dfrac{2}{5}\right)\left(\dfrac{3}{7}\right) = -\dfrac{6}{35}$

**15.** $\left(-\dfrac{4}{12}\right)\left(-\dfrac{3}{23}\right) = \dfrac{1}{23}$

**17.** $-64 \div 8 = -8$

**19.** $\dfrac{48}{-6} = -8$

**21.** $\dfrac{-120}{-20} = 6$

**23.** $\dfrac{3}{10} \div \left(-\dfrac{9}{20}\right) = \dfrac{3}{10} \times \left(-\dfrac{20}{9}\right)$

$\qquad\qquad = -\dfrac{2}{3}$

**25.** $\dfrac{-\frac{4}{5}}{-\frac{7}{10}} = -\dfrac{4}{5} \div \left(-\dfrac{7}{10}\right)$

$\phantom{\dfrac{-\frac{4}{5}}{-\frac{7}{10}}} = -\dfrac{4}{5} \times \left(-\dfrac{10}{7}\right)$

$\phantom{\dfrac{-\frac{4}{5}}{-\frac{7}{10}}} = \dfrac{8}{7}$

$\phantom{\dfrac{-\frac{4}{5}}{-\frac{7}{10}}} = 1\dfrac{1}{7}$

**27.** $50.28 \div (-6) = -8.38$

**29.** $\dfrac{-55.8}{-9} = 6.2$

**31.** $3(-6)(-4) = -18(-4)$

$\phantom{3(-6)(-4)} = 72$

**33.** $(-8)(4)(-6) = -32(-6)$

$\phantom{(-8)(4)(-6)} = 192$

**35.** $2(-8)(3)\left(-\dfrac{1}{3}\right) = -16(3)\left(-\dfrac{1}{3}\right)$

$\phantom{2(-8)(3)\left(-\dfrac{1}{3}\right)} = -48\left(-\dfrac{1}{3}\right)$

$\phantom{2(-8)(3)\left(-\dfrac{1}{3}\right)} = 18$

**37.** $(-5)(2)(-1)(-3)$

$\phantom{(-5)(2)} = -10(-1)(-3)$

$\phantom{(-5)(2)} = 10(-3)$

$\phantom{(-5)(2)} = -30$

**39.** $8(-3)(-5)(0)(-2)$

$\phantom{8(-3)} = -24(-5)(0)(-2)$

$\phantom{8(-3)} = 120(0)(-2)$

$\phantom{8(-3)} = 0(-2)$

$\phantom{8(-3)} = 0$

**41.** $\left(-\dfrac{2}{3}\right)\left(-\dfrac{3}{4}\right)\left(-\dfrac{5}{6}\right) = \left(\dfrac{1}{2}\right)\left(-\dfrac{5}{6}\right)$

$\phantom{\left(-\dfrac{2}{3}\right)\left(-\dfrac{3}{4}\right)\left(-\dfrac{5}{6}\right)} = -\dfrac{5}{12}$

**43.** $4(\$1.25) = \$5.00$

**45.** $\dfrac{-12° - 14° - 3° + 5° + 8° - 1° - 10° - 23°}{8}$

$\phantom{XX} = \dfrac{-12° + (-14°) + (-3°) + 5° + 8° + (-1°) + (-10°) + (-23°)}{8}$

$\phantom{XX} = -\dfrac{50°}{8}$

$\phantom{XX} = -\dfrac{25°}{4}$

$\phantom{XX} = -6.25°$

**47.** $7(-10) = -70$
Dropped 70 feet.

**49.** $17(2) = 34 \text{ or } +34$

**51.** $4(1) + 2(-1) = 4 + (-2)$

$\phantom{4(1) + 2(-1)} = 2 \text{ or } +2$

**53.** $-8(-1) = 8 \text{ or } +8$

**55.** $2(-1) + (-2) + 2(2)$

$\phantom{2(-1)} = -2 + (-2) + 4$

$\phantom{2(-1)} = -4 + 4$

$\phantom{2(-1)} = 0$

At par

**57.** It has to be 0.

**59.** The product of three negative numbers is a negative number which makes $b$ a negative number.

## Cumulative Review Problems

**61.** $A = bh = (15 \text{ in.})(6 \text{ in.})$
$= 90$ square inches

## 9.4   Exercises

**1.** $16 + 32 \div (-4) = 16 + (-8)$
$= 8$

**3.** $24 \div (-3) + 16 \div (-4)$
$= -8 + (-4)$
$= -12$

**5.** $3(-4) + 5(-2) - (-3)$
$= -12 + (-10) + 3$
$= -22 + 3$
$= -19$

**7.** $-54 \div 6 + 9 = -9 + 9$
$= 0$

**9.** $5 - 30 \div 3 = 5 - 10$
$= 5 + (-10)$
$= -5$

**11.** $36 \div 12(-2) = 3(-2)$
$= -6$

**13.** $3(-4) + 6(-2) - 3$
$= -12 + (-12) + (-3)$
$= -24 + (-3)$
$= -27$

**15.** $11(-6) - 3(12) = -66 - 36$
$= -66 + (-36)$
$= -102$

**17.** $16 - 4(8) + 18 \div (-9)$
$= 16 - 32 + (-2)$
$= 16 + (-32) + (-2)$
$= -16 + (-2)$
$= -18$

**19.** $\dfrac{8 + 6 - 12}{3 - 6 + 5} = \dfrac{2}{2} = 1$

**21.** $\dfrac{6(-2) + 4}{6 - 3 - 5} = \dfrac{-12 + 4}{6 - 3 - 5}$
$= \dfrac{-8}{-2}$
$= 4$

**23.** $\dfrac{-16 \div (-2)}{3(-4) - 4} = \dfrac{8}{-12 - 4}$
$= \dfrac{8}{-16}$
$= -\dfrac{1}{2}$

**25.** $\dfrac{4(-5) \div 4 + 7}{20 \div (-10)} = \dfrac{-20 \div 4 + 7}{-2}$
$= \dfrac{-5 + 7}{-2}$
$= \dfrac{2}{-2}$
$= -1$

**27.** $\dfrac{12 \div 3 + (-2)(2)}{9 - 9 \div (-3)} = \dfrac{4 + (-4)}{9 - (-3)}$
$= \dfrac{4 + (-4)}{9 + 3}$
$= \dfrac{0}{12}$
$= 0$

**29.** $3(2-6)+4^2 = 3(-4)+4^2$
$$= 3(-4)+16$$
$$= -12+16$$
$$= 4$$

**31.** $12 \div (-6) + (7-2)^3$
$$= 12 \div (-6) + 5^3$$
$$= 12 \div (-6) + 125$$
$$= -2 + 125$$
$$= 123$$

**33.** $\dfrac{3}{16} - \dfrac{3}{8} + \dfrac{1}{2}\left(-\dfrac{1}{2}\right)$
$$= \dfrac{3}{16} - \dfrac{3}{8} + \left(-\dfrac{1}{4}\right)$$
$$= \dfrac{3}{16} + \left(-\dfrac{6}{16}\right) + \left(-\dfrac{4}{16}\right)$$
$$= -\dfrac{7}{16}$$

**35.** $\left(\dfrac{3}{5}\right)^2 - \dfrac{3}{5}\left(\dfrac{1}{3}\right) = \dfrac{9}{25} - \dfrac{3}{5}\left(\dfrac{1}{3}\right)$
$$= \dfrac{9}{25} - \dfrac{1}{5}$$
$$= \dfrac{9}{25} + \left(-\dfrac{5}{25}\right)$$
$$= \dfrac{4}{25}$$

**37.** $(1.2)^2 - 3.6(-1.5)$
$$= 1.44 - 3.6(-1.5)$$
$$= 1.44 + 5.4$$
$$= 6.84$$

**39.** $\dfrac{-13° + (-14°) + (-20°)}{3}$
$$= \dfrac{-47°}{3}$$
$$\approx -15.7°\text{F}$$

**41.** $[14° + (-1°) + (-13°) + (-14°) + (-20°) + (-16°)] \div 8$
$$= \dfrac{-50°}{6}$$
$$= \dfrac{-25°}{3}$$
$$\approx -8.3°\text{F}$$

**43.** $[-14° + (-20°) + (-16°) + (-2°) + 19° + 33° + 39° + 38° + 31° + 14° + (-1°) + (-13°)] \div 12$
$$= \dfrac{108°}{12} = 9°\text{F}$$

**45.**   $-12° + 3° = -9°\text{C}$
$$-12° + 5(3°) = -12° + 15°$$
$$= 3°\text{C}$$

## Cumulative Review Problems

**47.** $3840 \text{ m} = \dfrac{3840}{1000} \text{ km} = 3.84 \text{ km}$

**49.** $A = \dfrac{h(b+B)}{2}$
$$= \dfrac{(14 \text{ in.})(23 \text{ in.} + 37 \text{ in.})}{2}$$
$$= \dfrac{(14 \text{ in.})(60 \text{ in.})}{2}$$
$$= \dfrac{840 \text{ in.}^2}{2}$$
$$= 420 \text{ square inches}$$

## 9.5   Exercises

**1.** $120 = 1.2 \times 10^2$

**3.** $10,000 = 1 \times 10^4$

**5.** $26,300 = 2.63 \times 10^4$

7. $288,000 = 2.88 \times 10^5$

9. $4,632,000 = 4.632 \times 10^6$

11. $12,000,000 = 1.2 \times 10^7$

13. $0.0931 = 9.31 \times 10^{-2}$

15. $0.00279 = 2.79 \times 10^{-3}$

17. $0.4 = 4 \times 10^{-1}$

19. $0.000016 = 1.6 \times 10^{-5}$

21. $0.00000531 = 5.31 \times 10^{-6}$

23. $0.00000018 = 1.8 \times 10^{-7}$

25. $5.36 \times 10^4 = 53,600$

27. $6.2 \times 10^{-2} = 0.062$

29. $3.71 \times 10^{-1} = 0.371$

31. $9 \times 10^{11} = 900,000,000,000$

33. $3.862 \times 10^{-8} = 0.00000003862$

35. $4.6 \times 10^{12} = 4,600,000,000,000$

37. $6.721 \times 10^{10} = 67,210,000,000$

39. $5,878,000,000,000 = 5.878 \times 10^{12}$ miles

41. $6,000,000,000 = 6 \times 10^9$ people

43. $1.25 \times 10^{13} = 12,500,000,000,000$ insects

45. $7.5 \times 10^{-5} = 0.000075$ centimeters

47. $1.4 \times 10^{10} = 14,000,000,000$ tons

49. $\begin{array}{r} 3.38 \times 10^7 \\ + \; 5.63 \times 10^7 \\ \hline 9.01 \times 10^7 \text{ dollars} \end{array}$

51. $\begin{array}{r} 5.87 \times 10^{21} \\ + \; 4.81 \times 10^{21} \\ \hline 10.68 \times 10^{21} = 1.068 \times 10^{22} \text{ tons} \end{array}$

53. $\begin{array}{r} 4.00 \times 10^8 \\ + \; 3.76 \times 10^7 \\ \hline \end{array}$

$\begin{array}{r} 40.00 \times 10^7 \\ + \; 3.76 \times 10^7 \\ \hline 36.24 \times 10^7 = 3.624 \times 10^8 \text{ feet} \end{array}$

55. $\begin{array}{r} 1.76 \times 10^7 \\ - \; 1.16 \times 10^7 \\ \hline 0.60 \times 10^7 = 6 \times 10^6 \text{ sq miles} \end{array}$

57. $\begin{aligned} 10.5 \,(5.88 \text{ trillion}) &= 61.74 \text{ trillion} \\ &= 61.74 \times 10^{12} \\ &= 6.174 \times 10^{13} \text{ miles} \end{aligned}$

## Cumulative Review Problems

59. $\begin{array}{r} 7.63 \\ \times \; 2.18 \\ \hline 6104 \\ 763 \\ 15\,26 \\ \hline 16.6334 \end{array}$

61. $\begin{aligned} \text{Cost} &= 4\,(2 \times 9 + 2 \times 13) \\ &= 4\,(18 + 26) \\ &= 4\,(44) \\ &= \$176 \end{aligned}$

## Putting Your Skills To Work

1. $1 \times 10^{-4} = 100 \times 10^{-6}$. Since a 3 foot thick wall can absorb $1 \times 10^{-6}$ curies per liter, then $1 \times 10^{-4}$ curies per liter will require a $3 \times 100 = 300$ foot thick wall.

3. 1 year $= 365 \times 24 \times 60$
   $\qquad = 525{,}600$ minutes
   This gives 525,600 liters.
   $525{,}600 \times 10^{-6} = 0.5256$ curies

## Chapter 9 Review Problems

1. $-20 + 5 = -15$

2. $-18 + 4 = -14$

3. $-3.6 + (-5.2) = -8.8$

4. $-7.6 + (-1.2) = -8.8$

5. $-\dfrac{1}{5} + \left(-\dfrac{1}{3}\right) = -\dfrac{3}{15} + \left(-\dfrac{5}{15}\right)$
   $\qquad\qquad\qquad = -\dfrac{8}{15}$

6. $-\dfrac{2}{7} + \dfrac{5}{4} = -\dfrac{4}{14} + \dfrac{5}{14}$
   $\qquad\qquad\quad = \dfrac{1}{14}$

7. $20 + (-14) = 6$

8. $-80 + 60 = -20$

9. $7 + (-2) + 9 + (-3)$
   $\quad = 5 + 9 + (-3)$
   $\quad = 14 + (-3)$
   $\quad = 11$

10. $6 + (-3) + 8 + (-9)$
    $\quad = 3 + 8 + (-9)$
    $\quad = 11 + (-9)$
    $\quad = 2$

11. $-36 - (-21) = -36 + 21$
    $\qquad\qquad\quad = -15$

12. $-21 - (-28) = -21 + 28$
    $\qquad\qquad\quad = 7$

13. $12 - (-7) = 12 + 7$
    $\qquad\qquad = 19$

14. $14 - (-3) = 14 + 3$
    $\qquad\qquad = 17$

15. $1.6 - 3.2 = 1.6 + (-3.2)$
    $\qquad\qquad = -1.6$

16. $-5.2 - 7.1 = -5.2 + (-7.1)$
    $\qquad\qquad\quad = -12.3$

17. $-\dfrac{2}{5} - \left(-\dfrac{1}{3}\right) = -\dfrac{2}{5} + \dfrac{1}{3}$
    $\qquad\qquad\qquad = -\dfrac{6}{15} + \dfrac{5}{15}$
    $\qquad\qquad\qquad = -\dfrac{1}{15}$

**18.** $\dfrac{1}{6} - \left(-\dfrac{5}{12}\right) = \dfrac{1}{6} + \dfrac{5}{12}$

$\qquad\qquad = \dfrac{2}{12} + \dfrac{5}{12}$

$\qquad\qquad = \dfrac{7}{12}$

**19.** $5 - (-2) - (-6) = 5 + 2 + 6$

$\qquad\qquad\qquad = 7 + 6$

$\qquad\qquad\qquad = 13$

**20.** $-15 - (-3) + 9 = -15 + 3 + 9$

$\qquad\qquad\qquad\quad = -12 + 9$

$\qquad\qquad\qquad\quad = -3$

**21.** $9 - 8 - 6 - 4$

$\quad = 9 + (-8) + (-6) + (-4)$

$\quad = 1 + (-6) + (-4)$

$\quad = -5 + (-4)$

$\quad = -9$

**22.** $-7 - 8 - (-3) = -7 + (-8) + 3$

$\qquad\quad -15 + 3 = -12$

**23.** $\left(-\dfrac{2}{7}\right)\left(-\dfrac{1}{5}\right) = \dfrac{2}{35}$

**24.** $\left(-\dfrac{6}{15}\right)\left(\dfrac{5}{12}\right) = -\dfrac{1}{6}$

**25.** $(5.2)(-1.5) = -7.8$

**26.** $(-3.6)(-1.2) = 4.32$

**27.** $-60 \div (-20) = 3$

**28.** $-18 \div (-3) = 6$

**29.** $\dfrac{-36}{4} = -9$

**30.** $\dfrac{-60}{12} = -5$

**31.** $\dfrac{-13.2}{-2.2} = 6$

**32.** $\dfrac{48}{-3.2} = -15$

**33.** $\dfrac{-\frac{2}{5}}{\frac{4}{7}} = -\dfrac{2}{5} \div \dfrac{4}{7}$

$\qquad = -\dfrac{2}{5} \times \dfrac{7}{4}$

$\qquad = -\dfrac{7}{10}$

**34.** $\dfrac{-\frac{1}{3}}{-\frac{7}{9}} = -\dfrac{1}{3} \div \left(-\dfrac{7}{9}\right)$

$\qquad = -\dfrac{1}{3} \times \left(-\dfrac{9}{7}\right)$

$\qquad = \dfrac{3}{7}$

**35.** $3(-5)(-2) = -15(-2)$

$\qquad\qquad\quad = 30$

**36.** $6(-8)(-1) = -48(-1)$

$\qquad\qquad\quad = 48$

**37.** $8 - (-30) \div 6 = 8 - (-5)$

$\qquad\qquad\qquad = 8 + 5$

$\qquad\qquad\qquad = 13$

**38.** $26 + (-28) \div 4 = 26 - 7$

$\qquad\qquad\qquad\quad = 19$

**39.** $2(-6) + 3(-4) - (-13)$
$$= -12 + (-12) + 13$$
$$= -24 + 13$$
$$= -11$$

**40.** $-49 \div (-7) + 3(-2) = 7 + (-6)$
$$= 1$$

**41.** $36 \div (-12) + 50 \div (-25)$
$$= -3 + (-2)$$
$$= -5$$

**42.** $21 - (-30) \div 15 = 21 + 2$
$$= 23$$

**43.** $50 \div 25(-4) = 2(-4)$
$$= -8$$

**44.** $-80 \div 20(-3) = -4(-3)$
$$= 12$$

**45.** $9(-9) + 9 = -81 + 9$
$$= -72$$

**46.** $\dfrac{5 - 9 + 2}{3 - 5} = \dfrac{-2}{-2}$
$$= 1$$

**47.** $\dfrac{4(-6) + 8 - 2}{15 - 7 + 2} = \dfrac{-24 + 8 - 2}{15 - 7 + 2}$
$$= \dfrac{-18}{10}$$
$$= -\dfrac{9}{5}$$
$$= -1\dfrac{4}{5}$$

**48.** $\dfrac{20 \div (-5) - (-6)}{(2)(-2)(-5)} = \dfrac{-4 + 6}{20}$
$$= \dfrac{2}{20}$$
$$= \dfrac{1}{10}$$

**49.** $\dfrac{6 - (-3) - 2}{5 - 2 \div (-1)} = \dfrac{6 + 3 - 2}{5 + 2}$
$$= \dfrac{7}{7}$$
$$= 1$$

**50.** $-3 + 4(2 - 6)^2 \div (-2)$
$$= -3 + 4(-4)^2 \div (-2)$$
$$= -3 + 4(16) \div (-2)$$
$$= -3 + 64 \div (-2)$$
$$= -3 + (-32)$$
$$= -35$$

**51.** $-6(12 - 15) + 2^3$
$$= -6(-3) + 2^3$$
$$= -6(-3) + 8$$
$$= 18 + 8$$
$$= 26$$

**52.** $-50 \div (-10) + (5 - 3)^4$
$$= -50 \div (-10) + 2^4$$
$$= -50 \div (-10) + 16$$
$$= 5 + 16$$
$$= 21$$

**53.** $\dfrac{1}{2} - \dfrac{1}{6} + \dfrac{2}{3}\left(\dfrac{3}{7}\right)$

$= \dfrac{1}{2} + \left(-\dfrac{1}{6}\right) + \dfrac{2}{7}$

$= \dfrac{21}{42} + \left(-\dfrac{7}{42}\right) + \dfrac{12}{42}$

$= \dfrac{26}{42}$

$= \dfrac{13}{21}$

**54.** $\left(\dfrac{2}{3}\right)^2 - \dfrac{3}{8}\left(\dfrac{8}{5}\right) = \dfrac{4}{9} - \dfrac{3}{8}\left(\dfrac{8}{5}\right)$

$= \dfrac{4}{9} - \dfrac{3}{5}$

$= \dfrac{20}{45} - \dfrac{27}{45}$

$= -\dfrac{7}{45}$

**55.** $(0.8)^2 - 3.2\,(-1.6)$
$= 0.64 - 3.2\,(-1.6)$
$= 0.64 + 5.12$
$= 5.76$

**56.** $1.4\,(4.7 - 4.9) - 12.8 \div (-0.2)$
$= 1.4\,(-0.2) - 12.8 \div (-0.2)$
$= -0.28 + 64$
$= 63.72$

**57.** $4160 = 4.16 \times 10^3$

**58.** $3{,}700{,}000 = 3.7 \times 10^6$

**59.** $218{,}000 = 2.18 \times 10^5$

**60.** $0.007 = 7 \times 10^{-3}$

**61.** $0.0000218 = 2.18 \times 10^{-5}$

**62.** $0.00000763 = 7.63 \times 10^{-6}$

**63.** $1.89 \times 10^4 = 18{,}900$

**64.** $3.76 \times 10^3 = 3760$

**65.** $7.52 \times 10^{-2} = 0.0752$

**66.** $6.61 \times 10^{-3} = 0.0061$

**67.** $9 \times 10^{-7} = 0.0000009$

**68.** $8 \times 10^{-8} = 0.00000008$

**69.** $5.36 \times 10^{-4} = 0.000536$

**70.** $1.98 \times 10^{-5} = 0.0000198$

**71.** $\begin{array}{r} 5.26 \times 10^{11} \\ +\ 3.18 \times 10^{11} \\ \hline 8.44 \times 10^{11} \end{array}$

**72.** $\begin{array}{r} 7.79 \times 10^{15} \\ +\ 1.93 \times 10^{15} \\ \hline 9.72 \times 10^{15} \end{array}$

**73.** $\begin{array}{r} 3.42 \times 10^{14} \\ -\ 1.98 \times 10^{14} \\ \hline 1.44 \times 10^{14} \end{array}$

**74.** $\begin{array}{r} 1.76 \times 10^{26} \\ -\ 1.08 \times 10^{26} \\ \hline 0.68 \times 10^{26} = 6.8 \times 10^{25} \end{array}$

**75.** $123{,}120{,}000{,}000{,}000$
$= 1.2312 \times 10^{14}$ drops

**76.** $5.983 \times 10^{24}$ kilograms

**77.** $5280 \times 2,500,000,000$

$\qquad = 13,200,000,000,000$

$\qquad = 1.32 \times 10^{13}$ feet

**78.** $1 \times 10^{-21}$

**79.** $\qquad 1.67$ yg $= 1.67 \times 10^{-24}$ grams

$\quad 0.00091$ yg $= 9.1 \times 10^{-28}$ grams

**80.** $0.000000000000001$

**81.** $384.4 \times 10^6 = 384,400,000$ meters

**82.** $-5 + 6 + (-7) = 1 + (-7)$

$\qquad\qquad\qquad = -6$ yards

Total loss of 6 yards.

**83.** $785 - (-98) = 785 + 98$

$\qquad\qquad\qquad = 883$ feet

**84.** $-\$18 + (-\$20) + \$40 = -\$38 + \$40$

$\qquad\qquad\qquad\qquad = \$2$

Balance is $2.

**85.** $\dfrac{-16° + (-18°) + (-5°) + 3° + (-12°)}{5}$

$\quad = \dfrac{-48°}{5}$

$\quad = -9.6°$

**86.** $2(-1) + (-2) + 4(1) + 2$

$\qquad = -2 + (-2) + 4 + 2$

$\qquad = -4 + 4 + 2$

$\qquad = 0 + 2$

$\qquad = 2$

2 points above par

## Chapter 9 Test

**1.** $-26 + 15 = -11$

**2.** $-31 + (-12) = -43$

**3.** $12.8 + (-8.9) = 3.9$

**4.** $-3 + (-6) + 7 + (-4) = -9 + 7 + (-4)$

$\qquad\qquad\qquad\qquad = -2 + (-4)$

$\qquad\qquad\qquad\qquad = -6$

**5.** $10 + (-7) + 3 + (-9) = 3 + 3 + (-9)$

$\qquad\qquad\qquad\qquad = 6 + (-9)$

$\qquad\qquad\qquad\qquad = -3$

**6.** $-\dfrac{1}{4} + \left(-\dfrac{5}{8}\right) = -\dfrac{2}{8} + \left(-\dfrac{5}{8}\right)$

$\qquad\qquad\qquad = -\dfrac{7}{8}$

**7.** $-32 - 6 = -32 + (-6)$

$\qquad\quad = -38$

**8.** $23 - 18 = 23 + (-18)$

$\qquad\quad = 5$

**9.** $\dfrac{4}{5} - \left(-\dfrac{1}{3}\right) = \dfrac{4}{5} + \dfrac{1}{3}$

$\qquad\qquad = \dfrac{12}{15} + \dfrac{5}{15}$

$\qquad\qquad = \dfrac{17}{15}$

$\qquad\qquad = 1\dfrac{2}{15}$

**10.** $-50 - (-7) = -50 + 7$

$\qquad\qquad = -43$

11. $-2.5 - (-6.5) = -2.5 + 6.5$
$$= 4$$

12. $4.8 - 2.7 = 4.8 + (-2.7)$
$$= 2.1$$

13. $\dfrac{1}{12} - \left(-\dfrac{5}{6}\right) = \dfrac{1}{12} + \dfrac{5}{6}$
$$= \dfrac{1}{12} + \dfrac{10}{12}$$
$$= \dfrac{11}{12}$$

14. $23 - (-23) = 23 + 23$
$$= 46$$

15. $(-20)(-6) = 120$

16. $27 \div \left(-\dfrac{3}{4}\right) = 27 \times \left(-\dfrac{4}{3}\right)$
$$= -36$$

17. $-40 \div (-4) = 10$

18. $(-9)(-1)(-2)(4)\left(\dfrac{1}{4}\right)$
$$= 9(-2)(4)\left(\dfrac{1}{4}\right)$$
$$= -18(4)\left(\dfrac{1}{4}\right)$$
$$= -72\left(\dfrac{1}{4}\right)$$
$$= -18$$

19. $\dfrac{-39}{-13} = 3$

20. $\dfrac{-\frac{3}{5}}{\frac{6}{7}} = -\dfrac{3}{5} \div \dfrac{6}{7}$
$$= -\dfrac{3}{5} \times \dfrac{7}{6}$$
$$= -\dfrac{7}{10}$$

21. $(-7)(-2)(4) = 14(4)$
$$= 56$$

22. $96 \div (-3) = -32$

23. $7 - 2(-5) = 7 + 10$
$$= 17$$

24. $(-42) \div (-7) + 8 = 6 + 8$
$$= 14$$

25. $18 \div (-3) + 24 \div (-12) = -6 + (-2)$
$$= -8$$

26. $8(-5) + 6(-8 - 6 + 12)$
$$= 8(-5) + 6(-2)$$
$$= -40 + (-12)$$
$$= -52$$

27. $1.3 + (-9.5) + 2.5 + (-1.5)$
$$= 1.3 + (-9.5) + 2.5 + (-1.5)$$
$$= -8.2 + 2.5 + (-1.5)$$
$$= -5.7 + (-1.5)$$
$$= -7.2$$

28. $-48 \div (-6) - 7(-2)^2$
$$= -48 \div (-6) - 7(4)$$
$$= 8 + (-28)$$
$$= -20$$

29. $\dfrac{3+8-5}{(-4)(6)+(-6)(3)} = \dfrac{3+8-5}{-24+(-18)}$

$= \dfrac{6}{-42}$

$= -\dfrac{1}{7}$

30. $\dfrac{5+28 \div (-4)}{7-(-5)} = \dfrac{5+(-7)}{7+5}$

$= \dfrac{-2}{12}$

$= -\dfrac{1}{6}$

31. $80{,}540 = 8.054 \times 10^4$

32. $0.000007 = 7 \times 10^{-6}$

33. $9.36 \times 10^{-5} = 0.0000936$

34. $7.2 \times 10^4 = 72{,}000$

35. $\dfrac{-14° + (-8°) + (-5°) + 7° + (-11°)}{5}$

$= \dfrac{-31°}{5}$

$= -6.2°$

36. $\begin{array}{r} 2 \times 5.8 \times 10^{-5} \\ + \; 2 \times 7.8 \times 10^{-5} \\ \hline \end{array}$

$\begin{array}{r} 11.6 \times 10^{-5} \\ + \; 15.6 \times 10^{-5} \\ \hline 27.2 \times 10^{-5} \end{array}$

or $2.72 \times 10^{-4}$ meters

---

# Chapters 1–9 Cumulative Test

1. $\begin{array}{r} 28{,}981 \\ - \; 16{,}598 \\ \hline 12{,}383 \end{array}$

2. $\begin{array}{r} 127 \\ 36)\overline{4572} \\ \underline{36} \\ 97 \\ \underline{72} \\ 252 \\ \underline{252} \\ 0 \end{array}$

3. $\begin{array}{r} 3\frac{1}{4} \\ + \; 8\frac{2}{3} \\ \hline \end{array} \qquad \begin{array}{r} 3\frac{3}{12} \\ + \; 8\frac{8}{12} \\ \hline 11\frac{11}{12} \end{array}$

4. $1\dfrac{5}{6} \times 2\dfrac{1}{2} = \dfrac{11}{6} \times \dfrac{5}{2}$

$= \dfrac{55}{12}$

$= 4\dfrac{7}{12}$

5. $9.812456$ rounds to $9.812$

6. $\begin{array}{r} 5.820 \\ 38.964 \\ 0.571 \\ 9.305 \\ + \; 8.800 \\ \hline 63.460 \text{ or } 63.46 \end{array}$

7. $\begin{array}{r} 12.89 \\ \times \; 5.12 \\ \hline 2578 \\ 1\,289 \\ 64\,45 \\ \hline 65.9968 \end{array}$

8. $\dfrac{n}{8} = \dfrac{56}{7}$

$n \times 7 = 8 \times 56$

$n \times 7 = 448$

$\dfrac{n \times 7}{7} = \dfrac{448}{7}$

$n = 64$

9. $\dfrac{n \text{ defects}}{2808 \text{ parts}} = \dfrac{7 \text{ defects}}{156 \text{ parts}}$

$156 \times n = 7 \times 2808$

$\dfrac{156 \times n}{156} = \dfrac{19{,}656}{156}$

$n = 126 \text{ defects}$

10. $n = 0.8\% \times 38$

$n = 0.008 \times 38$

$n = 0.304$

11. 12% of what number is 480?

$\dfrac{12}{100} = \dfrac{480}{b}$

$12 \times b = 100 \times 480$

$\dfrac{12 \times b}{12} = \dfrac{48{,}000}{12}$

$b = 4000$

12. 94 km = 94,000 m (move 3 places right)

13. $180 \text{ in.} \times \dfrac{1 \text{ yd}}{36 \text{ in.}} = 5 \text{ yd}$

14. $A = \pi r^2 = (3.14)(5)^2 = 78.5$

78.5 square meters

15. a. 300 students age 23–25

   b.   500 students age 20–22
        300 students age 23–25
        200 students age 26-28
        100 students age 29–31
        $\overline{1100 \text{ students over age 19}}$

16. $\sqrt{36} + \sqrt{49} = 6 + 7$

$= 13$

17. $-1.2 + (-3.5) = -4.7$

18. $-\dfrac{1}{4} + \dfrac{2}{3} = -\dfrac{3}{12} + \dfrac{8}{12}$

$= \dfrac{5}{12}$

19. $7 - 8 = 7 + (-18)$

$= -11$

20. $-8 - (-3) = -8 + 3$

$= -5$

21. $5(-3)(-1)(-2)(2) = -15(-1)(-2)(2)$

$= 15(-2)(2)$

$= -30(2)$

$= -60$

22. $\dfrac{-\frac{3}{7}}{-\frac{5}{14}} = \left(-\dfrac{3}{7}\right)\left(-\dfrac{14}{5}\right) = \dfrac{6}{5} = 1\dfrac{1}{5}$

23. $6 - 3(-4) = 6 - (-12)$

$= 6 + 12$

$= 18$

24. $(-20) \div (-2) + (-6) = 10 + (-6)$

$= 4$

25. $\dfrac{(-2)(-1) + (-4)(-3)}{1 + (-4)(2)} = \dfrac{2 + 12}{1 + (-8)}$

$= \dfrac{14}{-7}$

$= -2$

26. $\dfrac{(-11)(-8) \div 22}{1 - 7(-2)} = \dfrac{88 \div 22}{1 + 14}$

$= \dfrac{4}{15}$

27. $579{,}863 = 5.79863 \times 10^5$

**28.** $0.00078 = 7.8 \times 10^{-4}$

**29.** $3.85 \times 10^{7} = 38{,}500{,}000$

**30.** $7 \times 10^{-5} = 0.00007$

# INTRODUCTION TO ALGEBRA

## Pretest Chapter 10

**1.** $23x - 40x = -17x$

**2.** $-8y + 12y - 3y = y$

**3.** $6a - 5b - 9a + 7b$
$= 6a - 9a - 5b + 7b$
$= -3a + 2b$

**4.** $5x - y + 2 - 17x - 3y + 8$
$= 5x - 17x - y - 3y + 2 + 8$
$= -12x - 4y + 10$

**5.** $7x - 14 + 5y + 8 - 7y + 9x$
$= 7x + 9x + 5y - 7y - 14 + 8$
$= 16x - 2y - 6$

**6.** $4a - 7b + 3c - 5b = 4a - 7b - 5b + 3c$
$= 4a - 12b + 3c$

**7.** $(6)(7x - 3y) = (6)(7x) - (6)(3y)$
$= 42x - 18y$

**8.** $(-3)(a + 5b - 1)$
$= (-3)(a) + (-3)(5b) + (-3)(-1)$
$= -3a - 15b + 3$

**9.** $(-2)(1.5a + 3b - 6c - 5)$
$= (-2)(1.5a) + (-2)(3b)$
$\quad + (-2)(-6c) + (-2)(-5)$
$= -3a - 6b + 12c + 10$

**10.** $(5)(2x - y) - (3)(3x + y)$
$= 10x - 5y - 9x - 3y$
$= x - 8y$

**11.** $\qquad 5 + x = 42$
$5 + x + (-5) = 42 + (-5)$
$\qquad x = 37$

**12.** $\qquad x + 2.5 = 6$
$x + 2.5 + (-2.5) = 6 + (-2.5)$
$\qquad x = 3.5$

**13.** $\qquad x - \dfrac{5}{8} = \dfrac{1}{4}$
$x - \dfrac{5}{8} + \dfrac{5}{8} = \dfrac{1}{4} + \dfrac{5}{8}$
$\qquad x = \dfrac{1}{4} + \dfrac{5}{8}$
$\qquad = \dfrac{2}{8} + \dfrac{5}{8}$
$\qquad = \dfrac{7}{8}$

**14.** $7x = -56$
$\dfrac{7x}{7} = \dfrac{-56}{7}$
$x = -8$

**15.** $5.4x = 27$
$\dfrac{5.4x}{5.4} = \dfrac{27}{5.4}$
$x = 5$

**16.**
$$\frac{3}{5}x = \frac{9}{10}$$
$$\frac{5}{3} \cdot \frac{3}{5}x = \frac{5}{3} \cdot \frac{9}{10}$$
$$x = \frac{3}{2}$$
$$= 1\frac{1}{2}$$

**17.**
$$5x - 9 = 26$$
$$5x - 9 + 9 = 26 + 9$$
$$5x = 35$$
$$\frac{5x}{5} = \frac{35}{5}$$
$$x = 7$$

**18.**
$$12 - 3x = 7x - 4$$
$$12 - 3x + 3x = 7x + 3x - 4$$
$$12 = 10x - 4$$
$$12 + 4 = 10x - 4 + 4$$
$$16 = 10x$$
$$\frac{16}{10} = \frac{10x}{10}$$
$$\frac{16}{10} = x \text{ or } x = \frac{8}{5} = 1\frac{3}{5}$$

**19.**
$$5(x - 1) = 7 - 3(x - 4)$$
$$5x - 5 = 7 - 3x + 12$$
$$5x - 5 = -3x + 19$$
$$5x + 3x - 5 = -3x + 3x + 19$$
$$8x - 5 = 19$$
$$8x - 5 + 5 = 19 + 5$$
$$8x = 24$$
$$\frac{8x}{8} = \frac{24}{8}$$
$$x = 3$$

**20.**
$$3x + 7 = 5(5 - x)$$
$$3x + 7 = 25 - 5x$$
$$3x + 5x + 7 = 25 - 5x + 5x$$
$$8x + 7 = 25$$
$$8x + 7 + (-7) = 25 + (-7)$$
$$8x = 18$$
$$\frac{8x}{8} = \frac{18}{8}$$
$$x = \frac{18}{8} = \frac{9}{4} = 2\frac{1}{4}$$

**21.** $c = p + 9$

**22.** $l = 2w + 5$

**23.**
$$a = \text{height of Mt. Ararat}$$
$$a - 1758 = \text{height of Mt. Hood}$$

**24.**
$$x = \text{width}$$
$$2x - 3 = \text{length}$$
$$2(x) + 2(2x - 3) = 108$$
$$2x + 4x - 6 = 108$$
$$6x - 6 = 108$$
$$6x - 6 + 6 = 108 + 6$$
$$6x = 114$$
$$\frac{6x}{6} = \frac{114}{6}$$
$$x = 19$$

$$\text{width} = 19 \text{ meters}$$
$$\text{length} = 2(19) - 3 = 35 \text{ meters}$$

**25.**
$$x = \text{length of one piece}$$
$$x + 2.5 = \text{length of other piece}$$

$$x + x + 2.5 = 18$$
$$2x + 2.5 = 18$$
$$2x + 2.5 + (-2.5) = 18 + (-2.5)$$
$$2x = 15.5$$
$$\frac{2x}{2} = \frac{15.5}{2}$$
$$x = 7.75$$

one piece $= 7.75$ feet

other piece $= 7.75 + 2.5 = 10.25$ feet

---

## 10.1   Exercises

**1.** A variable is a symbol, usually a letter of the alphabet, that stands for a number.

**3.** $G = 5xy$: variables are $G$, $x$, $y$.

**5.** $p = \dfrac{4ab}{3}$: variables are $p$, $a$, $b$.

**7.** $r = 3 \times m + 5 \times n$
$r = 3m + 5n$

**9.** $H = 2 \times a - 3 \times b$
$H = 2a - 3b$

**11.** $-16x + 26x = 10x$

**13.** $2x - 8x + 5x = -x$

**15.** $-\dfrac{1}{2}x + \dfrac{3}{4}x + \dfrac{1}{12}x$

$= -\dfrac{6}{12}x + \dfrac{9}{12}x + \dfrac{1}{12}x$

$= \dfrac{4}{12}x$

$= \dfrac{1}{3}x$

**17.** $x + 3x + 8 - 7 = 4x + 1$

**19.** $1.3x + 10 - 2.4x - 3.6$
$= 1.3x - 2.4x + 10 - 3.6$
$= -1.1x + 6.4$

**21.** $-13x + 7 - 19x - 10$
$= -13x - 19x + 7 - 10$
$= -32x - 3$

**23.** $16x + 9y - 11 + 21x - 17y + 30$
$= 16x + 21x + 9y - 17y - 11 + 30$
$= 37x - 8y + 19$

**25.** $5a - 3b - c + 8a - 2b - 6c$
$= 5a + 8a - 3b - 2b - c - 6c$
$= 13a - 5b - 7c$

**27.** $\dfrac{1}{2}x + \dfrac{1}{7}y - \dfrac{3}{4}x + \dfrac{5}{21}y - \dfrac{5}{6}x$

$= \dfrac{1}{2}x - \dfrac{3}{4}x - \dfrac{5}{6}x + \dfrac{1}{7}y + \dfrac{5}{21}y$

$= \dfrac{6}{12}x - \dfrac{9}{12}x - \dfrac{10}{12}x + \dfrac{3}{21}y + \dfrac{5}{21}y$

$= -\dfrac{13}{12}x + \dfrac{8}{21}y$

**29.** $7.3x + 1.7x + 4 - 6.4x - 5.6x - 10$
$= 7.3x + 1.7x - 6.4x - 5.6x + 4 - 10$
$= -3x - 6$

**31.** $-7.6n + 1.2 + 11.2m - 3.7n - 8.1m$
$= -7.6n - 3.5n + 11.2m - 8.1m + 1.2$
$= -11.1n + 3.1m + 1.2$

**33.** Perimeter
$= 4x - 2 + 3x + 6 + 5x - 3$
$= 4x + 3x + 5x - 2 + 6 - 3$
$= 12x + 1$

## Cumulative Review Problems

**35.** $\dfrac{n}{6} = \dfrac{12}{15}$

$n \times 15 = 6 \times 12$

$\dfrac{n \times 15}{15} = \dfrac{72}{15}$

$n = 4.8$

**37.** $6n = 18$

$\dfrac{6n}{6} = \dfrac{18}{6}$

$n = 3$

## 10.2  Exercises

**1.** variable

**3.** $3x$ and $x$, $2y$ and $-3y$

**5.** $9(3x - 2) = 9(3x) + 9(-2)$
$= 27x - 18$

**7.** $(-2)(x + y) = (-2)(x) + (-2)(y)$
$= -2x - 2y$

**9.** $(-7)(1.5x - 3y)$
$= (-7)(1.5x) + (-7)(-3y)$
$= -10.5x + 21y$

**11.** $(-3x + 7y)(-10)$
$= (-3x)(-10) + (7y)(-10)$
$= 30x - 70y$

**13.** $4(p + 9q - 10)$
$= 4(p) + 4(9q) + 4(-10)$
$= 4p + 36q - 40$

**15.** $3\left(\dfrac{1}{5}x + \dfrac{2}{3}y - \dfrac{1}{4}\right)$

$= 3\left(\dfrac{1}{5}x\right) + 3\left(\dfrac{2}{3}y\right) + 3\left(-\dfrac{1}{4}\right)$

$= \dfrac{3}{5}x + 2y - \dfrac{3}{4}$

**17.** $15(-12a + 2.2b + 6.7)$
$= 15(-12a) + 15(2.2b) + 15(6.7)$
$= -180a + 33b + 100.5$

**19.** $(8a + 12b - 9c - 5)(4)$
$= (8a)(4) + (12b)(4)$
$+ (-9c)(4) + (-5)(4)$
$= 32a + 48b - 36c - 20$

**21.** $(-2)(1.3x - 8.5y - 5z + 12)$
$= (-2)(1.3x) + (-2)(-8.5y)$
$+ (-2)(-5z) + (-2)(12)$
$= -2.6x + 17y + 10z - 24$

**23.** $\dfrac{1}{2}\left(2x - 3y + 4z - \dfrac{1}{2}\right)$

$= \dfrac{1}{2}(2x) + \dfrac{1}{2}(-3y) + \dfrac{1}{2}(4z)$

$+ \dfrac{1}{2}\left(-\dfrac{1}{2}\right)$

$= x - \dfrac{3}{2}y + 2z - \dfrac{1}{4}$

**25.** $p = 2(l + w) = 2l + 2w$

**27.** $A = \dfrac{h(B + b)}{2}$

$A = \dfrac{hB + hb}{2}$

**29.** $4(5x - 1) + 7(x - 5)$
$= 20x - 4 + 7x - 35$
$= 27x - 39$

**31.** $6\,(3x + 2y) - 4\,(x + 7)$
$$= 18x + 12y - 4x - 28$$
$$= 14x + 12 - 28$$

**33.** $1.5\,(x + 2.2y) + 3\,(2.2x + 1.6y)$
$$= 1.5x + 3.3y + 6.6x + 4.8y$$
$$= 8.1x + 8.1y$$

**35.** $2\,(3b + c - 2a) - 5\,(a - 2c + 5b)$
$$= 6b + 2c - 4a - 5a + 10c - 25b$$
$$= -9a - 19b + 12c$$

**37.**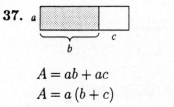

$A = ab + ac$
$A = a\,(b + c)$
Hence, $a\,(b + c) = ab + ac$.

**39.** $x$

$A = xy + xw + xz$
$A = x\,(y + w + z)$
Hence, $x\,(y + w + z) = xy + xw + xz$.

## Cumulative Review Problems

**41.** $C = \pi d = (3.14)\,(12 \text{ in.}) \approx 37.7 \text{ inches}$

**43.** $A = \dfrac{h\,(b + B)}{2} = \dfrac{7 \text{ cm}\,(9 \text{ cm} + 13 \text{ cm})}{2}$
$$= 77 \text{ sq cm}$$

**45.** $16{,}440 - 1110 = 15{,}330$ pounds
$$\frac{15{,}330}{2} = 7665 \text{ pounds}$$
Nereid: 7665 pounds
Auriole: $7665 + 1110 = 8775$ pounds

---

## 10.3   Exercises

**1.** equation

**3.** opposite

**5.** $\qquad y - 5 = 9$
$$y - 5 + 5 = 9 + 5$$
$$y = 14$$

**7.** $\qquad x + 6 = 15$
$$x + 6 + (-6) = 15 + (-6)$$
$$x = 9$$

**9.** $\qquad x + 16 = -2$
$$x + 16 + (-16) = -2 + (-16)$$
$$x = -18$$

**11.** $\qquad 14 + x = -11$
$$14 + (-14) + x = -11 + (-14)$$
$$x = -25$$

**13.** $\qquad -12 + x = 7$
$$-12 + 12 + x = 7 + 12$$
$$x = 19$$

**15.** $\qquad 5.2 = x - 4.6$
$$5.2 + 4.6 = x - 4.6 + 4.6$$
$$9.8 = x$$

**17.** $\qquad x + 3.7 = -5$
$$x + 3.7 + (-3.7) = -5 + (-3.7)$$
$$x = -8.7$$

**19.** $\qquad x - 25.2 = -12$
$$x - 25.2 + 25.2 = -12 + 25.2$$
$$x = 13.2$$

**21.**
$$x + \frac{1}{4} = \frac{3}{4}$$
$$x + \frac{1}{4} + \left(-\frac{1}{4}\right) = \frac{3}{4} + \left(-\frac{1}{4}\right)$$
$$x = \frac{2}{4}$$
$$= \frac{1}{2}$$

**23.**
$$x - \frac{3}{5} = \frac{2}{5}$$
$$x - \frac{3}{5} + \frac{3}{5} = \frac{2}{5} + \frac{3}{5}$$
$$x = \frac{5}{5}$$
$$= 1$$

**25.**
$$x + \frac{2}{3} = -\frac{5}{6}$$
$$x + \frac{2}{3} + \left(-\frac{2}{3}\right) = -\frac{5}{6} + \left(-\frac{2}{3}\right)$$
$$x = -\frac{5}{6} + \left(-\frac{4}{6}\right)$$
$$= -\frac{9}{6}$$
$$= -\frac{3}{2}$$
$$= -1\frac{1}{2}$$

**27.**
$$\frac{1}{5} + y = -\frac{2}{3}$$
$$\frac{1}{5} + \left(-\frac{1}{5}\right) + y = -\frac{2}{3} + \left(-\frac{1}{5}\right)$$
$$y = -\frac{10}{15} + \left(-\frac{3}{15}\right)$$
$$= -\frac{13}{15}$$

**29.**
$$3x - 5 = 2x + 9$$
$$3x + (-2x) - 5 = 2x + (-2x) + 9$$
$$x - 5 = 9$$
$$x - 5 + 5 = 9 + 5$$
$$x = 14$$

**31.**
$$2x - 7 = x - 19$$
$$2x + (-x) - 7 = x + (-x) - 19$$
$$x - 7 = -19$$
$$x - 7 + 7 = -19 + 7$$
$$x = -12$$

**33.**
$$7x - 9 = 6x - 7$$
$$7x + (-6x) - 9 = 6x + (-6x) - 7$$
$$x - 9 = -7$$
$$x - 9 + 9 = -7 + 9$$
$$x = 2$$

**35.**
$$18x + 28 = 17x + 19$$
$$18x + (-17x) + 28 = 17 + (-17x) + 19$$
$$x + 28 = 19$$
$$x + 28 + (-28) = 19 + (-28)$$
$$x = -9$$

**37.**
$$y + \frac{1}{2} = 6$$
$$y - \frac{1}{2} + \frac{1}{2} = 6 + \frac{1}{2}$$
$$y = 6\frac{1}{2}$$

**39.**
$$5 = z + 13$$
$$5 + (-13) = z + 13 + (-13)$$
$$-8 = z$$

**41.**
$$-5.9 + y = -4.7$$
$$-5.9 + 5.9 + y = -4.7 + 5.9$$
$$y = 1.2$$

**43.**
$$2x - 1 = x + 5$$
$$2x + (-x) - 1 = x + (-x) + 5$$
$$x - 1 = 5$$
$$x - 1 + 1 = 5 + 1$$
$$x = 6$$

**45.** To solve the equation $3x = 12$, divide both sides of the equation by 3 so that $x$ stands alone.

---

## Cumulative Review Problems

**47.** $7(2x + 3y) - 3(5x - 1)$
$$= 7(2x) + 7(3y) + (-3)(5x)$$
$$+ (-3)(-1)$$
$$= 14x + 21y - 15x + 3$$
$$= 14x - 15x + 21y + 3$$
$$= -x + 21y + 3$$

**49.** $V = \dfrac{1}{3}(8 \text{ ft})(3 \text{ ft})(4 \text{ ft})$
$$= 32 \text{ cubic feet}$$

---

## 10.4    Exercises

**1.** Answers may vary. To maintain the balance, whatever you do to one side of the scale, you need to do the same thing to the other side of the scale.

**3.** $\dfrac{4}{3}$

**5.** $4x = 36$
$$\dfrac{4x}{4} = \dfrac{36}{4}$$
$$x = 9$$

**7.** $7y = -28$
$$\dfrac{7y}{7} = \dfrac{-28}{7}$$
$$y = -4$$

**9.** $-9x = 16$
$$\dfrac{-9x}{-9} = \dfrac{16}{-9}$$
$$x = -\dfrac{16}{9}$$

**11.** $-5x = -40$
$$\dfrac{-5x}{-5} = \dfrac{-40}{-5}$$
$$x = 8$$

**13.** $-64 = -4m$
$$\dfrac{-64}{-4} = \dfrac{-4m}{-4}$$
$$16 = m$$

**15.** $0.6x = 6$
$$\dfrac{0.6x}{0.6} = \dfrac{6}{0.6}$$
$$x = 10$$

**17.** $5.5z = 9.9$
$$\dfrac{5.5z}{5.5} = \dfrac{9.9}{5.5}$$
$$z = 1.8$$

**19.** $-0.5x = 6.75$
$$\dfrac{-0.5x}{-0.5} = \dfrac{6.75}{-0.5}$$
$$x = -13.5$$

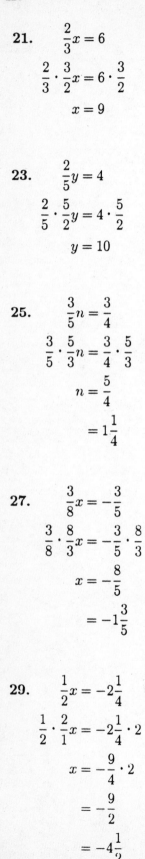

**21.** $\dfrac{2}{3}x = 6$

$\dfrac{2}{3} \cdot \dfrac{3}{2}x = 6 \cdot \dfrac{3}{2}$

$x = 9$

**23.** $\dfrac{2}{5}y = 4$

$\dfrac{2}{5} \cdot \dfrac{5}{2}y = 4 \cdot \dfrac{5}{2}$

$y = 10$

**25.** $\dfrac{3}{5}n = \dfrac{3}{4}$

$\dfrac{3}{5} \cdot \dfrac{5}{3}n = \dfrac{3}{4} \cdot \dfrac{5}{3}$

$n = \dfrac{5}{4}$

$= 1\dfrac{1}{4}$

**27.** $\dfrac{3}{8}x = -\dfrac{3}{5}$

$\dfrac{3}{8} \cdot \dfrac{8}{3}x = -\dfrac{3}{5} \cdot \dfrac{8}{3}$

$x = -\dfrac{8}{5}$

$= -1\dfrac{3}{5}$

**29.** $\dfrac{1}{2}x = -2\dfrac{1}{4}$

$\dfrac{1}{2} \cdot \dfrac{2}{1}x = -2\dfrac{1}{4} \cdot 2$

$x = -\dfrac{9}{4} \cdot 2$

$= -\dfrac{9}{2}$

$= -4\dfrac{1}{2}$

**31.** $1\dfrac{1}{4}z = 10$

$\dfrac{5}{4}z = 10$

$\dfrac{5}{4} \cdot \dfrac{4}{5}z = 10 \cdot \dfrac{4}{5}$

$z = 8$

**33.** First, undo the multiplication.

$4.5 \div 0.5 = 9$

Now divide.

$9 \div 0.5 = 18$

18 is the correct answer.

---

## Cumulative Review Problems

**35.** $6 - 3x + 5y + 7x - 12y$

$= -3x + 7x + 5y - 12y + 6$

$= 4x - 7y + 6$

**37.** $\dfrac{6.5}{68.4} \approx 0.095 = 9.5\%$

---

## 10.5   Exercises

**1.** You want to obtain the $x$-term all by itself on one side of the equation. So you want to remove the $-6$ from the left-hand side of the equation. Therefore, you would add the opposite of $-6$. This means you would add 6 to both sides.

**3.** $3 - 4\,(2) \overset{?}{=} 5 - 3\,(2)$

$3 - 8 \overset{?}{=} 5 - 6$

$-5 \neq -1$   No

**5.** $8\left(\dfrac{1}{2}\right) - 2 \overset{?}{=} 10 - 16\left(\dfrac{1}{2}\right)$

$$4 - 2 \overset{?}{=} 10 - 8$$

$$2 = 2 \quad \text{Yes}$$

**7.** $\qquad 12x - 30 = 6$

$$12x - 30 + 30 = 6 + 30$$

$$12x = 36$$

$$\dfrac{12x}{12} = \dfrac{36}{12}$$

$$x = 3$$

**9.** $\qquad 9x - 3 = -7$

$$9x - 3 + 3 = -7 + 3$$

$$9x = -4$$

$$\dfrac{9x}{9} = \dfrac{-4}{9}$$

$$x = -\dfrac{4}{9}$$

**11.** $\qquad -9x = 3x - 10$

$$-9x + (-3x) = 3x + (-3x) - 10$$

$$-12x = -10$$

$$\dfrac{-12x}{-12} = \dfrac{-10}{-12}$$

$$x = \dfrac{10}{12}$$

$$= \dfrac{5}{6}$$

**13.** $\qquad 18 - 2x = 4x + 6$

$$18 - 2x + 2x = 4x + 2x + 6$$

$$18 = 6x + 6$$

$$18 + (-6) = 6x + 6 + (-6)$$

$$12 = 6x$$

$$\dfrac{12}{6} = \dfrac{6x}{6}$$

$$2 = x$$

**15.** $\qquad 8 + x = 3x - 6$

$$8 + x + (-x) = 3x + (-x) - 6$$

$$8 = 2x - 6$$

$$8 + 6 = 2x - 6 + 6$$

$$14 = 2x$$

$$\dfrac{14}{2} = \dfrac{2x}{2}$$

$$7 = x$$

**17.** $\qquad 7 + 3x = 6x - 8$

$$7 + 3x + (-3x) = 6x + (-3x) - 8$$

$$7 = 3x - 8$$

$$7 + 8 = 3x - 8 + 8$$

$$15 = 3x$$

$$\dfrac{15}{3} = \dfrac{3x}{3}$$

$$5 = x$$

**19.** $\qquad 5 + 2y = 7 + 5y$

$$5 + 2y + (-2y) = 7 + 5y + (-2y)$$

$$5 = 7 + 3y$$

$$5 + (-7) = 7 + (-7) + 3y$$

$$-2 = 3y$$

$$\dfrac{-2}{3} = \dfrac{3y}{3}$$

$$-\dfrac{2}{3} = y$$

**21.** $\qquad 2x + 6 = -8 - 12x$

$$2x + 12x + 6 = -8 - 12x + 12x$$

$$14x + 6 = -8$$

$$14x + 6 + (-6) = -8 + (-6)$$

$$14x = -14$$

$$\dfrac{14x}{14} = \dfrac{-14}{14}$$

$$x = -1$$

**23.**
$$-10 + 6y + 2 = 3y - 26$$
$$-8 + 6y = 3y - 26$$
$$-8 + 6y + (-3y) = 3y + (-3y) - 26$$
$$-8 + 3y = -26$$
$$-8 + 8 + 3y = -26 + 8$$
$$3y = -18$$
$$\frac{3y}{3} = \frac{-18}{3}$$
$$y = -6$$

**29.**
$$-30 - 12y + 18 = -24y + 13 + 7y$$
$$-12 - 12y = -17y + 13$$
$$-12 - 12y + 17y = -17y + 17y + 13$$
$$-12 + 5y = 13$$
$$-12 + 12 + 5y = 13 + 12$$
$$5y = 25$$
$$\frac{5y}{5} = \frac{25}{5}$$
$$y = 5$$

**25.**
$$12 + 4y - 7 = 6y - 9$$
$$5 + 4y = 6y - 9$$
$$5 + 4y + (-6y) = 6y + (-6y) - 9$$
$$5 - 2y = -9$$
$$5 + (-5) - 2y = -9 + (-5)$$
$$-2y = -14$$
$$\frac{-2y}{-2} = \frac{-14}{-2}$$
$$y = 7$$

**31.**
$$5(x + 4) = 4x + 15$$
$$5x + 20 = 4x + 15$$
$$5x + (-4x) + 20 = 4x + (-4x) + 15$$
$$x + 20 = 15$$
$$x + 20 + (-20) = 15 + (-20)$$
$$x = -5$$

**33.**
$$7(y - 1) = 4(y + 2) + 18$$
$$7y - 7 = 4y + 8 + 18$$
$$7y - 7 = 4y + 26$$
$$7y + (-4y) - 7 = 4y + (-4y) + 26$$
$$3y - 7 = 26$$
$$3y - 7 + 7 = 26 + 7$$
$$3y = 33$$
$$\frac{3y}{3} = \frac{33}{3}$$
$$y = 11$$

**27.**
$$\frac{3}{4}x + 2 = -10$$
$$\frac{3}{4}x + 2 + (-2) = -10 + (-2)$$
$$\frac{3}{4}x = -12$$
$$\frac{3}{4} \cdot \frac{4}{3}x = -12 \cdot \frac{4}{3}$$
$$x = -16$$

**35.**
$$7x - 3(x - 6) = 2(x - 3) + 8$$
$$7x - 3x + 18 = 2x - 6 + 8$$
$$4x + 18 = 2x + 2$$
$$4x + (-2x) + 18 = 2x + (-2x) + 2$$
$$2x + 18 = 2$$
$$2x + 18 + (-18) = 2 + (-18)$$
$$2x = -16$$
$$\frac{2x}{2} = \frac{-16}{2}$$
$$x = -8$$

**37.**
$$7x - 16 = 3(x + 2)$$
$$7x - 16 = 3x + 6$$
$$7x + (-3x) - 16 = 3x + (-3x) + 6$$
$$4x - 16 = 6$$
$$4x - 16 + 16 = 6 + 16$$
$$4x = 22$$
$$\frac{4x}{4} = \frac{22}{4}$$
$$x = \frac{22}{4} = \frac{11}{2} = 5\frac{1}{2}$$
$$= 5.5$$

**39.** 
$$7x - 6(x + 3) = -2(x - 4)$$
$$7x - 6x - 18 = -2x + 8$$
$$x - 18 = -2x + 8$$
$$x + 2x - 18 = -2x + 2x + 8$$
$$3x - 18 = 8$$
$$3x - 18 + 18 = 8 + 18$$
$$3x = 26$$
$$\frac{3x}{3} = \frac{26}{3}$$
$$x = \frac{26}{3}$$
$$= 8\frac{2}{3}$$

**41. a.**
$$7 + 3x = 6x - 8$$
$$7 + 3x + (-6x) = 6x + (-6x) - 8$$
$$7 - 3x = -8$$
$$7 + (-7) - 3x = -8 + (-7)$$
$$-3x = -15$$
$$\frac{-3x}{-3} = \frac{-15}{-3}$$
$$x = 5$$

**b.**
$$7 + 3x = 6x - 8$$
$$7 + 3x + (-3x) = 6x + (-3x) - 8$$
$$7 = 3x - 8$$
$$7 + 8 = 3x - 8 + 8$$
$$15 = 3x$$
$$\frac{15}{3} = \frac{3x}{3}$$
$$5 = x$$

**c.** If you prefer to collect $x$-terms on the left side, then method (a) is easier. If you prefer to have a positive coefficient on the $x$-term, then method (b) is easier. But either method will give you the correct answer.

## Cumulative Review Problems

**43.**
$$V = \frac{4\pi r^3}{3}$$
$$= \frac{4(3.14)(46 \text{ cm})^3}{3}$$
$$= \frac{4(3.14)(97{,}336 \text{ cu cm})}{3}$$
$$\approx 407{,}513.4 \text{ cu cm}$$

**45.** 
$$640 - 300 - 150$$
$$= 640 + (-300) + (-150)$$
$$= 340 + (-150)$$
$$= 190 \text{ dealers}$$

## Putting Your Skills To Work

**1.**
$$
\begin{array}{r}
112{,}000 \\
810{,}000 \\
1{,}318{,}000 \\
+\ 3{,}951{,}000 \\
\hline
6{,}191{,}000
\end{array}
$$

$$\frac{3{,}951{,}000}{6{,}191{,}000} \approx 0.638 = 6.38\%$$

**3.** With degree:

percent is $\dfrac{3{,}951{,}000}{10{,}215{,}000} \approx 0.39 = 39\%$

Without degree:

| women | men |
|---|---|
| 112,000 | 567,000 |
| 810,000 | 3,217,000 |
| + 1,318,000 | + 4,223,000 |
| 2,240,000 | 8,007,000 |

percent is $\dfrac{2{,}240{,}000}{8{,}007{,}000} \approx 0.28 = 28\%$

## 10.6   Exercises

**1.** $h = 34 + r$

**3.** $b = n - 107$

**5.** $d = e - 5$

**7.** $l = 2w + 7$

**9.** $s = 3f - 2$

**11.** $n = 3000 + 3s$

**13.** $j + s = 26$

**15.** $ht = 500$

**17.** $j = $ Julie's mileage
$j + 386 = $ Barbara's mileage

**19.** $b = $ measure of angle $B$
$b - 46 = $ measure of angle $A$

**21.** $w = $ height of Mount Whitney
$w + 4430 = $ height of Mount Everest

**23.** $l = $ Lisa's tips
$l + 12 = $ Sam's tips
$l - 6 = $ Brenda's tips

**25.** $w = $ width
$w + 7 = $ height
$2w - 1 = $ length

**27.** $x = $ first angle
$2x = $ second angle
$x - 14 = $ third angle

**29.** $1.08s = $ speed of Caravan
$1.08(s + 10) = $ speed of Lexus

## Cumulative Review Problems

**31.** $-6 - (-7)(2) = -6 - (-14)$
$= -6 + 14$
$= 8$

**33.** $\dfrac{135}{349} \approx 0.387$

$\dfrac{112}{0.360} \approx 311$

$275 \times 0.385 \approx 106$

## 10.7   Exercises

**1.**    $x = $ length of shorter piece
$x + 5.5 = $ length of longer piece

$$x + x + 5.5 = 16$$
$$2x = 10.5$$
$$\frac{2x}{2} = \frac{10.5}{2}$$
$$x = 5.25$$

shorter piece $= 5.25$ feet
longer piece $= 5.25 + 5.5 = 10.75$ feet

**3.**    $x = $ Jack
$x - 140 = $ Jodie

$$x + x - 140 = 670$$
$$2x - 140 = 670$$
$$2x = 810$$
$$\frac{2x}{2} = \frac{810}{2}$$
$$x = 405$$

Jack can leg-press 405 lbs.
Jodie can leg-press $405 - 140 = 265$ lbs.

**5.**    $x = $ number of cars in November
$x + 84 = $ number of cars in May
$x - 43 = $ number of cars in July

$$x + x + 83 + x - 43 = 398$$
$$3x + 40 = 398$$
$$3x = 358$$
$$x \approx 119$$

119 cars in November
$119 + 84 = 203$ cars in May
$119 - 43 = 76$ cars in July

**7.**    $x = $ longer piece
$x - 47 = $ shorter piece

$$x + x - 4.7 = 12$$
$$2x - 4.7 = 12$$
$$2x = 16.7$$
$$\frac{2x}{2} = \frac{16.7}{2}$$
$$x = 8.35$$

longer piece $= 8.35$ feet
shorter piece $= 8.35 - 4.7 = 3.65$ feet

**9.**    $x = $ width of board
$2x - 4 = $ length of board

$$2(x) + 2(2x - 4) = 76$$
$$2x + 4x - 8 = 76$$
$$6x - 8 = 76$$
$$6x = 84$$
$$\frac{6x}{6} = \frac{84}{6}$$
$$x = 14$$

width $= 14$ in.
length $= 2(14) - 4$
$= 28 - 4$
$= 24$ in.

**11.**    $x = $ length of the first side
$x + 20 = $ length of the second side
$x - 4 = $ length of the third side

$$x + x + 20 + x - 4 = 199$$
$$3x + 16 = 199$$
$$3x = 183$$
$$x = 61$$

first side $= 61$ mm
second side $= 61 + 20 = 81$ mm
third side $= 61 - 4 = 75$ mm

**13.**      $x = $ length of the first side

$2x = $ length of the second side

$x + 12 = $ length of the third side

$$x + 2x + x + 12 = 44$$

$$4x + 12 = 44$$

$$4x = 32$$

$$\frac{4x}{4} = \frac{32}{4}$$

$$x = 8$$

first side $= 8$ cm

second side $= 2\,(8) = 16$ cm

third side $= 8 + 12 = 20$ cm

**15.**      $x = $ no. of degrees in angle $A$

$3x = $ no. of degrees in angle $B$

$x + 40 = $ no. of degrees in angle $C$

$$x + 3x + x + 40 = 180$$

$$5x + 40 = 180$$

$$5x = 140$$

$$\frac{5x}{5} = \frac{140}{5}$$

$$x = 28$$

Angle $A = 28°$

Angle $B = 3\,(28) = 84°$

Angle $C = 28 + 40 = 68°$

**17.**      $x = $ total sales

$0.05x = $ commission

$$0.05x + 1200 = 5000$$

$$0.05x = 3800$$

$$\frac{0.05x}{0.05} = \frac{3800}{0.05}$$

$$x = 76{,}000$$

total sales $= \$76{,}000$

**19.**      $x = $ yearly rent

$0.12x = $ commission

$$100 + 0.12x = 820$$

$$0.12x = 720$$

$$\frac{0.12x}{0.12} = \frac{720}{0.12}$$

$$x = 6000$$

yearly rent $= \$6000$

**21.**      $x = $ length of the adult section

$x + 6.2 = $ length of the children section

$$x + x + 6.2 = 32$$

$$2x + 6.2 = 32$$

$$2x = 25.8$$

$$\frac{2x}{2} = \frac{25.8}{2}$$

$$x = 12.9$$

adult section $= 12.9$ ft

children section $= 12.9 + 6.2 = 19.1$ ft

**23.**      $x = $ cost of first program

$2x - 20 = $ cost of second program

$3\,(2x - 20) + 17 = $ cost of third program

$$x + 2x - 20 + 3\,(2x - 20) + 17 = 570.33$$

$$x + 2x - 20 + 6x - 60 + 17 = 570.33$$

$$9x - 63 = 570.33$$

$$9x = 633.33$$

$$\frac{9x}{9} = \frac{633.33}{9}$$

$$x = 70.37$$

first program $= \$70.37$

second program $= 2\,(70.37) - 20$

$= \$120.74$

third program $= 3\,(120.74) + 17$

$= \$379.22$

## Cumulative Review Problems

**25.** $\dfrac{n}{100} = \dfrac{12}{20}$

$n \times 20 = 100 \times 12$

$\dfrac{n \times 20}{20} = \dfrac{1200}{20}$

$n = 60 \text{ or } 60\%$

**27.** $C = \pi d$

$\approx (3.14)(37.4)$

$= 117.436$

$\approx 117.4 \text{ meters}$

## Chapter 10 Review Problems

**1.** $-8a + 6 - 5a - 3 = -8a - 5a + 6 - 3$

$\qquad\qquad\qquad\qquad = -13a + 3$

**2.** $\dfrac{3}{4}x + \dfrac{2}{3} + \dfrac{1}{8}x + \dfrac{1}{4}$

$= \dfrac{3}{4}x + \dfrac{1}{8}x + \dfrac{2}{3} + \dfrac{1}{4}$

$= \dfrac{6}{8}x + \dfrac{1}{8}x + \dfrac{8}{12} + \dfrac{3}{12}$

$= \dfrac{7}{8}x + \dfrac{11}{12}$

**3.** $5x + 2y - 7x - 9y$

$= 5x - 7x + 2y - 9y$

$= -2x - 7y$

**4.** $3x - 7y + 8x + 2y$

$= 3x + 8x - 7y + 2y$

$= 11x - 5y$

**5.** $5x - 9y - 12 - 6x - 3y + 18$

$= 5x - 6x - 9y - 3y - 12 + 18$

$= -x - 12y + 6$

**6.** $7x - 2y - 20 - 5x - 8y + 13$

$= 7x - 5x - 2y - 8y - 20 + 13$

$= 2x - 10y - 7$

**7.** $-3(5x + y) = -3(5x) + (-3)(y)$

$\qquad\qquad\quad = -15x - 3y$

**8.** $-4(2x + 3y)$

$= -4(2x) + (-4)(3y)$

$= -8x - 12y$

**9.** $9(x - 3y + 4)$

$= 9(x) + 9(-3y) + 9(4)$

$= 9x - 27y + 36$

**10.** $3(2x - 6y - 1)$

$= 3(2x) + 3(-6y) + 3(-1)$

$= 6x - 18y - 3$

**11.** $-8(3a - 5b - c)$

$= (-8)(3a) + (-8)(-5b) + (-8)(-c)$

$= -24a + 40b + 8c$

**12.** $-9(2a - 8b - c)$

$= (-9)(2a) + (-9)(-8b) + (-9)(-c)$

$= -18a + 72b + 9c$

**13.** $5(1.2x + 3y - 5.5)$

$= 5(1.2x) + 5(3y) + 5(-5.5)$

$= 6x + 15y - 27.5$

**14.** $6(1.4x - 2y + 3.4)$

$= 6(1.4x) + 6(-2y) + 6(3.4)$

$= 8.4x - 12y + 20.4$

**15.** $2(x + 3y) - 4(x - 2y)$
$$= 2x + 6y - 4x + 8y$$
$$= -2x + 14y$$

**16.** $2(5x - y) - 3(x + 2y)$
$$= 10x - 2y - 3x - 6y$$
$$= 7x - 8y$$

**17.** $-2(a + b) - 3(2a + 8)$
$$= -2a - 2b - 6a - 24$$
$$= -8a - 2b - 24$$

**18.** $-4(a - 2b) + 3(5 - a)$
$$= -4a + 8b + 15 - 3a$$
$$= -7a + 8b + 15$$

**19.**
$$x - 3 = 9$$
$$x + (-3) + 3 = 9 + 3$$
$$x = 12$$

**20.**
$$-8 = x - 12$$
$$-8 + 12 = x - 12 + 12$$
$$4 = x$$

**21.**
$$x + 8.3 = 20$$
$$x + 8.3 + (-8.3) = 20 + (-8.3)$$
$$x = 11.7$$

**22.**
$$2.4 = x - 5$$
$$2.4 + 5 = x - 5 + 5$$
$$7.4 = x$$

**23.**
$$3.1 + x = -9$$
$$3.1 + (-3.1) + x = -9 + (-3.1)$$
$$x = -12.1$$

**24.**
$$x - 7 = 5.8$$
$$x - 7 + 7 = 5.8 + 7$$
$$x = 12.8$$

**25.**
$$x - \frac{3}{4} = 2$$
$$x + \left(-\frac{3}{4}\right) + \frac{3}{4} = 2 + \frac{3}{4}$$
$$x = \frac{11}{4} = 2\frac{3}{4}$$

**26.**
$$x + \frac{1}{2} = 3\frac{3}{4}$$
$$x + \frac{1}{2} + \left(-\frac{1}{2}\right) = 3\frac{3}{4} + \left(-\frac{1}{2}\right)$$
$$x = 3\frac{3}{4} + \left(-\frac{2}{4}\right)$$
$$x = 3\frac{1}{4}$$

**27.**
$$x + \frac{3}{8} = \frac{1}{2}$$
$$x + \frac{3}{8} + \left(-\frac{3}{8}\right) = \frac{1}{2} + \left(-\frac{3}{8}\right)$$
$$x = \frac{4}{8} + \left(-\frac{3}{8}\right)$$
$$= \frac{1}{8}$$

**28.**
$$x - \frac{5}{6} = \frac{2}{3}$$
$$x - \frac{5}{6} + \frac{5}{6} = \frac{2}{3} + \frac{5}{6}$$
$$x = \frac{4}{6} + \frac{5}{6}$$
$$x = \frac{9}{6}$$
$$= \frac{3}{2}$$
$$= 1\frac{1}{2}$$

**29.**
$$2x + 20 = 25 + x$$
$$2x + (-x) + 20 = 25 + x + (-x)$$
$$x + 20 = 25$$
$$x + 20 + (-20) = 25 + (-20)$$
$$x = 5$$

**30.**
$$5x - 3 = 4x - 15$$
$$5x + (-4x) - 3 = 4x + (-4x) - 154$$
$$x - 3 = -15$$
$$x - 3 + 3 = -15 + 3$$
$$x = -12$$

**31.** $8x = -20$
$$\frac{8x}{8} = \frac{-20}{8}$$
$$x = \frac{-20}{8} = -\frac{5}{2}$$
$$= -2\frac{1}{2}$$

**32.** $-12y = 60$
$$\frac{-12y}{-12} = \frac{60}{-12}$$
$$y = -5$$

**33.** $1.5x = 9$
$$\frac{1.5x}{1.5} = \frac{9}{1.5}$$
$$x = 6$$

**34.** $1.8y = 12.6$
$$\frac{1.8y}{1.8} = \frac{12.6}{1.8}$$
$$y = 7$$

**35.** $-7.2x = 36$
$$\frac{-7.2x}{-7.2} = \frac{36}{-7.2}$$
$$x = -5$$

**36.** $6x = 1.5$
$$\frac{6x}{6} = \frac{1.5}{6}$$
$$x = 0.25$$

**37.** $\dfrac{3}{4}x = 6$
$$\frac{4}{3} \cdot \frac{3}{4}x = \frac{4}{3} \cdot 6$$
$$x = 8$$

**38.** $\dfrac{2}{3}x = \dfrac{5}{9}$
$$\frac{3}{2} \cdot \frac{2}{3}x = \frac{3}{2} \cdot \frac{5}{9}$$
$$x = \frac{5}{6}$$

**39.**
$$3 - 2x = 9 - 8x$$
$$3 - 2x + 8x = 9 - 8x + 8x$$
$$3 + 6x = 9$$
$$3 + (-3) + 6x = 9 + (-3)$$
$$6x = 6$$
$$\frac{6x}{6} = \frac{6}{6}$$
$$x = 1$$

**40.**
$$8 - 6x = -7 - 3x$$
$$8 - 6x + 6x = -7 - 3x + 6x$$
$$8 = -7 + 3x$$
$$8 + 7 = -7 + 7 + 3x$$
$$15 = 3x$$
$$\frac{15}{3} = \frac{3x}{3}$$
$$5 = x$$

**44.**
$$4 + 3x - 8 = 12 + 5x + 4$$
$$3x - 4 = 5x + 16$$
$$3x + (-3x) - 4 = 5x + (-3x) + 16$$
$$-4 = 2x + 16$$
$$-4 + (-16) = 2x + 16 + (-16)$$
$$-20 = 2x$$
$$\frac{-20}{2} = \frac{2x}{2}$$
$$-10 = x$$

**41.**
$$10 + x = 3x - 6$$
$$10 + x + (-x) = 3x + (-x) - 6$$
$$10 = 2x - 6$$
$$10 + 6 = 2x - 6 + 6$$
$$16 = 2x$$
$$\frac{16}{2} = \frac{2x}{2}$$
$$8 = x$$

**45.**
$$5x - 2 = 27$$
$$5x - 2 + 2 = 27 + 2$$
$$5x = 29$$
$$\frac{5x}{5} = \frac{29}{5}$$
$$x = \frac{29}{5}$$
$$= 5\frac{4}{5}$$

**42.**
$$8x - 7 = 5x + 8$$
$$8x + (-5x) - 7 = 5x + (-5x) + 8$$
$$3x - 7 = 8$$
$$3x - 7 + 7 = 8 + 7$$
$$3x = 15$$
$$\frac{3x}{3} = \frac{15}{3}$$
$$x = 5$$

**46.**
$$2(3x - 4) = 7 - 2x + 5x$$
$$6x - 8 = 7 + 3x$$
$$6x + (-3x) - 8 = 7 + 3x + (-3x)$$
$$3x - 8 = 7$$
$$3x - 8 + 8 = 7 + 8$$
$$3x = 15$$
$$\frac{3x}{3} = \frac{15}{3}$$
$$x = 5$$

**43.**
$$9x - 3x + 18 = 36$$
$$6x + 18 = 36$$
$$6x + 18 + (-18) = 36 + (-18)$$
$$6x = 18$$
$$\frac{6x}{6} = \frac{18}{6}$$
$$x = 3$$

**47.**
$$5 + 2y + 5(y - 3) = 6(y + 1)$$
$$5 + 2y + 5y - 15 = 6y + 6$$
$$7y - 10 = 6y + 6$$
$$7y + (-6y) - 10 = 6y + (-6y) + 6$$
$$y - 10 = 6$$
$$y - 10 + 10 = 6 + 10$$
$$y = 16$$

**48.**
$$3 + 5\,(y + 4) = 4\,(y - 2) + 3$$
$$3 + 5y + 20 = 4y - 8 + 3$$
$$5y + 23 = 4y - 5$$
$$5y + (-4y) + 23 = 4y + (-4y) - 5$$
$$y + 23 = -5$$
$$y + 23 + (-23) = -5 + (-23)$$
$$y = -28$$

**49.** $w = c + 3000$

**50.** $a = m - 18$ or $m = a + 18$

**51.** $A = 3B$

**52.** $l = 2w - 3$

**53.**
$$r = \text{Roberto's salary}$$
$$r + 2050 = \text{Michael's salary}$$

**54.**
$$x = \text{length of the first side}$$
$$2x = \text{length of the second side}$$

**55.**
$$c = \text{number of Connie's courses}$$
$$c - 6 = \text{number of Nancy's courses}$$

**56.**
$$b = \text{number of books in old library}$$
$$2b + 450 = \text{number of books in new library}$$

**57.**
$$x = \text{length of one piece}$$
$$x + 6.5 = \text{length of other piece}$$
$$x + x + 6.5 = 60$$
$$2x + 6.5 = 60$$
$$2x + 6.5 + (-6.5) = 60 + (-6.5)$$
$$2x = 53.5$$
$$\frac{2x}{2} = \frac{53.5}{2}$$
$$x = 26.75$$

one piece $= 26.75$ feet
other piece $= 26.75 + 6.5 = 32.25$ feet

**58.**
$$x = \text{old employee's salary}$$
$$x - 28 = \text{new employee's salary}$$
$$x + x - 28 = 412$$
$$2x - 28 = 412$$
$$2x - 28 + 28 = 412 + 28$$
$$2x = 440$$
$$\frac{2x}{2} = \frac{440}{2}$$
$$x = 220$$

old employee $= \$220$
new employee $= 220 - 28 = \$192$

**59.**
$$x = \#\text{ customers in February}$$
$$2x = \#\text{ customers in March}$$
$$x + 3000 = \#\text{ customers in April}$$
$$x + 2x + x + 3000 = 45{,}200$$
$$4x + 3000 = 45{,}200$$
$$4x + 3000 + (-3000) = 45{,}200 + (-3000)$$
$$4x = 42{,}200$$
$$\frac{4x}{4} = \frac{42{,}200}{4}$$
$$x = 10{,}550$$

February $= 10{,}550$
March $= 2\,(10{,}550) = 21{,}100$
April $= 10{,}550 + 3000 = 13{,}550$

**60.**
$$x = \text{miles on Thursday}$$
$$x + 106 = \text{miles on Friday}$$
$$x - 39 = \text{miles on Saturday}$$

$$x + x + 106 + x - 39 = 856$$
$$3x + 67 = 856$$
$$3x + 67 + (-67) = 856 + (-67)$$
$$3x = 789$$
$$\frac{3x}{3} = \frac{789}{3}$$
$$x = 263$$

Thursday = 263 miles
Friday = 263 + 106 = 369 miles
Saturday = 263 − 39 = 224 miles

**61.**
$$w = \text{width}$$
$$2x - 3 = \text{length}$$

$$72 = 2w + 2(2w - 3)$$
$$72 = 2w + 4w - 6$$
$$72 = 6w - 6$$
$$72 + 6 = 6w + (-6) + 6$$
$$78 = 6w$$
$$\frac{78}{6} = \frac{6w}{6}$$
$$w = 13$$

width = 13 in.
length = 2(13) − 3 = 23 in.

**62.**
$$x = \text{width}$$
$$3x + 2 = \text{length}$$

$$2(x) + 2(3x + 2) = 180$$
$$2x + 6x + 4 = 180$$
$$8x + 4 = 180$$
$$8x + 4 + (-4) = 180 + (-4)$$
$$8x = 176$$
$$\frac{8x}{9} = \frac{176}{8}$$
$$x = 22$$

width = 22 m
length = 3(22) + 2 = 68 m

**63.**
$$x = \text{length of the first side}$$
$$x + 7 = \text{length of the second side}$$
$$x - 4 = \text{length of the third side}$$

$$x + x + 7 + x - 4 = 99$$
$$3x + 3 = 99$$
$$3x + 3 + (-3) = 99 + (-3)$$
$$3x = 96$$
$$\frac{3x}{3} = \frac{96}{3}$$
$$x = 32$$

first side = 32 m
second side = 32 + 7 = 39 m
third side = 32 − 4 = 28 m

**64.**
$$x + 74 = \text{angle } A$$
$$x = \text{angle } B$$
$$3x = \text{angle } C$$

$$x + 74 + x + 3x = 180$$
$$5x + 74 = 180$$
$$5x + 74 + (-74) = 180 + (-74)$$
$$5x = 106$$
$$\frac{5x}{5} = \frac{106}{5}$$
$$x = 21.2$$

angle $A = 21.2 + 74 = 95.2°$
angle $B = 21.2°$
angle $C = 3(21.2) = 63.6°$

**65.** $x$ = cost of a chair
345 − 275 = 70 is the cost for 2 chairs.

$$2x = 70$$
$$\frac{2x}{2} = \frac{70}{2}$$
$$x = 35$$

chair = $35
table = 275 − 4(35)
= 275 − 140 = $135

**66.**
$$x = \text{length}$$
$$x - 44 = \text{width}$$
$$2(x) + 2(x - 44) = 288$$
$$2x + 2x - 88 = 288$$
$$4x - 88 = 288$$
$$4x - 88 + 88 = 288 + 88$$
$$4x = 376$$
$$\frac{4x}{4} = \frac{376}{4}$$
$$x = 94$$

length = 94 feet
width = 94 − 44 = 50 feet

**67.**
$$x = \text{miles for second day}$$
$$x + 88 = \text{miles for first day}$$
$$x + x + 88 = 760$$
$$2x + 88 = 760$$
$$2x + 88 + (-88) = 760 + (-88)$$
$$2x = 672$$
$$\frac{2x}{x} = \frac{672}{2}$$
$$x = 336$$

second day = 336 miles
first day = 336 + 88 = 424 miles

**68.**
$$x = \text{first week}$$
$$x + 156 = \text{second week}$$
$$x - 142 = \text{third week}$$
$$x + x + 156 + x - 142 = 800$$
$$3x + 14 = 800$$
$$3x + 14 + (-14) = 800 + (-14)$$
$$3x = 786$$
$$\frac{3x}{3} = \frac{786}{3}$$
$$x = 262$$

first week = 262
second week = 262 + 156 = 418
third week = 262 − 142 = 120

**69.**
$$x = \text{number of first-class passengers}$$
$$4x = \text{number of coach passengers}$$
$$x + 4x = 760$$
$$5x = 760$$
$$\frac{5x}{5} = \frac{760}{5}$$
$$x = 152$$

first class = 152
coach = 4(150) = 608

**70.**
$$x = \text{cost of the furniture}$$
$$0.08x = \text{commission}$$
$$0.08x + 1500 = 3050$$
$$0.08x + 1500 + (-1500) = 3050 + (-1500)$$
$$0.08x = 1550$$
$$\frac{0.08x}{0.08} = \frac{1550}{0.08}$$
$$x = 19{,}375$$

furniture = $19,375

## Chapter 10 Test

**1.** $5a - 11a = -6a$

**2.** $\dfrac{1}{3}x + \dfrac{5}{8}y - \dfrac{1}{5}x + \dfrac{1}{2}y$

$$= \frac{1}{3}x - \frac{1}{5}x + \frac{5}{8}y + \frac{1}{2}y$$

$$= \frac{5}{15}x - \frac{3}{15}x + \frac{5}{8}y + \frac{4}{8}y$$

$$= \frac{2}{15}x + \frac{9}{8}y$$

**3.** $-3x + 7y - 8x - 5y$

$\quad = -3x - 8x + 7y - 5y$

$\quad = -11x + 2y$

**4.** $6a - 5b + 7 - 5a - 3b + 4$

$\quad = 6a - 5a - 5b - 3b + 7 + 4$

$\quad = a - 8b + 11$

**5.** $7x - 8y + 2z - 9z + 8y$

$\quad = 7x - 8y + 8y + 2z - 9z$

$\quad = 7x - 7z$

**6.** $x + 5y - 6 - 5x - 7y + 11$

$\quad = x - 5x + 5y - 7y - 6 + 11$

$\quad = -4x - 2y + 5$

**7.** $5(12x - 5y) = 5(12x) + 5(-5y)$

$\qquad\qquad\quad = 60x - 25y$

**8.** $-4(2x - 3y + 7)$

$\quad = (-4)(2x) + (-4)(-3y) + (-4)(7)$

$\quad = -8x + 12y - 28$

**9.** $-1.5(3a - 2b + c - 8)$

$\quad = (-1.5)(3a) + (-1.5)(-2b)$

$\qquad + (-1.5)(c) + (-1.5)(-8)$

$\quad = -4.5a + 3b - 1.5c + 12$

**10.** $2(-3a + 2b) - 5(1 - 2b)$

$\quad = -6a + 4b - 5a + 10b$

$\quad = -11a + 14b$

**11.**
$$-5 - 3x = 19$$
$$-5 + 5 - 3x = 19 + 5$$
$$-3x = 24$$
$$\frac{-3x}{-3} = \frac{24}{-3}$$
$$x = -8$$

**12.**
$$-3(4 - x) = 5(6 + 2x)$$
$$-12 + 3x = 30 + 10x$$
$$-12 + 3x + (-3x) = 30 + 10x + (-3x)$$
$$-12 = 30 + 7x$$
$$-12 + (-30) = 30 + (-30) + 7x$$
$$-42 = 7x$$
$$\frac{-42}{7} = \frac{7x}{7}$$
$$-6 = x$$

**13.**
$$-5x + 9 = -4x - 6$$
$$-5x + 5x + 9 = -4x + 5x - 6$$
$$9 = x - 6$$
$$9 + 6 = x - 6 + 6$$
$$15 = x$$

**14.**
$$8x - 2 - x = 3x - 9 - 10x$$
$$7x - 2 = -7x - 9$$
$$7x + 7x - 2 = -7x + 7x - 9$$
$$14x - 2 = -9$$
$$14x - 2 + 2 = -9 + 2$$
$$14x = -7$$
$$\frac{14x}{14} = \frac{-7}{14}$$
$$x = -\frac{1}{2}$$

**15.**
$$2x - 5 + 7x = 4x - 1 + x$$
$$9x - 5 = 5x - 1$$
$$9x + (-5x) - 5 = 5x + (-5x) - 1$$
$$4x - 5 = -1$$
$$4x - 5 + 5 = -1 + 5$$
$$4x = 4$$
$$\frac{4x}{4} = \frac{4}{4}$$
$$x = 1$$

**16.**
$$3 - (x + 2) = 5 + (3)(x + 2)$$
$$3 - x - 2 = 5 + 3x + 6$$
$$-x + 1 = 3x = 11$$
$$-x + x + 1 = 3x + x + 11$$
$$1 = 4x + 11$$
$$1 + (-11) = 4x + 11 + (-11)$$
$$-10 = 4x$$
$$\frac{-10}{4} = \frac{4x}{4}$$
$$-\frac{5}{2} = x \text{ or } x = -2\frac{1}{2}$$

**17.** $s = f + 15$

**18.** $n = s - 15{,}000$

**19.** $\frac{1}{2}s$ = first angle

$s$ = second angle

$2s$ = third angle

**20.** $\quad w$ = width

$2w - 5$ = length

**21.** $\quad x$ = acres in Prentice farm

$3x$ = acres in Smithfield farm

$$x + 3x = 348$$
$$4x = 348$$
$$\frac{4x}{4} = \frac{348}{4}$$
$$x = 87$$

Prentice farm = 87 acres

Smithfield farm = 261 acres

**22.** $\quad x$ = Marcia's earnings

$x - 1500$ = Sam's earnings

$$x + x - 1500 = 46{,}500$$
$$2x - 1500 = 46{,}500$$
$$2x - 1500 + 1500 = 46{,}500 + 1500$$
$$2x = 48{,}000$$
$$\frac{2x}{2} = \frac{48{,}000}{2}$$
$$x = 24{,}000$$

Marcia = $24,000

Sam = 24,000 − 1500 = $22,500

**23.** $\quad x$ = miles on Monday

$x + 56$ = miles on Tuesday

$x - 14$ = miles on Wednesday

$$x + x + 56 + x - 14 = 975$$
$$3x + 42 = 975$$
$$3x + 42 + (-42) = 975 + (-42)$$
$$3x = 933$$
$$\frac{3x}{3} = \frac{933}{3}$$
$$x = 311$$

Monday = 311 miles

Tuesday = 311 + 56 = 367 miles

Wednesday = 311 − 14 = 297 miles

**24.** $\quad x$ = length

$\frac{1}{2}x + 8$ = width

$$2\,(x) + 2\left(\frac{1}{2}x + 8\right) = 118$$

$$2x + x + 16 = 118$$

$$3x + 16 = 118$$

$$3x + 16 + (-16) = 118 + (-16)$$

$$3x = 102$$

$$\frac{3x}{3} = \frac{102}{3}$$

$$x = 34$$

length = 34 feet

width $= \frac{1}{2}\,(34) + 8 = 25$ feet

---

## Chapters 1–10 Cumulative Test

**1.**
```
  456
   89
  123
+  79
  747
```

**2.**
```
    309
  ×  35
  1 545
  9 27
 10,815
```

**3.** 45,678,934 rounds to 45,678,900

**4.** $\dfrac{5}{12} \div \dfrac{1}{6} = \dfrac{5}{12} \times \dfrac{6}{1}$

$$= \frac{5}{2}$$

$$= 2\frac{1}{2}$$

**5.** $3\dfrac{1}{4} \times 2\dfrac{1}{2} = \dfrac{13}{4} \times \dfrac{5}{2}$

$$= \frac{65}{8}$$

$$= 8\frac{1}{8}$$

**6.**
```
  0.0078
×    9.3
    234
    702
0.07254
```

**7.**
```
  34,007.090
-  3,456.789
  30,550.301
```

**8.** $\dfrac{9}{n} = \dfrac{40.5}{72}$

$$9 \times 72 = n \times 40.5$$

$$648 = n \times 40.5$$

$$\frac{648}{40.5} = \frac{n \times 40.5}{40.5}$$

$$16 = n$$

**9.** What is 28.5% of $5600?

$$28.5\% \times 5600 = n$$

$$0.285 \times 5600 = n$$

$$n = \$1596$$

**10.** $34\% \times n = 1870$

$$0.34 \times n = 1870$$

$$\frac{0.34 \times n}{0.34} = \frac{1870}{0.34}$$

$$n = 5500$$

**11.** 345 mm = 0.345 m

**12.** 10 ft $\times \dfrac{12 \text{ in.}}{1 \text{ ft}} = 120$ in.

**13.** $c = \pi d$

$$= 3.14\,(12)$$

$$\approx 37.7 \text{ yd}$$

**14.**  $A = \dfrac{bh}{2}$

$\phantom{A} = \dfrac{(13)\,(22)}{2}$

$\phantom{A} = 143$

143 sq m

**15.**  $4 - 8 + 12 - 32 - 7$

$= 4 + (-8) + 12 + (-32) + (-7)$

$= -4 + 12 + (32) + (-7)$

$= 8 + (-32) + (-7)$

$= -24 + (-7)$

$= -31$

**16.**  $(5)\,(-2)\,(3)\,(-1) = (-10)\,(3)\,(-1)$

$\phantom{(5)\,(-2)\,(3)\,(-1)} = (-30)\,(-1)$

$\phantom{(5)\,(-2)\,(3)\,(-1)} = 30$

**17.**  $\dfrac{1}{2}a + \dfrac{1}{7}b + \dfrac{1}{4}a - \dfrac{3}{14}b$

$= \dfrac{1}{2}a + \dfrac{1}{4}a + \dfrac{1}{7}b - \dfrac{3}{14}b$

$= \dfrac{2}{4}a + \dfrac{1}{4}a + \dfrac{2}{14}b - \dfrac{3}{14}b$

$= \dfrac{3}{4}a - \dfrac{1}{14}b$

**18.**  $-4x + 5y - 9 - 2x - 3y + 12$

$= -4x - 2x + 5y - 3y - 9 + 12$

$= -6x + 2y + 3$

**19.**  $-7\,(-3x + y - 8)$

$= -7\,(-3x) + (-7)\,(y) + (-7)\,(-8)$

$= 21x - 7y + 56$

**20.**  $2\,(3x - 4y) - 8\,(x + 2y)$

$= 6x - 8y - 8x - 16y$

$= -2x - 24y$

**21.**

$$5x - 5 = 7x - 13$$
$$5x + (-7x) - 5 = 7x + (-7x) - 13$$
$$-2x - 5 = -13$$
$$-2x - 5 + 5 = -13 + 5$$
$$-2x = -8$$
$$\frac{-2x}{-2} = \frac{-8}{-2}$$
$$x = 4$$

**22.**

$$7 - 9y - 12 = 3y + 5 - 8y$$
$$-9y - 5 = -5y + 5$$
$$-9y + 9y - 5 = -5y + 9y + 5$$
$$-5 = 4y + 5$$
$$-5 + (-5) = 4y + 5 + (-5)$$
$$-10 = 4y$$
$$\frac{-10}{4} = \frac{4y}{4}$$
$$-\frac{5}{2} = y \text{ or } y = -2\frac{1}{2}$$

**23.**

$$x - 2 + 5x + 3 = 183 - x$$
$$6x + 1 = 183 - x$$
$$6x + x + 1 = 183 - x + x$$
$$7x + 1 = 183$$
$$7x + 1 + (-1) = 183 + (-1)$$
$$7x = 182$$
$$\frac{7x}{7} = \frac{182}{7}$$
$$x = 26$$

**24.**

$$9\left(2x+8\right)=20-\left(x+5\right)$$
$$18+72=20-x-5$$
$$18x+72=-x+15$$
$$18x+x+72=-x+x+15$$
$$19x+72=15$$
$$19x+72+(-72)=15+(-72)$$
$$19x=-57$$
$$\frac{19x}{19}=\frac{-57}{19}$$
$$x=-3$$

$$2\left(x\right)+2\left(2x+8\right)=98$$
$$2x+4x+16=98$$
$$6x+16=98$$
$$6x+16+(-16)=98+(-16)$$
$$6x=82$$
$$\frac{6x}{6}=\frac{82}{6}$$
$$x=\frac{41}{3}$$
$$=13\frac{2}{3}$$

$$\text{width}=13\frac{2}{3}\text{ feet}$$

**25.**

$$p=\text{weight of printer}$$
$$p+322=\text{weight of computer}$$

$$\text{length}=2x+8=2\left(\frac{41}{3}\right)+8$$
$$=\frac{82}{3}+\frac{24}{3}=\frac{106}{3}$$
$$=35\frac{1}{3}\text{ feet}$$

**26.**

$$f=\text{fall enrollment}$$
$$f-87=\text{summer enrollment}$$

**27.**

$$x=\text{miles driven on Thursday}$$
$$x+48=\text{miles driven on Friday}$$
$$x-95=\text{miles driven on Saturday}$$
$$x+x+48+x-95=1081$$
$$3x-47=1081$$
$$3x+(-47)+47=1081+47$$
$$3x=1128$$
$$\frac{3x}{3}=\frac{1128}{3}$$
$$x=376$$

$$\text{Thursday}=376\text{ miles}$$
$$\text{Friday}=376+48=424\text{ miles}$$
$$\text{Saturday}=376-95=281\text{ miles}$$

**28.**

$$x=\text{width}$$
$$2x+8=\text{length}$$

# PRACTICE FINAL EXAM

**1.** 82,367 = Eighty-two thousand, three hundred sixty-seven

**2.**
$$\begin{array}{r} 13,428 \\ +\ 16,905 \\ \hline 30,333 \end{array}$$

**3.** $19 + 23 + 16 + 45 + 70 = 173$

**4.**
$$\begin{array}{r} 89,071 \\ -\ 54,968 \\ \hline 34,103 \end{array}$$

**5.**
$$\begin{array}{r} 78 \\ \times\ 54 \\ \hline 312 \\ 390 \\ \hline 4212 \end{array}$$

**6.**
$$\begin{array}{r} 2035 \\ \times\ 107 \\ \hline 14\ 245 \\ 203\ 50 \\ \hline 217,745 \end{array}$$

**7.**
$$\begin{array}{r} 158 \\ 7\overline{)1106} \\ \underline{7}\phantom{000} \\ 40\phantom{0} \\ \underline{35}\phantom{0} \\ 56 \\ \underline{56} \\ 0 \end{array}$$

**8.**
$$\begin{array}{r} 606 \\ 26\overline{)15,756} \\ \underline{15\ 6}\phantom{00} \\ 156 \\ \underline{156} \\ 0 \end{array}$$

**9.** $3^4 + 20 \div 4 \times 2 + 5^2 = 81 + 10 + 25$
$$= 116$$

**10.** $512 \div 16 = 32$

32 mi/gal

**11.** $\dfrac{14}{30} = \dfrac{14 \div 2}{30 \div 2} = \dfrac{7}{15}$

**12.** $3\dfrac{9}{11} = \dfrac{3 \times 11 + 9}{11} = \dfrac{42}{11}$

**13.** $\dfrac{1}{10} + \dfrac{3}{4} + \dfrac{4}{5} = \dfrac{1}{10} \times \dfrac{2}{2} + \dfrac{3}{4} \times \dfrac{5}{5} + \dfrac{4}{5} \times \dfrac{4}{4}$
$$= \dfrac{2}{20} + \dfrac{15}{20} + \dfrac{16}{20}$$
$$= \dfrac{33}{20}$$
$$= 1\dfrac{13}{20}$$

**14.** $2\dfrac{1}{3} + 3\dfrac{3}{5} = 2\dfrac{5}{15} + 3\dfrac{9}{15}$
$$= 5\dfrac{14}{15}$$

**15.**
$$\begin{array}{cc} 4\frac{5}{7} & 4\frac{10}{14} \\ -\ 2\frac{1}{2} & -\ 2\frac{7}{14} \\ \hline & 2\frac{3}{14} \end{array}$$

**16.** $1\dfrac{1}{4} \times 3\dfrac{1}{5} = \dfrac{5}{4} \times \dfrac{16}{5}$
$$= \dfrac{5 \cdot 4 \cdot 4}{4 \cdot 5}$$
$$= 4$$

**17.** $\dfrac{7}{9} \div \dfrac{5}{18} = \dfrac{7}{9} \times \dfrac{18}{5}$

$\phantom{\dfrac{7}{9} \div \dfrac{5}{18}} = \dfrac{14}{5}$

$\phantom{\dfrac{7}{9} \div \dfrac{5}{18}} = 2\dfrac{4}{5}$

**18.** $\dfrac{5\frac{1}{2}}{3\frac{1}{4}} = \dfrac{\frac{11}{2}}{\frac{13}{4}}$

$\phantom{\dfrac{5\frac{1}{2}}{3\frac{1}{4}}} = \dfrac{11}{2} \times \dfrac{4}{13}$

$\phantom{\dfrac{5\frac{1}{2}}{3\frac{1}{4}}} = \dfrac{22}{13}$

$\phantom{\dfrac{5\frac{1}{2}}{3\frac{1}{4}}} = 1\dfrac{9}{13}$

**19.** $1\dfrac{1}{2} + 3\dfrac{1}{4} + 2\dfrac{1}{10} = 1\dfrac{10}{20} + 3\dfrac{5}{20} + 2\dfrac{2}{20}$

$\phantom{1\dfrac{1}{2} + 3\dfrac{1}{4} + 2\dfrac{1}{10}} = 6\dfrac{17}{20}$ miles

**20.** $11\dfrac{2}{3} \div 2\dfrac{1}{3} = \dfrac{35}{3} \div \dfrac{7}{3}$

$\phantom{11\dfrac{2}{3} \div 2\dfrac{1}{3}} = \dfrac{35}{3} \cdot \dfrac{3}{7}$

$\phantom{11\dfrac{2}{3} \div 2\dfrac{1}{3}} = 5$

5 packages

**21.** $\dfrac{719}{1000} = 0.719$

**22.** $0.86 = \dfrac{86}{100} = \dfrac{43}{50}$

**23.** $0.315 > 0.309$

**24.** $506.3782 \approx 506.38$

**25.** $9.6 + 3.82 + 1.05 + 7.3 = 21.77$

**26.**
$$\begin{array}{r} 3.610 \\ -\ 2.853 \\ \hline 0.757 \end{array}$$

**27.**
$$\begin{array}{r} 1.23 \\ \times\ 0.4 \\ \hline 0.492 \end{array}$$

**28.**
$$\begin{array}{r}
3.69\phantom{0} \\
0.24\overline{)\,0.8856} \\
\underline{72}\phantom{00} \\
165\phantom{0} \\
\underline{144}\phantom{0} \\
216 \\
\underline{216} \\
0
\end{array}$$

**29.**
$$\begin{array}{r}
0.8125 \\
16\overline{)\,13.0000} \\
\underline{12\ 8}\phantom{000} \\
20\phantom{00} \\
\underline{16}\phantom{00} \\
40\phantom{0} \\
\underline{32}\phantom{0} \\
80 \\
\underline{80} \\
0
\end{array}$$

$\dfrac{13}{16} = 0.8125$

**30.** $0.7 + (0.2)^3 - 0.08\,(0.03)$

$\phantom{0.7} = 0.7 + 0.008 - 0.0024$

$\phantom{0.7} = 0.708 - 0.0024$

$\phantom{0.7} = 0.7056$

**31.** $\dfrac{7000}{215} = \dfrac{7000 \div 5}{215 \div 5}$

$\phantom{\dfrac{7000}{215}} = \dfrac{1400 \text{ students}}{43 \text{ faculty}}$

**32.** $\dfrac{12}{15} = \dfrac{17}{21}$

$12 \times 21 \overset{?}{=} 15 \times 17$

$252 \neq 255$  False

**33.**
$$\frac{5}{9} = \frac{n}{17}$$
$$5 \times 17 = 9 \times n$$
$$\frac{85}{9} = \frac{9 \times n}{9}$$
$$n \approx 9.4$$

**34.**
$$\frac{3}{n} = \frac{7}{18}$$
$$3 \times 18 = 7 \times n$$
$$\frac{54}{7} = 7 \times n$$
$$7.7 \approx n$$

**35.**
$$\frac{n}{12} = \frac{5}{4}$$
$$n \times 4 = 12 \times 5$$
$$\frac{n \times 4}{4} = \frac{60}{4}$$
$$n = 15$$

**36.**
$$\frac{n}{7} = \frac{36}{28}$$
$$n \times 28 = 7 \times 36$$
$$\frac{n \times 28}{28} = \frac{252}{28}$$
$$n = 9$$

**37.**
$$\frac{2000}{3} = \frac{n}{5}$$
$$2000 \times 5 = 3 \times n$$
$$\frac{10,000}{3} = \frac{3 \times n}{3}$$
$$n \approx \$3333.33$$

**38.**
$$\frac{200}{6} = \frac{325}{n}$$
$$200 \times n = 6 \times 325$$
$$\frac{200 \times n}{200} = \frac{1950}{200}$$
$$n = 9.75$$

9.75 inches

**39.**
$$\frac{68}{5} = \frac{4000}{n}$$
$$68 \times n = 5 \times 4000$$
$$\frac{68 \times n}{68} = \frac{20,000}{68}$$
$$n \approx \$294.12 \text{ withheld}$$

**40.**
$$\frac{18}{1.2} = \frac{24}{n}$$
$$18 \times n = 1.2 \times 24$$
$$\frac{18 \times n}{18} = \frac{28.8}{18}$$
$$n = 1.6$$

1.6 lb of butter

**41.** $0.0063 = 0.63\%$

**42.**
$$\frac{17}{80} = \frac{n}{100}$$
$$17 \times 100 = 80 \times n$$
$$\frac{1700}{80} = \frac{80 \times n}{80}$$
$$21.25 = n$$

21.25%

**43.** $164\% = 1.64$

**44.** $300 \times n = 52$

$$\frac{300 \times n}{300} = \frac{52}{300}$$

$$n \approx 0.173$$

17.3%

**45.** 6.3% of 4800

$$6.3\% \times 4800 = n$$

$$0.063 \times 4800 = n$$

$$n = 302.4$$

**46.**
$$\frac{58}{100} = \frac{145}{b}$$

$$58 \times b = 100 \times 145$$

$$\frac{58 \times b}{58} = \frac{14{,}500}{58}$$

$$b = 250$$

**47.** 126% of 3400

$$126\% \times 3400 = n$$

$$1.26 \times 3400 = n$$

$$n = 4284$$

**48.** $11{,}800 - 0.08\,(11{,}800) = 11{,}800 - 944$
$$= 10{,}856$$

$10,856

**49.**
$$\frac{28}{100} = \frac{1260}{b}$$

$$28 \times b = 100 \times 1260$$

$$\frac{28 \times b}{28} = \frac{126{,}000}{28}$$

$$b = 4500$$

4500 students

**50.** Difference $= 11.28 - 8.40 = 2.88$

$$\text{percent} = \frac{2.88}{8.40}$$

$$\approx 0.343 = 34.3\%$$

34.3%

**51.** $17 \text{ qt} \times \dfrac{1 \text{ gallon}}{4 \text{ quarts}} = 4.25 \text{ gal}$

**52.** $3.25 \text{ tons} \times \dfrac{2000 \text{ lb}}{1 \text{ ton}} = 6500 \text{ lb}$

**53.** $16 \text{ ft} \times \dfrac{12 \text{ in.}}{1 \text{ ft}} = 192 \text{ in.}$

**54.** $5.6 \text{ km} = 5600 \text{ m}$

**55.** $6.98 \text{ g} = 0.0698 \text{ kg}$

**56.** $2.48 \text{ ml} = 0.00248 \text{ L}$

**57.** $12 \text{ mi} \times \dfrac{1.61 \text{ km}}{1 \text{ mi}} = 19.32 \text{ km}$

**58.** $0.00063182 = 6.3182 \times 10^{-4}$

**59.** $126{,}400{,}000{,}000 = 1.264 \times 10^{11}$

**60.** $0.623 \text{ cm} + 0.74 \text{ cm} + 0.0428 \text{ mm}$
$$= 0.623 \text{ cm} + 0.74 \text{ cm} + 0.00428 \text{ cm}$$
$$= 1.36728 \text{ cm thick}$$

**61.** $P = 2l + 2w$
$$= 2\,(6) + 2\,(1.2)$$
$$= 12 + 2.4$$
$$= 14.4 \text{ m}$$

**62.** $P = 82 + 13 + 98 + 13$
$$= 206 \text{ cm}$$

**63.** $A = \dfrac{bh}{2} = \dfrac{6\,(1.8)}{2} = 5.4$ sq ft

**64.** $A = \dfrac{h\,(b + B)}{2}$

$\quad = \dfrac{7.5\,(8 + 12)}{2}$

$\quad = \dfrac{7.5\,(20)}{2}$

$\quad = 75$

75 sq m

**65.** $A = \pi r^2 = 3.14\,(6)^2 \approx 113.04$ sq m

**66.** $C = 2\pi r$

$\quad C = 2\,(3.14)\left(\dfrac{18}{2}\right)$

$\quad = 56.52$

56.52 m

**67.** $V = \dfrac{\pi r^2 h}{3}$

$\quad = \dfrac{\pi\,(4)^2\,(10)}{3}$

$\quad \approx 167.46$ cu cm

**68.** $V = \dfrac{12\,(19)\,(2.7)}{3}$

$\quad = 205.2$

205.2 cu ft

**69.** Total area

$\quad =$ Area of square $+$ Area of triangle

$\quad = s^2 + \dfrac{bh}{2}$

$\quad = (5)^2 + \dfrac{3\,(5)}{2}$

$\quad = 32.5$ sq m

**70.** $\quad \dfrac{n}{130} = \dfrac{30}{120}$

$\quad n \times 120 = 30 \times 130$

$\quad \dfrac{n \times 120}{120} = \dfrac{3900}{120}$

$\quad n = 32.5$

**71.** 8 million dollars in profit

**72.** $9 - 8 = 1$

one million dollars

**73.** $50°$F

**74.** From 1990 to 2000

**75.** 600 students are between 17-22 years old.

**76.** $10 + 4 = 14$

1400 students

**77.** Mean $= \dfrac{8 + 12 + 16 + 17 + 20 + 22}{6}$

$\quad \approx 15.83$

Median $= \dfrac{16 + 17}{2} = 16.5$

**78.** $\sqrt{49} + \sqrt{81} = 7 + 9$

$\quad = 16$

**79.** $\sqrt{123} \approx 11.091$

**80.** hypotenuse $= \sqrt{9^2 + 12^2}$

$\quad = \sqrt{81 + 44}$

$\quad = \sqrt{225}$

$\quad = 15$

15 ft

**81.** $-8 + (-2) + (-3) = -10 + (-3)$
$$= -13$$

**82.** $-\dfrac{1}{4} + \dfrac{3}{8} = -\dfrac{1}{4} \cdot \dfrac{2}{2} + \dfrac{3}{8}$
$$= -\dfrac{2}{8} + \dfrac{3}{8}$$
$$= \dfrac{-2 + 3}{8}$$
$$= \dfrac{1}{8}$$

**83.** $9 - 12 = 9 + (-12)$
$$= -3$$

**84.** $-20 - (-3) = -20 + 3$
$$= -17$$

**85.** $2(-3)(4)(-1) = -6(4)(-1)$
$$= -24(-1)$$
$$= 24$$

**86.** $-\dfrac{2}{3} \div \dfrac{1}{4} = -\dfrac{2}{3} \cdot \dfrac{4}{1}$
$$= -\dfrac{8}{3}$$
$$= -2\dfrac{2}{3}$$

**87.** $(-16) \div (-2) + (-4) = 8 + (-4)$
$$= 4$$

**88.** $12 - 3(-5) = 12 + 15$
$$= 27$$

**89.** $7 - (-3) + 12 \div (-6) = 7 + 3 + (-2)$
$$= 10 + (-2)$$
$$= 8$$

**90.** $\dfrac{(-3)(-1) + (-4)(2)}{(0)(6) + (-5)(2)} = \dfrac{3 + (-8)}{0 + (-10)}$
$$= \dfrac{-5}{-10}$$
$$= \dfrac{1}{2}$$

**91.** $5x - 3y - 8x - 4y$
$$= 5x - 8x - 3y - 4y$$
$$= -3x - 7y$$

**92.** $5 + 2a - 8b - 12 - 6a - 9b$
$$= 5 + (-12) + 2a + (-6a)$$
$$\quad + (-8b) + (-9b)$$
$$= -7 - 4a - 17b$$

**93.** $-2(x - 3y - 5)$
$$= -2(x) + (-2)(-3y) + (-2)(-5)$$
$$= -2x + 6y + 10$$

**94.** $-2(4x + 2) - 3(x + 3y)$
$$= -8x - 4 - 3x - 9y$$
$$= -11x - 9y - 4$$

**95.** $5 - 4x = -3$
$$5 + (-5) - 4x = -3 + (-5)$$
$$-4x = -8$$
$$\dfrac{-4x}{-4} = \dfrac{-8}{-4}$$
$$x = 2$$

**96.**

$$5 - 2(x - 3) = 15$$
$$5 - 2x + 6 = 15$$
$$-2x + 11 = 15$$
$$-2x + 11 + (-11) = 15 + (-11)$$
$$-2x = 4$$
$$\frac{-2x}{-2} = \frac{4}{-2}$$
$$x = -2$$

$$x + (x + 12) + 2x = 452$$
$$4x + 12 = 452$$
$$4x + 12 + (-12) = 452 + (-12)$$
$$4x = 440$$
$$\frac{4x}{4} = \frac{440}{4}$$
$$x = 110$$

$$\text{math} = 110$$
$$\text{history} = 110 + 12 = 122$$
$$\text{psychology} = 2(110) = 220$$

**97.**

$$7 - 2x = 10 + 4x$$
$$7 - 2x + (-4x) = 10 + 4x + (-4x)$$
$$7 - 6x = 10$$
$$7 + (-7) - 6x = 10 + (-7)$$
$$-6x = 3$$
$$\frac{-6x}{-6} = \frac{3}{-6}$$
$$x = -\frac{3}{6}$$
$$= -\frac{1}{2}$$

**100.**

$$x = \text{width}$$
$$2x + 5 = \text{length}$$
$$2(x) + 2(2x + 5) = 106$$
$$2x + 4x + 10 = 106$$
$$6x + 10 = 106$$
$$6x + 10 + (-10) = 106 + (-10)$$
$$6x = 96$$
$$\frac{6x}{6} = \frac{96}{6}$$
$$x = 16$$

$$\text{width} = 16 \text{ m}$$
$$\text{length} = 2(16) + 5 = 37 \text{ m}$$

**98.**

$$-3(x + 4) = 2(x - 5)$$
$$-3x - 12 = 2x - 10$$
$$-3x + 3x - 12 = 2x + 3x - 10$$
$$-12 = 5x - 10$$
$$-12 + 10 = 5x - 10 + 10$$
$$-2 = 5x$$
$$\frac{-2}{5} = \frac{5x}{5}$$
$$-\frac{2}{5} = x$$

**99.**

$$x = \# \text{ of students taking math}$$
$$x + 12 = \# \text{ of students taking history}$$
$$2x = \# \text{ of students taking psychology}$$